AF356210

Emerging Research in Rotary Engines and Sustainability in Vehicle Engineering

Emerging Research in Rotary Engines and Sustainability in Vehicle Engineering

Editors

Cheng Shi
Jinxin Yang
Jianbing Gao
Peng Zhang

Basel • Beijing • Wuhan • Barcelona • Belgrade • Novi Sad • Cluj • Manchester

Editors

Cheng Shi
School of Vehicle and Energy
Yanshan University
Qinhuangdao
China

Jinxin Yang
College of Energy and Power
Engineering
Beijing University of
Technology
Beijing
China

Jianbing Gao
School of Mechanical
Engineering
Beijing Institute of
Technology
Beijing
China

Peng Zhang
School of Automobile
Chang'an University
Xi'an
China

Editorial Office
MDPI
St. Alban-Anlage 66
4052 Basel, Switzerland

This is a reprint of articles from the Special Issue published online in the open access journal *Sustainability* (ISSN 2071-1050) (available at: https://www.mdpi.com/journal/sustainability/special_issues/rotary_engines).

For citation purposes, cite each article independently as indicated on the article page online and as indicated below:

Lastname, A.A.; Lastname, B.B. Article Title. *Journal Name* **Year**, *Volume Number*, Page Range.

ISBN 978-3-7258-0026-1 (Hbk)
ISBN 978-3-7258-0025-4 (PDF)
doi.org/10.3390/books978-3-7258-0025-4

Contents

 *sustainability*

Article

An Explainable Prediction Model for Aerodynamic Noise of an Engine Turbocharger Compressor Using an Ensemble Learning and Shapley Additive Explanations Approach

Rong Huang [1], Jimin Ni [1], Pengli Qiao [1], Qiwei Wang [1,*], Xiuyong Shi [1] and Qi Yin [2]

[1] School of Automotive Studies, Tongji University, Shanghai 201804, China; huangrong0801@163.com (R.H.); njmwjyx@hotmail.com (J.N.); 2133533@tongji.edu.cn (P.Q.); shixy@tongji.edu.cn (X.S.)

[2] SAIC Motor, General Institute of Innovation Research and Development, Shanghai 201804, China; yinqi@saicmotor.com

* Correspondence: wangqiwei1987@hotmail.com; Tel./Fax: +86-021-6958-9980

Abstract: In the fields of environment and transportation, the aerodynamic noise emissions emitted from heavy-duty diesel engine turbocharger compressors are of great harm to the environment and human health, which needs to be addressed urgently. However, for the study of compressor aerodynamic noise, particularly at the full operating range, experimental or numerical simulation methods are costly or long-period, which do not match engineering requirements. To fill this gap, a method based on ensemble learning is proposed to predict aerodynamic noise. In this study, 10,773 datasets were collected to establish and normalize an aerodynamic noise dataset. Four ensemble learning algorithms (random forest, extreme gradient boosting, categorical boosting (CatBoost) and light gradient boosting machine) were applied to establish the mapping functions between the total sound pressure level (SPL) of the aerodynamic noise and the speed, mass flow rate, pressure ratio and frequency of the compressor. The results showed that, among the four models, the CatBoost model had the best prediction performance with a correlation coefficient and root mean square error of 0.984798 and 0.000628, respectively. In addition, the error between the predicted total SPL and the observed value was the smallest, at only 0.37%. Therefore, the method based on the CatBoost algorithm to predict aerodynamic noise is proposed. For different operating points of the compressor, the CatBoost model had high prediction accuracy. The noise contour cloud in the predicted MAP from the CatBoost model was better at characterizing the variation in the total SPL. The maximum and minimum total SPLs were 122.53 dB and 115.42 dB, respectively. To further interpret the model, an analysis conducted by applying the Shapley Additive Explanation algorithm showed that frequency significantly affected the SPL, while the speed, mass flow rate and pressure ratio had little effect on the SPL. Therefore, the proposed method based on the CatBoost algorithm could well predict aerodynamic noise emissions from a turbocharger compressor.

Keywords: turbocharger compressor; aerodynamic noise; ensemble learning; emission prediction model; Shapley Additive Explanation

Citation: Huang, R.; Ni, J.; Qiao, P.; Wang, Q.; Shi, X.; Yin, Q. An Explainable Prediction Model for Aerodynamic Noise of an Engine Turbocharger Compressor Using an Ensemble Learning and Shapley Additive Explanations Approach. *Sustainability* **2023**, *15*, 13405. https://doi.org/10.3390/su151813405

Academic Editors: Jinxin Yang, Jianbing Gao, Cheng Shi and Peng Zhang

Received: 24 July 2023
Revised: 29 August 2023
Accepted: 4 September 2023
Published: 7 September 2023

1. Introduction

As the problems of energy shortage and environmental pollution are becoming more and more prominent, reducing fuel consumption and pollutant emissions from road vehicles is one of the most important approaches to achieving environmentally and economically sustainable development [1–3]. Turbochargers are widely used in the transportation field because they increase engines' specific power and reduce gas emissions [4–7]. Unfortunately, the noise emissions generated by turbochargers become a non-negligible part of the engine noise source, hindering environmentally sustainable development to some extent [8,9]. In addition, due to the increase in output power requirements for diesel engines, the turbocharger pressure ratio increases, resulting in an increase in the compressor load and higher aerodynamic noise emissions [10,11]. The existing literature indicates that aerodynamic noise is considered to be the main noise source in turbochargers [12,13]. Aerodynamic noise mainly consists of discrete noise and broadband noise, which is generated by the turbulent motion between the airflow and the compressor components [14,15]. Due to the complexity of the turbulent motion, it is difficult to quantitatively describe the flow field and the resulting induced sound field during the operation of a turbocharger compressor by means of a complete mathematical analytical formula. Therefore, experimental or numerical simulation methods are often relied upon to obtain a realistic situation of the compressor aerodynamic noise.

Analyzing the aerodynamic noise distribution of a compressor is the basis for achieving aerodynamic noise emission control. Researchers have conducted numerous experimental studies on compressors' aerodynamic noise. Raitor et al. [16] studied the main noise sources of centrifugal compressors. The results indicated that blade passing frequency (BPF) noise, buzzsaw noise and tip clearance noise were the main noise sources. Figurella et al. [17] showed that in the compressor, BPF and its harmonic frequencies, discrete noise could be observed. Sun et al. [18] conducted experiments to investigate the influences of foam metal casing treatment on an axial flow compressor's aerodynamic noise. The results showed that the use of the foam metal casing treatment could reduce aerodynamic noise within a range of 0.18 dB~1.6 dB. Zhang et al. [19] used the experiment method to investigate the effect of differential tip clearances on the noise emissions of an axial compressor. The results showed that when the tip clearance was small, the sound pressure level (SPL) of the compressor was lowest. Furthermore, Galindo et al. [20] carried out experiments to study the influence of inlet geometry on an automotive turbocharger compressor noise. They found that the aerodynamic noise emissions and surge margin could significantly improve using a convergent-divergent nozzle. Therefore, the experimental approach is an effective way to study the aerodynamic noise of turbocharger compressors. However, the test bench operation and cost limitations make it difficult to carry out the measurement of SPL for compressor aerodynamic noise under arbitrary operating conditions. This brings challenges for reducing compressor noise emissions and promoting environmentally sustainable development.

With the gradual development of computational fluid dynamics (CFD), numerical simulation techniques for compressor noise coupled with CFD and computational aerodynamic acoustics methods have been widely used [21,22]. Liu et al. [23] calculated the unsteady flow field of a compressor and used the flow field results to obtain noise source information. In order to calculate centrifugal fan and axial compressor noises, the RANS method and the Ffowcs Williams and Hawkings (FW–H) equation were used by Khelladi et al. [24] and Laborderie et al. [25]. Karim et al. [12] conducted a CFD numerical simulation with the use of the Large Eddy Simulation approach to measure the pressure signal at the inlet and outlet of a compressor, and to calculate the SPL and spectral distribution. Lu et al. [26] conducted an experimental and simulation investigation on the aerodynamic noise of an axial compressor. They found that the main sound source areas were the rotor and stator. In addition, Zhang et al. [27] used multiple calculation methods to investigate the effect of an approximately solid surface wall on fan noise propagation. However, due to the large resource consumption and long computation time of the multi-dimensional and dense-grid numerical simulation of compressor noise, there are certain disadvantages in engineering applications.

From the above literature analysis, it is clear that traditional experimental measurements are costly, long-period and complicated to operate. The advent of numerical simulation methods has made it possible to obtain more detailed flow structures and richer flow field information than experiments at a lower cost. However, turbulence is a nonlinear mechanical system with a large number of degrees of freedom and an extremely wide range of scales. For models with complex geometrical shapes and high Reynolds number flows, even if they rely on numerical simulation by computer, they still need to perform very complicated calculations on a rapidly increasing number of grids, which consumes huge computational resources. Therefore, in order to save the costs of experiments or simulations and shorten their time cycles, data-driven methods are gradually becoming a focus of attention [28]. That is, by means of machine learning, key information can be extracted and "black box" models constructed based on sample data from experiments or numerical simulations.

Machine learning, as an interdisciplinary discipline, has received sustained attention from many scholars in recent years [29–31]. Ensemble learning algorithms based on decision trees, such as extreme gradient boosting (XGBoost) and random forest (RF), are widely used in the study of complex nonlinear models in the environmental field [32,33]. Furthermore, the recently proposed categorical boosting (CatBoost) and light gradient boosting machine (LightGBM) algorithms have gained attention due to their excellent performance on small datasets and their strong overfitting resistance [34,35]. However, these algorithms are based on decision trees and are often considered "black box" algorithms, making it difficult to know their prediction process. In recent years, researchers have introduced a number of techniques to explain machine learning algorithms. The partial dependence diagram (PDP), as a classical method to reveal the mean partial relationship of one or more features in the model results, has been adopted by many researchers [36]. However, the average marginal benefits calculated by PDPs may hide the variability among data. Therefore, the Shapley Additive Explanation (SHAP) method was introduced to overcome these problems. The SHAP method is a game theory-based model diagnosis method that can improve interpretability by calculating the importance value of each input feature on the prediction results [37,38]. In addition, the SHAP method offers the possibility to visualize and interpret the contribution of a feature value to the predicted results using SHAP values.

In the existing literature, the experimental and numerical simulation methods are the main approaches used to study the aerodynamic noise characteristics. However, they are costly and long-period at the full compressor operating range, which has some drawbacks in engineering applications. To fill this gap, based on compressor aerodynamic noise datasets, four ensemble learning methods (RF, XGBoost, CatBoost and LightGBM) and the SHAP algorithm were used to establish an interpretable compressor aerodynamic noise prediction model in this study. The model based on the CatBoost algorithm with the best predictive performance among the four models was selected through tenfold cross-validation to carry out the aerodynamic noise prediction, and the differences between the predicted results and the observed values were compared and analyzed. A MAP diagram of the aerodynamic noise at the full operating range is presented. Furthermore, in order to understand the prediction process of the proposed method, the SHAP algorithm was used to reveal the nonlinear relationship between the model input features and the predicted results. The interpretable prediction model proposed in this study could accurately evaluate the compressor aerodynamic noise under arbitrary operating conditions and provide data and theoretical support for realizing the control of noise emissions and contributing to environmentally sustainable development. Figure 1 shows the research framework of this study.

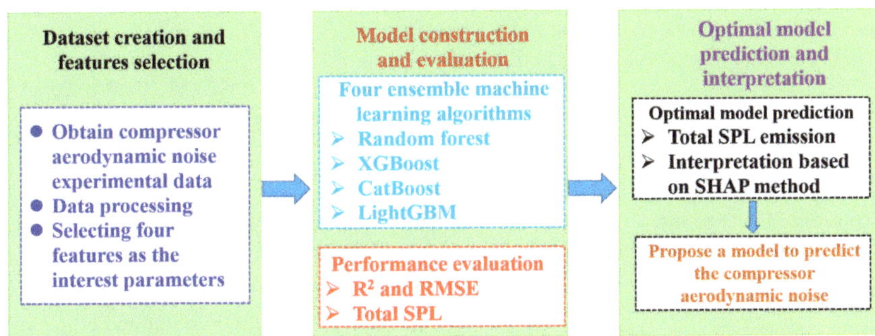

Figure 1. Investigation procedure.

2. Research Methodology

The interpretable prediction model building and analysis process for predicting compressor aerodynamic noise is shown in Figure 2. The compressor aerodynamic noise data were obtained from the experiments, and the data were processed to build the emission prediction model. Four ensemble machine learning methods (random forest (RF), extreme gradient boosting (XGBoost), categorical boosting (CatBoost) and light gradient boosting machine (LightGBM)) were used to construct the models, and then an interpretable algorithm of Shapley Additive Explanations (SHAP) was used to analyze the extent to which the input features contributed to the output results and provided explanations for the aerodynamic noise prediction process. The results of the study can provide decision-making for compressor aerodynamic noise emission control.

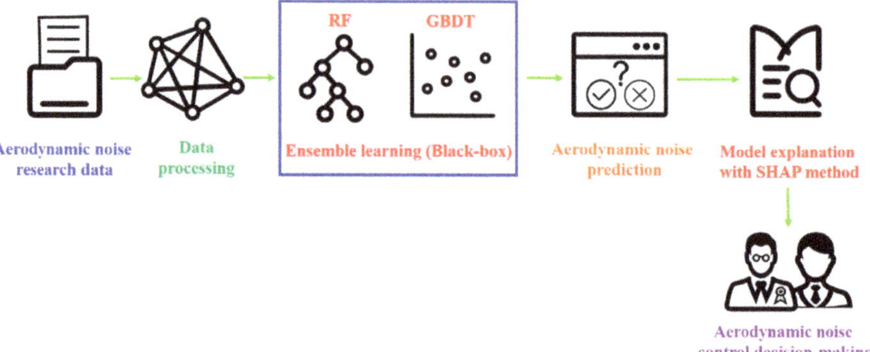

Figure 2. Explainable prediction model for compressor aerodynamic noise using Shapley Additive Explanations approach.

2.1. Experimental System and Method

Figure 3 shows the schematic of the turbocharger compressor aerodynamic noise test rig. In the compressor noise experimental system, the PCB-SN152495 type microphone was used to measure the sound pressure level (SPL) of aerodynamic noise, the PCB-HT356B21 type vibration sensor was used to test the vibration acceleration on the surface of the compressor worm shell and the SPL and vibration signals were collected and analyzed by the SIEMENS signal acquisition method. A detailed description of turbocharger test bench can be found in the literature [39,40], and the turbocharger test bench and aerodynamic noise test instruments are shown in Table 1. As can be seen from Table 1, the turbocharger performance and noise test rig consisted of four parts, which were the compressor section, turbine section, intake and exhaust piping and components, and noise test section, respectively. Table 2 lists the measuring ranges, accuracies and uncertainties of the aerodynamic noise test instruments [40].

Table 1. Distributions of turbocharger test bench and aerodynamic noise test instruments.

Compressor Section	Turbine Section	Intake and Exhaust Piping and Components	Noise Test Section
Compressor inlet flowmeter	Turbine	Automatic circulating valve	Vibration sensors
Compressor inlet pressure sensor	Turbine inlet flowmeter	Electric exhaust control valve	Microphone
Compressor inlet and outlet temperature sensors	Turbine inlet control valve	Electric trimming valve	Signal acquisition port
Speed sensor	–	Burner	Computer
Compressor	–	Air source and vent valve	–

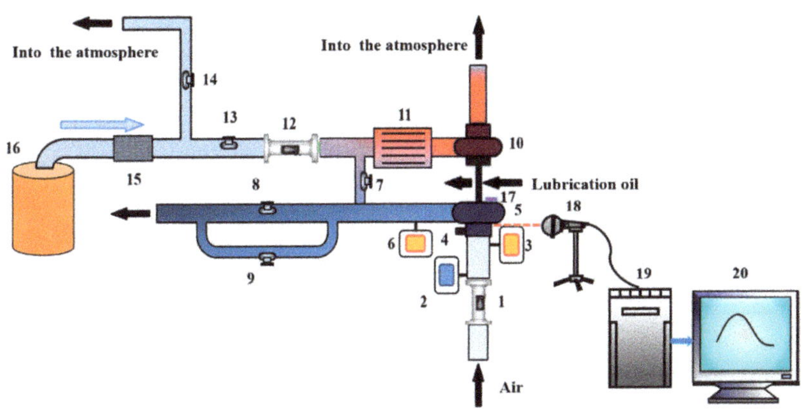

1. Compressor inlet flowmeter   2. Compressor inlet pressure sensor   3. Compressor inlet temperature sensor

4. Speed sensor   5. Compressor   6. Compressor outlet temperature sensor

7. Automatic circulating valve   8. Electric exhaust control valve   9. Electric trimming valve

10. Turbine   11. Burner   12. Turbine inlet flowmeter

13. Turbine inlet control valve   14. Air source vent valve   15. Filter

16. Air source   17. Vibration sensors   18. Microphone

19. Signal acquisition port   20. Computer

Figure 3. Schematic diagram of aerodynamic noise for compressor test setup [40].

Table 2. Measuring ranges, accuracies and uncertainties of instruments [40].

Instruments	Parameters	Measuring Range	Accuracy	Uncertainty
Tachometer	Speed	0~400,000 r/min	0.1 r/min	0.01 r/min
Microphone	Noise	15~165 dB	0.1 dB	±0.02 dB
Vibration sensor	Vibration	±490 m/s² pk	1%	±0.2 m/s²
Temperature sensor	Compressor inlet and outlet temperature	−200~400 °C	0.25 °C	±0.1 °C
Pressure sensor	Compressor inlet pressure	−175~35,000 Pa, −40~85 °C	0.05%	±0.01%
Pressure sensor	Compressor outlet pressure	0~700,000 Pa, −40~85 °C	0.05%	±0.01%
Temperature sensor	Turbine inlet and outlet temperature	−200~1372 °C	0.4%	±0.1 °C
Pressure sensor	Turbine inlet pressure	0~700,000 Pa, −40~85 °C	0.05%	±0.01%
Pressure sensor	Turbine outlet pressure	−175~35,000 Pa, −40~85 °C	0.05%	±0.01%
Flow meter	Compressor outlet mass flow rate	30 standard liters per minute (SLM)~3000 SLM, 0~70 °C	0.5%	±0.1% F.S

2.2. Research Object

The research object of this study was the turbocharger compressor of a heavy-duty diesel engine. The compressor structure using a splitter blade, the diffuser using a bladeless structure and the specific parameters are shown in Table 3. A detailed description of the specification dimensions of the compressor can be found in the literature [40].

Table 3. Compressor specification parameters [40].

Item	Value
Turbocharger compatibility	Heavy-duty diesel engine
Outlet diameter of impeller (mm)	94.4
Inlet diameter of impeller (mm)	66.46
Main blade number	7
Splitter blade number	7

Table 3. *Cont.*

Item	Value
Diffuser height (mm)	4.77
Design pressure ratio	4.5
Rated speed (r/min)	117,000
Outlet diameter of diffuser (mm)	166.15
Inlet diameter of diffuser (mm)	90
Type of cooling	Oil cooling
Mass flow rate range (kg/s)	0.08, 0.66

2.3. Dataset Creation

During the experiments, the JB/T 12332-2015 "Turbocharger Noise Test Method" standard was referenced to test the noise of the compressor [41]. In order to ensure the repeatability and accuracy of the aerodynamic noise experiments of the compressor, the laboratory environment and instruments needed to be measured and calibrated before the test started. The measurement methods and procedures are described in the literature [40]. In addition, the turbine, exhaust pipes and facilities of the turbocharger for the test were covered and soundproofed to ensure that the compressor inlet aerodynamic noise experiments were not affected by other noise sources. In the experiments, the SPLs of aerodynamic noise corresponding to different frequencies were recorded by adjusting the speed, pressure ratio and mass flow rate of the compressor. The formula for calculating the total SPL of aerodynamic noise is shown in Equation (1) [42]:

$$L_{total} = 10\log\left(\sum_{i=1}^{n} 10^{\frac{L_i}{10}}\right) \tag{1}$$

where L_{total} and L_i are the total SPL and SPL at a fixed frequency point, respectively. n is the number of frequency points.

In this study, a total of 10,773 sets of aerodynamic noise data were obtained from the experiments. The noise test points were determined based on a MAP diagram of the compressor performance, as shown in Figure 4. The noise test points included 21 operating points. In addition, the datasets collected in the experiments were obtained from a previous study [40]. The remaining operating points were the compressor performance distribution points.

The distribution of the dataset is shown in Table 4. From the table, it can be seen that the dataset covered a total of seven speed lines ranging from 60,000 r/min to 110,000 r/min, including three operating regions of the compressor: near-choke region, high-efficiency region and near-surge region.

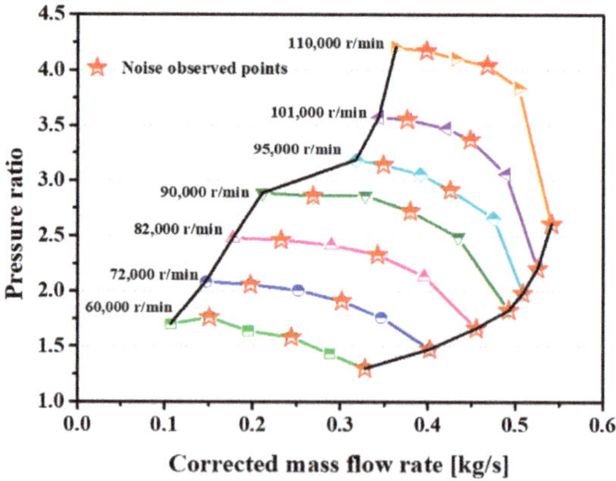

Figure 4. Observed operating points of aerodynamic noise [40].

Table 4. Dataset distributions of noise test points.

Speed (r/min)	Mass Flow Rate (kg/s)	Pressure Ratio	Frequency (Hz)	Number of Datasets
60,000	0.151	1.764	0~25,600 (interval of 50)	513
60,000	0.244	1.583	0~25,600 (interval of 50)	513
60,000	0.328	1.3	0~25,600 (interval of 50)	513
72,000	0.197	1.476	0~25,600 (interval of 50)	513
72,000	0.302	1.914	0~25,600 (interval of 50)	513
72,000	0.402	2.063	0~25,600 (interval of 50)	513
82,000	0.232	2.465	0~25,600 (interval of 50)	513
82,000	0.342	2.331	0~25,600 (interval of 50)	513
82,000	0.456	1.663	0~25,600 (interval of 50)	513
90,000	0.268	2.863	0~25,600 (interval of 50)	513
90,000	0.379	2.723	0~25,600 (interval of 50)	513
90,000	0.493	1.83	0~25,600 (interval of 50)	513
95,000	0.349	3.144	0~25,600 (interval of 50)	513
95,000	0.426	2.916	0~25,600 (interval of 50)	513
95,000	0.509	1.985	0~25,600 (interval of 50)	513
101,000	0.376	3.555	0~25,600 (interval of 50)	513
101,000	0.449	3.375	0~25,600 (interval of 50)	513
101,000	0.526	2.207	0~25,600 (interval of 50)	513
110,000	0.398	4.175	0~25,600 (interval of 50)	513
110,000	0.468	4.041	0~25,600 (interval of 50)	513
110,000	0.542	2.609	0~25,600 (interval of 50)	513

2.4. Model Building and Performance Evaluation

In this study, one traditional ensemble learning algorithm (RF) and three gradient boosting decision tree (GBDT) algorithms (XGBoost, CatBoost and LightGBM) were used to build a compressor aerodynamic noise emissions prediction model. Compared to complex deep learning models, using four ensemble models made it easier to capture the variation in parameters and variable interpretation within each model. For the ensemble learning component, the RF was a typical bagging algorithm that accomplished a classification task by voting and a regression task by averaging [43]. Specifically, the RF was a set of decision trees, and each tree was constructed using the best split for each node among a subset of predictors randomly chosen at that node. In the end, a simple majority vote was taken for prediction. The GBDT was a machine learning model for regression and classification, and its effective implementations included XGBoost. However, the efficiency and standardization of XGBoost was not satisfactory when feature dimensionality was high and the data size was large. Therefore, the CatBoost and LightGBM models were proposed, and these models were shown to significantly outperform other models in terms of accuracy for structured and tabulated data [44]. To be specific, CatBoost used greedy strategies to consider combinations to improve classification accuracy when constructing new split points for the current tree. Meanwhile LightGBM contained two novel techniques including Gradient-based One-Sided Sampling and Exclusive Feature Bundling [45].

The ensemble learning models in this study were all implemented based on scikit-learn and Python libraries. To ensure the accuracy of the models, each model uniformly used 80% of the dataset as the training set and 20% of the dataset as the validation set. The optimal model was obtained by adjusting the training strategy using GridSearch and tenfold cross-validation methods, in which the training set was randomly divided into ten copies and the ten subsets were traversed in turn, with the current subset used for testing and the remaining nine copies used for training. The performance of the prediction model was evaluated using the coefficient of determination (R^2) and the root mean square error (RMSE). The R^2 and RMSE were calculated as shown in Equations (2) and (3):

$$R^2 = 1 - \frac{\sum_1^N \left(y_p - y_o\right)^2}{\sum_1^N \left(\overline{y} - y_o\right)^2} \tag{2}$$

$$RMSE = \sqrt{\frac{\sum_1^N (y_p - y_o)^2}{N}} \tag{3}$$

where N is the sample size, y_p is the predicted value, y_o is the test observation and $\bar{y}$ is the average of y_o.

The setup parameters of the four models are listed in Table 5.

Table 5. The setup parameters of the four models.

Model	Number of Trees	Depth of Trees	Number of Leaf Nodes	Minimum Number of Samples for Leaf Nodes	Learning Rate	Sampling Ratio of Training Set
RF	1000	10	200	5	0.01	0.9
XGBoost	500	6	200	5	0.01	0.9
CatBoost	4000	10	200	5	0.01	0.9
LightGBM	3000	10	200	5	0.01	0.9

The distributions of the predicted operating points of the prediction models are shown in Figure 5. The remaining operating points were the aerodynamic noise test points of the compressor at different speed lines.

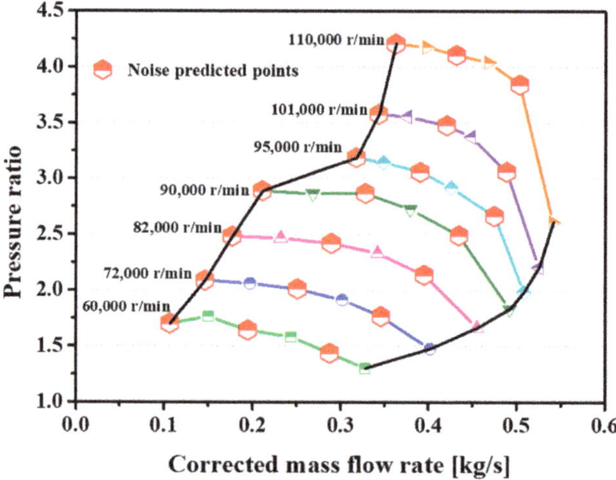

Figure 5. Predicted operating points of aerodynamic noise.

2.5. Model Interpretation

Ensemble learning models based on decision trees have often been considered as "black box" models. However, while establishing prediction models accurately, it is also necessary to explain how the prediction models work effectively. SHAP summary graphs obtained using the SHAP method have been shown to be effective in explaining the predicted results of decision tree models [38]. In a SHAP summary graph, the horizontal axis (x-axis) represents the SHAP value, and the magnitude of the value indicates the average marginal contribution of the input features to the model output. A SHAP value of less than 0 indicates a negative contribution; equal to 0, no contribution; and greater than 0, a positive contribution. A positive contribution means that the input features are highly important to the final predicted result, while the least important features result in a negative contribution. Each input feature was ranked from top to bottom according to its importance, with the top features contributing more to the predicted results of the model than the bottom features. The points representing the feature values were plotted horizontally, and the color of each point from low (blue) to high (red) represents the magnitude of the SHAP value for that feature [46].

3. Results and Discussion

3.1. Descriptive Statistics of Parameters Affecting Sound Pressure Level

3.1.1. Identification of Interest Parameters

In this study, two interest parameters included the compressor operating characteristics and the aerodynamic noise characteristics, which were introduced to explain their effects on the aerodynamic noise SPL. The two parameters are shown in Table 6. The parameters of compressor operating characteristics include speed, pressure ratio and mass flow rate. Related studies [47,48] have shown that compressor operating characteristics reflect the operating condition of the compressor and have an obvious impact on the SPL of aerodynamic noise. Aerodynamic noise characteristics refere to the frequencies corresponding to the SPL of aerodynamic noise. The SPL of aerodynamic noise varied for different frequencies. However, the coupling effect of these four characteristics (speed, pressure ratio, mass flow rate and frequency) on the SPL of the compressor was not well investigated, especially in terms of the contribution of each characteristic to the SPL, which was one of the focuses of this study.

Table 6. Specifications of four types of interest parameters.

Type	Parameters	Unit	Calculation Method
Compressor operating characteristics	Speed	r/min	Directly measured by tachometer
	Pressure ratio	–	Directly measured by pressure sensor and with the equation P_{out}/P_{in}
	Mass flow rate	kg/s	Directly measured by flow meter
Noise characteristics	Frequency	Hz	Directly measured by microphone

3.1.2. Distribution of Input Features

During the experiments, the compressor operating conditions were adjusted by changing the compressor speed, pressure ratio and mass flow rate, and the aerodynamic noise was measured. As can be seen from Figure 4 and Table 4, the compressor speed distribution ranged from 60,000 r/min to 110,000 r/min, the pressure ratio distribution ranged from 1.3 to 4.175, the mass flow rate distribution ranged from 0.151 kg/s to 0.542 kg/s and the frequency distribution ranged from 0 to 25,600 Hz. Therefore, by changing the speed, pressure ratio and mass flow rate within a certain range, the compressor was operated under different operating conditions, and then the aerodynamic noise was generated.

Different frequencies corresponded to different SPLs of aerodynamic noise. The aerodynamic noise characteristics of the compressor under various operating conditions are shown in Figure 6, and the experimental data were provided by a previous study [40]. From the figure, it can be seen that under the same operating conditions, the SPL of aerodynamic noise basically tended to decrease as the frequency increased. The frequency distribution ranged from 0 to 25,600 Hz, which shows that different frequencies had an effect on the SPL of aerodynamic noise.

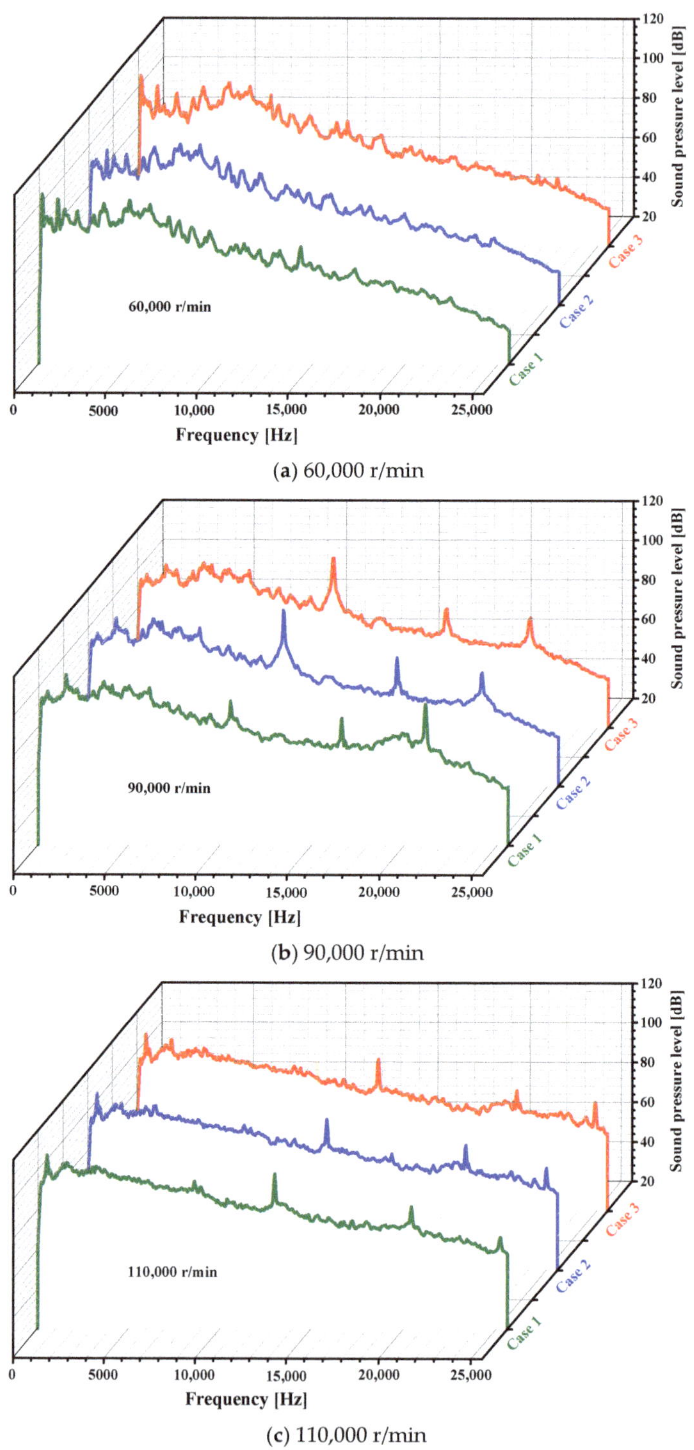

(**a**) 60,000 r/min

(**b**) 90,000 r/min

(**c**) 110,000 r/min

Figure 6. Aerodynamic noise characteristics of compressor under various operating conditions.

Therefore, speed, pressure ratio, mass flow rate and frequency were selected as the four features describing the aerodynamic noise generated during the operation of the compressor, and the output result was the SPL of the corresponding frequency. In this study, a total of 10,773 sets of valid data were collected, in which each set of data contained the SPL of aerodynamic noise and the four characteristic values affecting the SPL. To prevent the influence of the magnitude on the model training results, all the eigenvalues were normalized.

3.2. Importance of Input Features

The importance of the input features of the models was analyzed using the SHAP method. Figure 7 shows the results of ranking the importance of the input features for the four models. In the figure, the SHAP values of all features obtained by applying the SHAP method were within 0.18. Among the four models, the SHAP values of each feature were frequency > speed > mass flow rate > pressure ratio in descending order. Among all the features, frequency was the most important feature which affected the SPL of the aerodynamic noise, and its average SHAP value was above 0.16. This was because the SPLs of the aerodynamic noise corresponding to different frequencies were significantly different under the same compressor operating conditions (speed, mass flow rate and pressure ratio were the same), which made frequency have the greatest effect on the SPL. This result is consistent with that of Xu et al. [49]. Compared with the RF model, the SHAP values of speed in the three models of XGBoost, CatBoost and LightGBM were all above 0.02, and there was a significant difference with the third ranked mass flow rate. This indicates that the influence of speed was still larger in these three models. The above results show that among the four models, the frequency, speed, mass flow rate and pressure ratio had an influence on the output results of the prediction models and could be used as the input features.

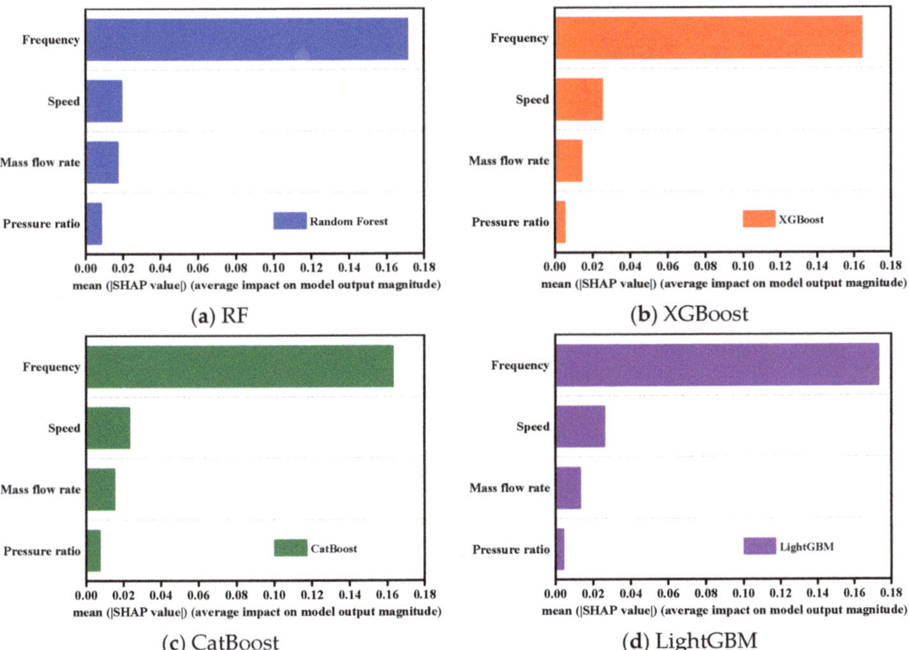

Figure 7. Analysis of the importance of the input features in different models.

3.3. Model Performance Comparison

Four prediction models of compressor aerodynamic noise were obtained by training and tenfold cross-validation with 8618 sets of a training dataset. The purpose of the tenfold cross-validation for prediction models was to select the optimal model parameters corresponding to the four models, thus improving the generalization ability of the models [43]. The R^2 and RMSE obtained for each calculation in the tenfold cross-validation are shown in Figure 8, and the average R^2 and average RMSE values from the tenfold cross-validation of the four models are shown in Figure 9. As can be seen from Figure 8, among the ten tests, the best prediction performance tests of RF, XGBoost, CatBoost and LightGBM were Test 5, Test 10, Test 6 and Test 3, respectively. In addition, from Figures 8 and 9, it can be seen that the R^2, RMSE and the mean R^2 and mean RMSE of the model training results in the tenfold cross-validation were close. Among them, the CatBoost model had the largest mean R^2 and the smallest mean RMSE with the values of 0.983579 and 0.000694, respectively. Therefore, for the four models, the optimal model was selected, respectively, for predictions adopting the tenfold cross-validation method.

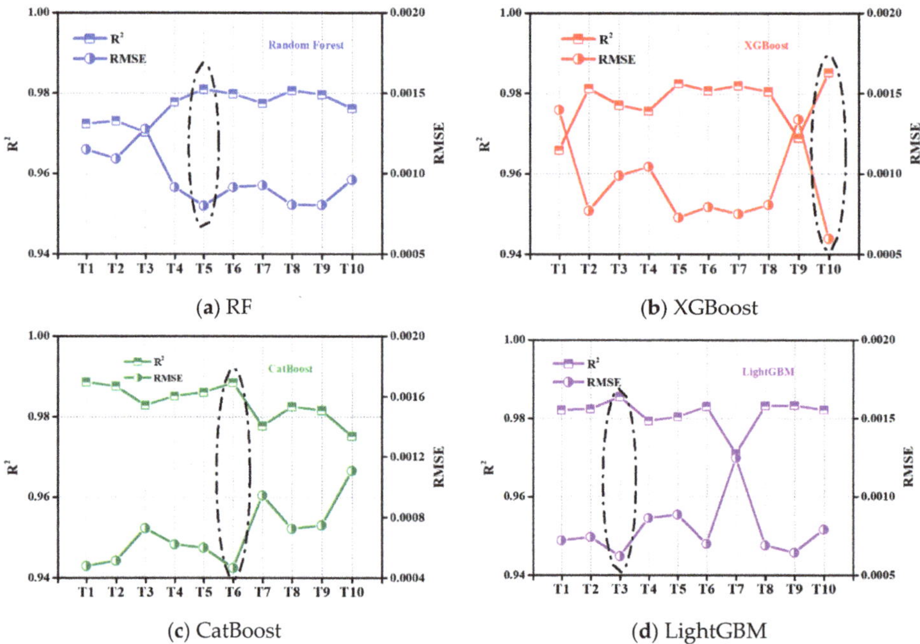

Figure 8. R^2 and RMSE values from the tenfold cross-validation of four models.

To determine the prediction performances of the models, the best models built by the four ensemble machine learning algorithms were applied to predict the 2155 datasets in the validation set, respectively. Figure 10 shows the R^2 and RMSE of the predicted results of the four models. It can be seen that overfitting was avoided in all four models. Among the four models, the largest R^2 and the smallest RMSE with the values of 0.984798 and 0.000628 can be seen in the CatBoost model, respectively, which indicates that the CatBoost-based model had the best predictive performance. Therefore, in this study, frequency, speed, mass flow rate and pressure ratios were used as the model input features, and the CatBoost algorithm was applied to build the compressor aerodynamic noise emission prediction model.

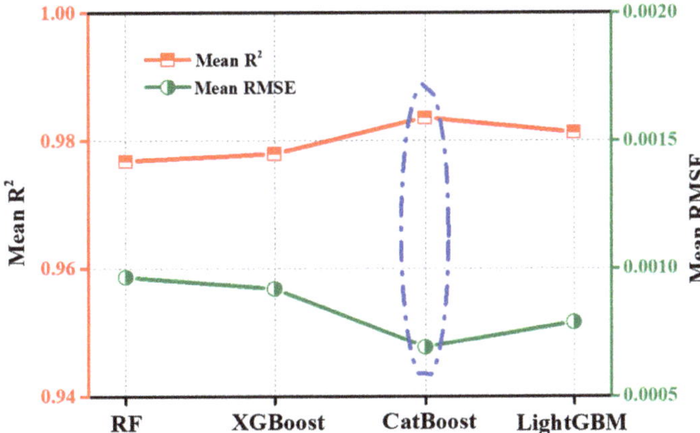

Figure 9. Comparison of mean R^2 and mean RMSE values of predicted results from the tenfold cross-validation of four models.

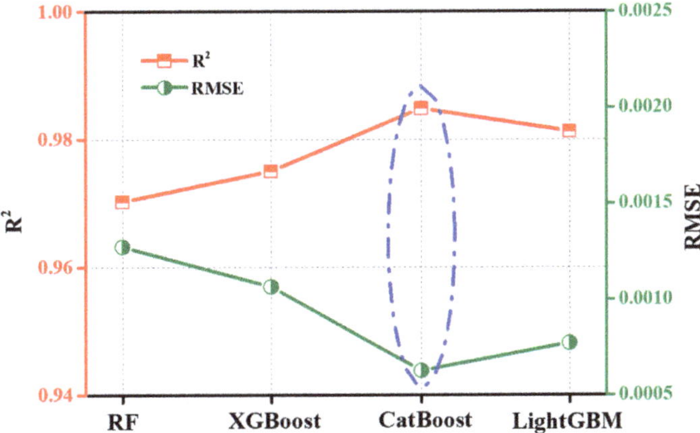

Figure 10. Comparison of R^2 and RMSE values of predicted results for four models.

Figure 11 shows the observed values and the predicted total SPL of aerodynamic noise for the four models. The predicted total SPL of aerodynamic noise based on the CatBoost algorithm had only 0.37% error compared with the observed value, which was the smallest error among the four models, indicating that the model established by applying the CatBoost algorithm had the highest prediction accuracy.

The predicted results of the CatBoost model based on 2155 sets of validation datasets were compared with the observed values. The comparison results for three randomly selected operating condition points are shown in Figure 12. The slanted straight line indicated the degree of fit between the predicted and observed values. In this study, 60,000 r/min, 90,000 r/min and 110,000 r/min were chosen to represent the low, medium and high speeds of the compressor, respectively. The CatBoost model had a high prediction accuracy under all three operating conditions. Compared with the medium and high speeds, the CatBoost model had the highest prediction accuracy under low-speed conditions (60,000 r/min) with an R^2 and RMSE of 0.997237 and 1.290883, respectively, indicating that the CatBoost model could accurately capture and predict the nonlinear relationship between the SPL of the compressor's aerodynamic noise and different input features.

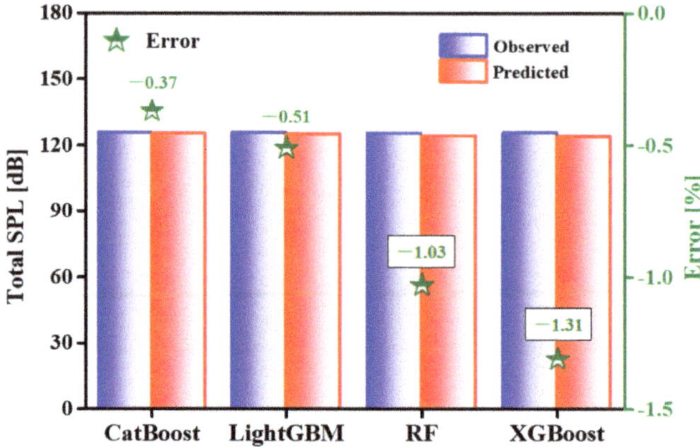

Figure 11. Comparison of total SPL of predicted and observed values for four models.

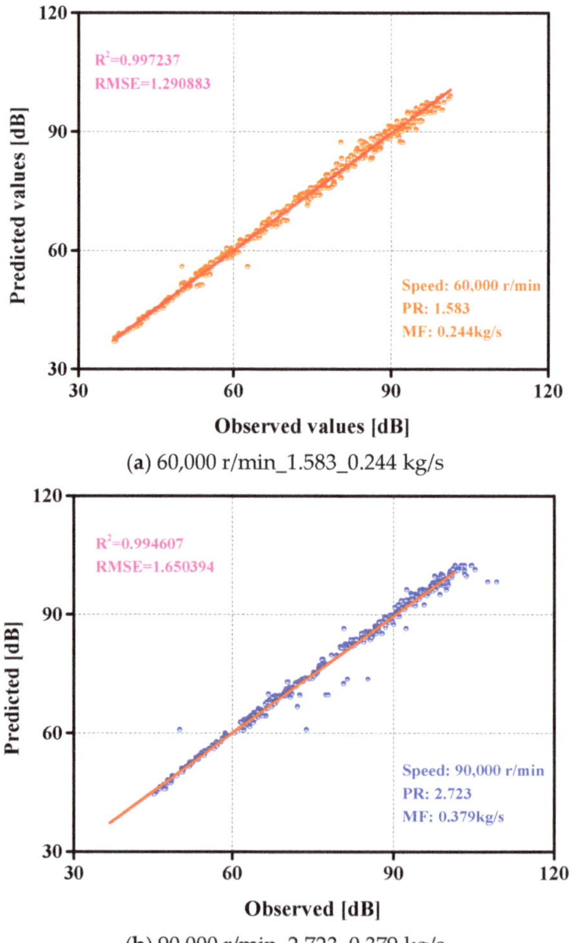

(**a**) 60,000 r/min_1.583_0.244 kg/s

(**b**) 90,000 r/min_2.723_0.379 kg/s

Figure 12. *Cont.*

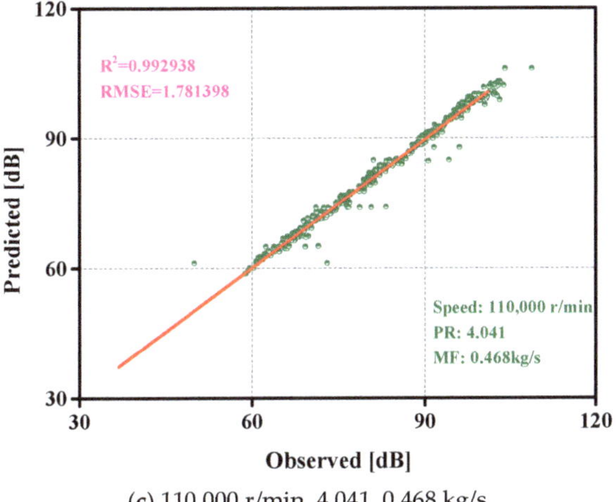

(**c**) 110,000 r/min_4.041_0.468 kg/s

Figure 12. The modeling results of the CatBoost model on the validation dataset.

3.4. Predicted Results and Model Interpretation of CatBoost Model

3.4.1. Predicted Results

Related studies have shown that the blade passing frequency (BPF) noise is one of the main noise components of a compressor's aerodynamic noise. The BPF is calculated as follows:

$$f_B = \frac{nZ}{60} \tag{4}$$

where n and Z are the compressor speed and the number of blade sets, respectively.

Figure 13 shows the predicted results of aerodynamic noise at the untested operating points in the MAP diagram of the compressor. The SPL of aerodynamic noise decreased with an increase in the frequency at different operating condition points, which was consistent with the trend in the observed values. It was further observed that for all predicted points, there was one peak at the BPF, and the peak was more obvious as the speed increased. This indicates that the model based on the CatBoost algorithm could well predict the acoustic information at specific frequencies.

From the above analysis, it can be seen that the prediction model based on the Cat-Boost algorithm could predict the aerodynamic noise for any operating conditions of the compressor and calculate the total SPL. Therefore, the aerodynamic noise MAP diagram could be given correspondingly while predicting the aerodynamic performance of the compressor. Figure 14 shows the noise MAP drawn directly using the observed values and the MAP drawn using the predicted results of the model. Among them, the noise MAP of the observed values consisted of the total SPL for the 21 test conditions, and the noise MAP predicted by the CatBoost model included the 21 observed values and the total SPLs predicted by 21 predicted points. As can be seen from the figure, the compressor aerodynamic noise increased with an increase in the compressor speed. At the same speed, the lowest total SPL of aerodynamic noise was found in the region of medium pressure ratio and medium mass flow rate. The MAP diagrams of aerodynamic noise predicted by the CatBoost model and the observed values are in good agreement, and the locations of the SPL contours are basically the same. In addition, compared with the experimental aerodynamic noise MAP, the noise contours in the predicted MAP are better at characterizing the changes in the total SPL. The maximum and minimum predicted total SPLs were 122.53 dB and 115.42 dB, respectively. Therefore, the comparison in Figure 14 further verifies the feasibility of the model built based on the CatBoost algorithm in the prediction of compressor aerodynamic noise, which could provide an accurate and usable numerical tool for the analysis of compressor aerodynamic noise prediction.

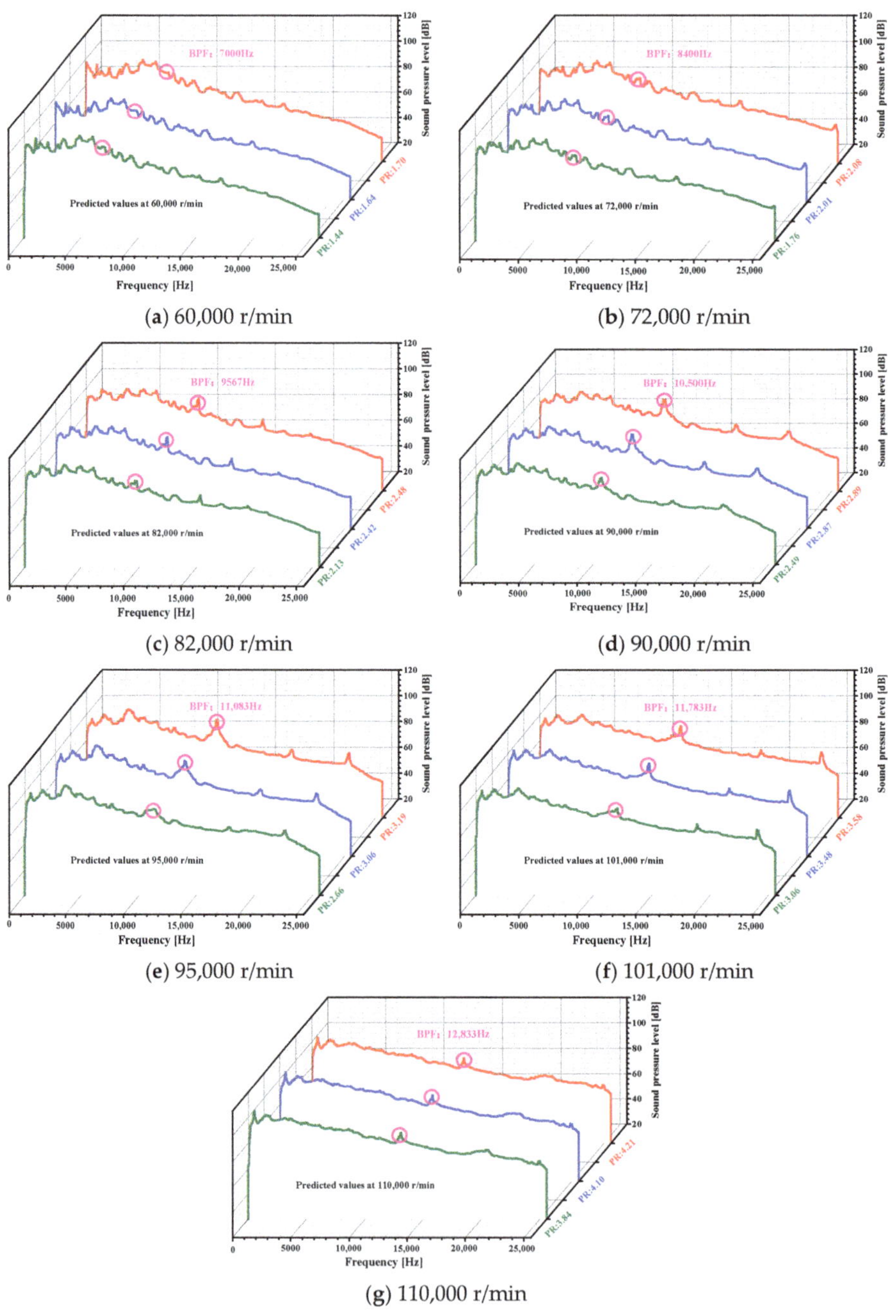

Figure 13. The modeling results of the CatBoost model of the predicted operating conditions.

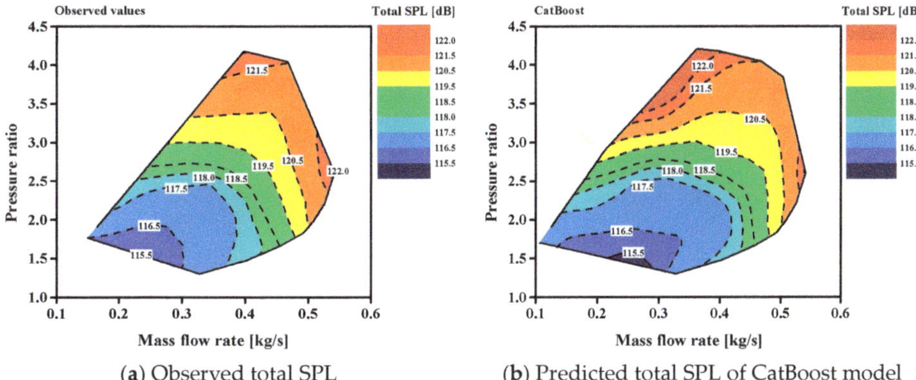

(a) Observed total SPL   **(b)** Predicted total SPL of CatBoost model

Figure 14. Comparison of experimental and CatBoost model predicted total SPL emission clouds for aerodynamic noise.

3.4.2. Interpretation of CatBoost Model Based on SHAP Method

The non-linear relationship between the four input features of the CatBoost-based aerodynamic noise prediction model and the SPL of the aerodynamic noise was revealed by the SHAP method. The results were extracted from a Python SHAP library. Figure 15 shows the effect of changing the input features on the SHAP value of the aerodynamic noise SPL. The color trends of the four input features show that the SPL of aerodynamic noise increased with an increase in the speed, mass flow rate and pressure ratio, and decreased with an increase in frequency. Among them, changing speed had the greatest effect on the change in SPL compared with the mass flow rate and pressure ratio. It was further observed that the SHAP values of the remaining three input features, except frequency, on the SPL of aerodynamic noise, were mainly concentrated around 0. This indicates that under similar operating conditions, the speed, mass flow rate and pressure ratio had less influence on the SPL of aerodynamic noise, while frequency could significantly affect the aerodynamic noise SPL of the compressor.

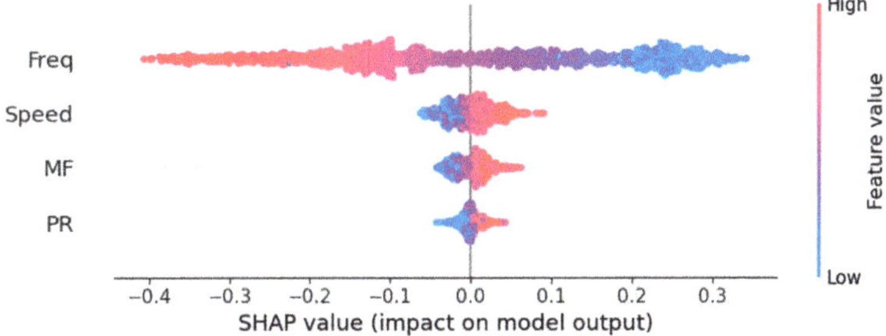

Figure 15. Relationship between the SHAP value and the values of different input features.

The SHAP method was used to further quantify the contribution of the four input features at each operating point of the compressor aerodynamic noise. The CatBoost model was applied to predict the SPL of the aerodynamic noise for one randomly selected data point from 10,773 datasets, and the SHAP method was used to calculate the contribution of the feature values. The calculation results were extracted from a Python SHAP library as shown in Figure 16. $E[f(x)]$ represents the average of the predicted results of all samples and $f(x)$ represents the predicted result of that point. The red color indicates that the feature led to an increase in the SPL, and the blue color indicates that the feature led to a decrease in the SPL. As can be seen from the figure, the frequency, speed and pressure ratio played a role in reducing the SPL of aerodynamic noise. At the same frequency, changing the compressor mass flow rate and speed had a greater effect on the SPL. Therefore, the SHAP method could effectively evaluate and quantify the influences of all features on the SPL during the operation of the compressor, which further increased the credibility of the prediction model for compressor aerodynamic noise based on the CatBoost algorithm.

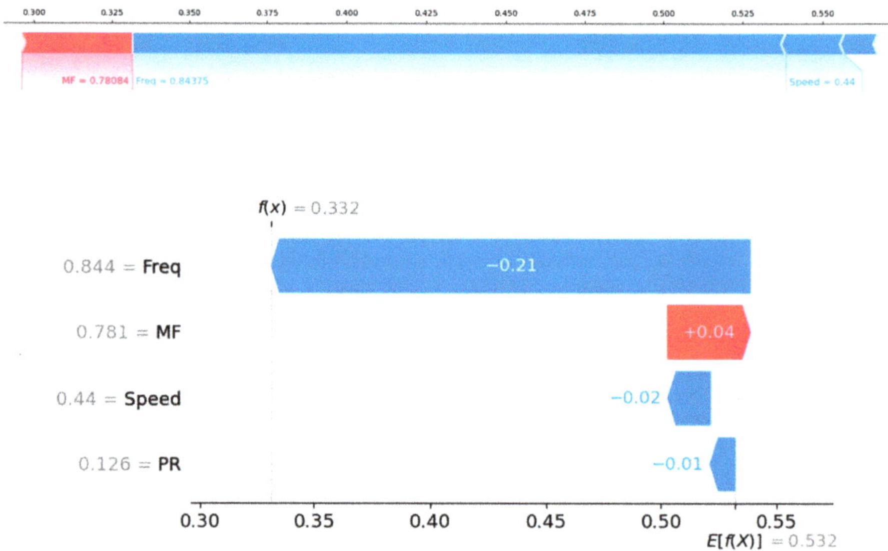

Figure 16. Interpretation of the features contribution of the compressor aerodynamic noise prediction model.

3.5. Comparative Analysis of the Results with the Existing Research Findings

In order to further emphasize the novelty of this study, a detailed comparison was conducted between this study and existing research findings, as shown in Table 7.

Table 7. Comparison of the investigation and a survey of the other existing literature.

Study	Research Object	Research Method	Main Conclusions
Broatch et al. [50]	Centrifugal compressor	Numerical simulation and experiments	A numerical model of a centrifugal compressor was presented to predict a presented peak pressure point.
Hu et al. [51]	Rotating vane compressor	Numerical simulation and experiments	A radiated noise prediction model of a rotary vane compressor was established.
Zhao et al. [52]	A commercial vehicle turbocharger compressor	Numerical simulation	A new one-dimensional prediction model was proposed to predict intake system noise.
Soulat et al. [53]	A single stage compressor	Numerical simulation	The effects of wake modelling on the prediction of broadband noise generated by the impingement of turbulent wakes on a stationary blade row were studied.
Sharma et al. [54]	A turbocharger compressor with ported shroud design	Numerical simulation and experiments	(1) Spectral signatures using statistical and scale-resolving turbulence modelling methods were obtained. (2) Rotating structures through the slot was found to potentially impact the acoustic and vibrational response.

Table 7. *Cont.*

Study	Research Object	Research Method	Main Conclusions
Wen et al. [55]	A centrifugal compressor	Numerical simulation and experiments	(1) A hybrid method based on Morhing acoustic analogy could accurately predict the acoustic radiation of a compressor. (2) Compared with TWPF, AWPF was a more significant source.
Li et al. [56]	A turbocharger compressor	Numerical simulation	(1) The compressor's static pressure values and the sound power values near the impeller outlet were the largest. (2) The noise levels of the inlets of cylindrical and cone types were smaller.
This investigation	A turbocharger centrifugal compressor from a heavy-duty diesel engine	Ensemble learning algorithms, SHAP method and experiments	(1) A prediction method of compressor aerodynamic noise was proposed using the CatBoost algorithm. (2) During the prediction process, the nonlinear relationships between the input features (speed, mass flow rate, pressure ratio and frequency) and the SPL were elaborated upon. (3) The predicted noise MAP was better at characterizing the variation in the total SPL for the aerodynamic noise.

As can been seen from Table 7, the existing literature focused on compressor noise characteristics analysis, and the research methods used in the studies included experimentation, numerical simulation, and a combination of experimental and numerical simulation. The findings mainly included aerodynamic noise characteristics of compressors at specific operating points. However, there were few studies on the aerodynamic noise prediction of centrifugal compressors for engine turbochargers under the entire operating range. In addition, the coupling effect between the influencing parameters (speed, mass flow rate, pressure ratio and frequency) and the total SPL of the compressor was not sufficiently analyzed, especially the contribution of each characteristic toward the SPL. Therefore, the innovation involved in this study was to propose a method that could accurately predict the aerodynamic noise of a turbocharger compressor under arbitrary working conditions based on ensemble learning. In addition, the SHAP algorithm was used to analyze the aerodynamic noise prediction process, which illustrated that the speed, mass flow rate and pressure ratio had little effect on the SPL of the aerodynamic noise, while frequency could significantly affect the SPL. The results of this study could provide a theoretical basis for reducing the aerodynamic noise emissions of compressors and have engineering guidance significance.

4. Conclusions

Environmentally sustainable development plays an important role in human health and social development. The analysis of the aerodynamic noise of turbocharger compressors is significant for reducing noise emissions. In order to accurately evaluate the aerodynamic noise emissions under arbitrary operating conditions of a heavy-duty diesel engine turbocharger compressor, aerodynamic noise experiments on turbocharger compressors and established datasets were conducted in this study. Four ensemble machine learning algorithms (random forest (RF), extreme gradient boosting (XGBoost), categorical boosting (CatBoost) and light gradient boosting machine (LightGBM)) were introduced to establish a compressor aerodynamic noise emission prediction model, and the SHAP algorithm was used to analyze the contribution of input features toward the model results. The main findings were as follows:

- In the compressor aerodynamic noise prediction model, the speed, pressure ratio, mass flow rate and frequency were the important input features. The degree of importance of the input features calculated based on the SHAP algorithm was frequency > speed > mass flow rate > pressure ratio in descending order. Compared with RF, the SHAP values of speed were above 0.02 in all three models of XGBoost, CatBoost and LightGBM, indicating that speed had some influence on the output results of the prediction models.
- The compressor aerodynamic noise model based on the CatBoost algorithm had the best prediction performance with the largest R^2 and the smallest RMSE with the values of 0.984798 and 0.000628, respectively. In addition, among the four models, the CatBoost model had the smallest error between the total SPL of aerodynamic noise and the observed value, which was only 0.37%.
- The CatBoost model had a high prediction accuracy at different operating points of the compressor. The predicted aerodynamic noise MAP from the CatBoost model and the experimental noise MAP were in good agreement, and the SPL contour locations were basically the same. In addition, compared with the experimental noise MAP, the predicted noise MAP was better at characterizing the variation in the total SPL of the aerodynamic noise.
- The analysis of the input characteristics of the prediction model based on the SHAP algorithm showed that the frequency and the SPL were negatively correlated. The speed, mass flow rate and pressure ratio and the SPL showed a positive correlation. In addition, the effects of the speed, mass flow rate and pressure ratio on the SPL were small, while frequency could significantly affect the SPL of the compressor.
- The prediction model of compressor aerodynamic noise established by applying the CatBoost algorithm could accurately evaluate aerodynamic noise under arbitrary operating conditions and provide data and theoretical support for realizing the control of aerodynamic noise emissions, contributing to environmentally sustainable development.

Author Contributions: Methodology, Q.W.; Software, P.Q.; Formal analysis, R.H. and X.S.; Investigation, R.H. and P.Q.; Resources, J.N. and Q.W.; Writing—original draft, R.H. and P.Q.; Writing—review & editing, R.H. and Q.Y.; Supervision, Q.W. and X.S.; Project administration, J.N.; Funding acquisition, J.N. and Q.W. All authors have read and agreed to the published version of the manuscript.

Funding: This research was funded by National Natural Science Foundation of China Youth Science Foundation project (Grant No. 22102116) and the State Key Laboratory of Internal Combustion Engine Reliability Open Subject Foundation of China (Grant No. skler-202114).

Institutional Review Board Statement: Not applicable.

Informed Consent Statement: Not applicable.

Data Availability Statement: The data presented in this study are available in the main text of the article.

Acknowledgments: The team of authors acknowledges anonymous reviewers for their feedback, which certainly improved the clarity and quality of this paper.

Conflicts of Interest: The authors declare that they have no known competing financial interest or personal relationship that could have appeared to influence the work reported in this paper.

References

1. Ahmed, S.A.; Zhou, S.; Zhu, Y.; Tsegay, A.S.; Feng, Y.; Ahmad, N.; Malik, A. Effects of Pig Manure and Corn Straw Generated Biogas and Methane Enriched Biogas on Performance and Emission Characteristics of Dual Fuel Diesel Engines. *Energies* **2020**, *13*, 889. [CrossRef]
2. Huang, R.; Ni, J.; Cheng, Z.; Wang, Q.; Shi, X.; Yao, X. Assessing the effects of ethanol additive and driving behaviors on fuel economy, particle number, and gaseous emissions of a GDI vehicle under real driving conditions. *Fuel* **2021**, *15*, 121642. [CrossRef]
3. Ahmed, S.A.; Zhou, S.; Zhu, Y.; Feng, Y.; Malik, A.; Ahmad, N. Influence of Injection Timing on Performance and Exhaust Emission of CI Engine Fuelled with Butanol-Diesel Using a 1D GT-Power Model. *Processes* **2019**, *7*, 299. [CrossRef]
4. Wang, J.; Shen, L.; Bi, L.; Lei, J. Influences of a variable nozzle turbocharger on the combustion and emissions of a light-duty diesel engine at different altitudes. *Fuel* **2023**, *349*, 128626. [CrossRef]
5. Bao, J.; Qu, P.; Wang, H.; Zhou, C.; Zhang, L.; Shi, C. Implementation of various bowl designs in an HPDI natural gas engine focused on performance and pollutant emissions. *Chemosphere* **2022**, *303*, 135275.
6. Shi, C.; Chai, S.; Di, L.; Ji, C.; Ge, Y.; Wang, H. Combined experimental-numerical analysis of hydrogen as a combustion enhancer applied to Wankel engine. *Energy* **2023**, *263*, 125896. [CrossRef]
7. Broatch, A.; Ruiz, S.; García-Tíscar, J.; Roig, F. On the influence of inlet elbow radius on recirculating backflow, whoosh noise and efficiency in turbocharger compressors. *Exp. Therm. Fluid* **2018**, *96*, 224–233.
8. Mao, Y.; Fan, C.; Zhang, Z.; Song, S.; Xu, C. Control of noise generated from centrifugal refrigeration compressor. *Mech. Syst. Signal Process.* **2021**, *152*, 107466. [CrossRef]
9. Chen, S.; Zuo, S.; Wu, Z.; Liu, C. Comprehensive vibro-acoustic characteristics and mathematical modeling of electric high-speed centrifugal compressor surge for fuel cell vehicles at various compressor speeds. *Mech. Syst. Signal Process.* **2022**, *178*, 109311.
10. Abom, M.; Kabral, R. Turbocharger noise—Generation and control. *SAE Tech. Pap.* **2014**. [CrossRef]
11. Sharma, S.; Broatch, A.; García-Tíscar, J.; Allport, J.; Nickson, A.K. Acoustic characterisation of a small high-speed centrifugal compressor with casing treatment: An experimental study. *Aerosp. Sci. Technol.* **2019**, *95*, 105518. [CrossRef]
12. Karim, A.; Miazgowicz, K.; Lizotte, B.; Abdelkrim, Z. Computational Aero-Acoustics Simulation of Compressor Whoosh Noise in Automotive Turbochargers. *SAE Tech. Pap.* **2013**. [CrossRef]
13. Wolfram, D.; Carolus, T. Experimental and numerical investigation of the unsteady flow field and tone generation in an isolated centrifugal fan impeller. *J. Sound Vib.* **2010**, *329*, 4380–4397. [CrossRef]
14. Zhang, Y.; Xu, S.; Wan, Y. Performance improvement of centrifugal compressors for fuel cell vehicles using the aerodynamic optimization and data mining methods. *Int. J. Hydrog. Energy* **2020**, *45*, 11276–11286. [CrossRef]
15. Chen, C.; Wang, Z.; Du, L.; Sun, D.; Sun, X. Simulating unsteady flows in a compressor using immersed boundary method with turbulent wall model. *Aerosp. Sci. Technol.* **2021**, *115*, 106834. [CrossRef]
16. Raitor, T.; Neise, W. Sound generation in centrifugal compressors. *J. Sound Vib.* **2008**, *314*, 738–756. [CrossRef]
17. Figurella, N.; Dehner, R.; Selamet, A.; Kevin, T.; Keith, M.; Robert, M. Noise at the mid to high flow range of a turbocharger compressor. *Noise Control Eng. J.* **2014**, *62*, 306–312. [CrossRef]
18. Sun, D.; Li, J.; Xu, R.; Dong, X.; Zhao, D.; Sun, X. Effects of the foam metal casing treatment on aerodynamic stability and aerocoustic noise in an axial flow compressor. *Aerosp. Sci. Technol.* **2021**, *115*, 106793.
19. Zhang, M.; Dong, X.; Jia, L.; Sun, D.; Sun, X. Effect of differential tip clearance on the performance and noise of an axial compressor. *Aerosp. Sci. Technol.* **2023**, *132*, 108070. [CrossRef]
20. Galindo, J.; Tiseira, A.; Navarro, R.; Tarí, D.; Meano, C.M. Effect of the inlet geometry on performance, surge margin and noise emission of an automotive turbocharger compressor. *Appl. Therm. Eng.* **2017**, *110*, 875–882. [CrossRef]
21. Stefano, F.; Stefano, P.; Giuseppe, C. CFD analysis of the acoustic behavior of a centrifugal compressor for high performance engine application. *Energy Procedia* **2014**, *45*, 759–768.
22. Sun, H.; Lee, S. Numerical prediction of centrifugal compressor noise. *J. Sound Vib.* **2004**, *269*, 421–430. [CrossRef]
23. Liu, Q.; Qi, D.; Mao, Y. Numerical calculation of centrifugal fan noise. *Proc. Inst. Mech. Eng.-Part C* **2006**, *220*, 1167–1177. [CrossRef]
24. Khelladi, S.; Kouidri, S.; Bakir, F.; Rey, R. Predicting tonal noise from a high rotational speed centrifugal fan. *J. Sound Vib.* **2008**, *313*, 113–133. [CrossRef]
25. Laborderie, J.; Soulat, L.; Moreau, S. Rediction of noise sources in axial compressor from URANS simulation. *J. Propuls. Power* **2014**, *30*, 1257–1271. [CrossRef]
26. Lu, H.; Xiao, Y.; Liu, Z.; Yuan, Y. Simulation and experimental research on aerodynamic noise of gas turbine 1.5-stage axial compressor. *Appl. Acoust.* **2022**, *192*, 108722. [CrossRef]
27. Zhang, J.; Chu, W.; Zhang, H.; Wu, Y.; Dong, X. Numerical and experimental investigations of the unsteady aerodynamics and aero-acoustics characteristics of a backward curved blade centrifugal fan. *Appl. Acoust.* **2016**, *110*, 256–267. [CrossRef]
28. Huang, H.; Lim, T.; Wu, J.; Ding, W.; Pang, J. Multitarget prediction and optimization of pure electric vehicle tire/road airborne noise sound quality based on a knowledge- and data-driven method. *Mech. Syst. Signal Process.* **2023**, *197*, 110361. [CrossRef]
29. Aygun, H.; Dursun, O.; Toraman, S. Machine learning based approach for forecasting emission parameters of mixed flow turbofan engine at high power modes. *Energy* **2023**, *271*, 127026. [CrossRef]
30. Barbado, A.; Corcho, S. Interpretable Machine Learning Models for Predicting and Explaining Vehicle Fuel Consumption Anomalies. *Eng. Appl. Artif. Intell.* **2020**, *115*, 105222. [CrossRef]

31. Cunha, B.Z.; Droz, C.; Zine, A.-M.; Foulard, S.; Ichchou, M. A Review of Machine Learning Methods Applied to Structural Dynamics and Vibroacoustic. *Mech. Syst. Signal Process.* **2023**, *200*, 110535. [CrossRef]
32. Wen, L.; Yuan, X. Forecasting CO_2 emissions in Chinas commercial department, through BP neural network based on random forest and PSO. *Sci. Total Environ.* **2020**, *718*, 137194. [CrossRef] [PubMed]
33. Ding, X.; Feng, C.; Yu, P.; Li, K.; Chen, X. Gradient boosting decision tree in the prediction of NO_X emission of waste incineration. *Energy* **2023**, *718*, 137194. [CrossRef]
34. Thongthammachart, T.; Araki, S.; Shimadera, H.; Matsuo, M.; Kondo, M. Incorporating Light Gradient Boosting Machine to land use regression model for estimating NO_2 and $PM_{2.5}$ levels in Kansai region, Japan. *Environ. Model. Softw.* **2022**, *155*, 105447. [CrossRef]
35. Nguyen, N.; Tong, K.T.; Lee, S.; Karamanli, A.; Vo, T.P. Prediction compressive strength of cement-based mortar containing metakaolin using explainable Categorical Gradient Boosting model. *Eng. Struct.* **2022**, *269*, 114768. [CrossRef]
36. Wang, D.; Thunéll, S.; Lindberg, U.; Jiang, L.; Trygg, J.; Tysklind, M.; Souihi, N. A machine learning framework to improve effluent quality control in wastewater treatment plants. *Sci. Total Environ.* **2021**, *784*, 147138. [CrossRef] [PubMed]
37. Amiri, S.; Mueller, M.; Hoque, S. Investigating the application of a commercial and residential energy consumption prediction model for urban Planning scenarios with Machine Learning and Shapley Additive explanation methods. *Energy Build.* **2023**, *287*, 112965. [CrossRef]
38. Lundberg, S.; Lee, S. Unified Approach to Interpreting Model Predictions. In Proceedings of the Advances in Neural Information Processing Systems 30 (NIPS 2017), Long Beach, CA, USA, 4–9 December 2017.
39. Huang, R.; Ni, J.; Fan, H.; Shi, X.; Wang, Q. Investigating a New Method-Based Internal Joint Operation Law for Optimizing the Performance of a Turbocharger Compressor. *Sustainability* **2023**, *15*, 990. [CrossRef]
40. Huang, R.; Ni, J.; Wang, Q.; Shi, X.; Yin, Q. Experimental and Mechanism Study of Aerodynamic Noise Emission Characteristics from a Turbocharger Compressor of Heavy-Duty Diesel Engine Based on Full Operating Range. *Sustainability* **2023**, *15*, 11300. [CrossRef]
41. Torregrosa, A.J.; Broatch, A.; Navarro, R.; García-Tíscar, J. Acoustic characterization of automotive turbocompressors. *Int. J. Eng. Res.* **2015**, *16*, 31–37. [CrossRef]
42. Liu, C.; Cao, Y.; Zhang, W.; Ming, P.; Liu, Y. Numerical and experimental investigations of centrifugal compressor BPF noise. *Appl. Acoust.* **2019**, *150*, 290–301. [CrossRef]
43. Wei, N.; Men, Z.; Ren, C.; Jia, Z.; Zhang, Y.; Jin, J.; Chang, J.; Lv, Z.; Guo, D.; Yang, Z.; et al. Applying machine learning to construct braking emission model for real-world road driving. *Environ. Int.* **2022**, *166*, 107386. [CrossRef] [PubMed]
44. Robinson, C.; Dilkina, B.; Hubbs, J.; Zhang, W.; Guhathakurta, S.; Brown, M.A.; Pendyala, R.M. Machine learning approaches for estimating commercial building energy consumption. *Appl. Energy* **2017**, *208*, 889–904. [CrossRef]
45. Ke, G.; Meng, Q.; Finley, T.; Wang, T.; Chen, W.; Ma, W.; Ye, Q.; Liu, T. LightGBM: A Highly Efficient Gradient Boosting Decision Tree. In Proceedings of the 31st Conference on Neural Information Processing Systems (NIPS 2017), Long Beach, CA, USA, 4–9 December 2017.
46. Xu, J.; Saleh, M.; Hatzopoulou, M. A machine learning approach capturing the effects of driving behaviour and driver characteristics on trip-level emissions. *Atmos. Environ.* **2020**, *224*, 117311. [CrossRef]
47. Zuo, S.; He, H.; Wu, X.; Wei, K. Numerical analysis on effects of inlet bent pipe's position on a centrifugal compressor's aerodynamic noise. *Appl. Mech. Mater.* **2014**, *633–634*, 1196–1201. [CrossRef]
48. He, Z.; Han, Y.; Chen, W.; Zhou, M.; Xing, Z. Noise control of a two-stage screw refrigeration compressor. *Appl. Acoust.* **2020**, *167*, 107383. [CrossRef]
49. Xu, C.; Zhou, H.; Mao, Y. Analysis of vibration and noise induced by unsteady flow inside a centrifugal compressor. *Aerosp. Sci. Technol.* **2020**, *107*, 106286. [CrossRef]
50. Broatch, A.; Galindo, J.; Navarro, R.; García-Tíscar, J. Methodology for experimental validation of a CFD model for predicting noise generation in centrifugal compressors. *Int. J. Heat Fluid Flow* **2014**, *50*, 134–144. [CrossRef]
51. Hu, X.; He, Z.; Yang, J.; Tao, P.; Hu, L.; Sun, S. Prediction and experimental study of radiated noise of rotating vane compressor under compound effects of multiple sources excitations. *Int. J. Refrig.* **2022**, *142*, 19–26. [CrossRef]
52. Zhao, Y.; Cai, Y.; Zhou, Z.; Xu, Z.; Chen, S. Research on Intake System Noise Prediction and Analysis for a Commercial Vehicle with Air Compressor Model. *SAE Tech. Pap.* **2023**. [CrossRef]
53. Soulat, L.; Moreau, S.; Posson, H. Wake model effects on the prediction of turbulence-interaction broadband noise in a realistic compressor stage. In Proceedings of the 41st AIAA Fluid Dynamics Conference and Exhibit 2011, Honolulu, HI, USA, 27–30 June 2011.
54. Sharma, S.; Broatch, A.; García-Tíscar, J.; Allport, J.M.; Nickson, A.K. Acoustic characteristics of a ported shroud turbocompressor operating at design conditions. *Int. J. Eng. Res.* **2020**, *21*, 1454–1468. [CrossRef]
55. Wen, H.; Wang, J.; Liu, H.; Qian, X.; Cui, Y.; Hong, L. Test and prediction on the acoustic radiation of centrifugal compressor using hybrid method based on Morhing acoustic analogy. *J. Ocean. Eng. Mar. Energy* **2023**, *9*, 193–201. [CrossRef]
56. Li, H.; Sun, Z.; Peng, X. Turbocharger Noise Prediction Using Broadband Noise Source Model. *J. Beijing Inst. Technol.* **2010**, *19*, 312–317.

sustainability

Article

Experimental and Mechanism Study of Aerodynamic Noise Emission Characteristics from a Turbocharger Compressor of Heavy-Duty Diesel Engine Based on Full Operating Range

Rong Huang [1], Jimin Ni [1], Qiwei Wang [1,*], Xiuyong Shi [1] and Qi Yin [2]

[1] School of Automotive Studies, Tongji University, Shanghai 201804, China; huangrong0801@163.com (R.H.); njmwjyx@hotmail.com (J.N.); shixy@tongji.edu.cn (X.S.)

[2] SAIC Motor, General Institute of Innovation Research and Development, Shanghai 201804, China; yinqi@saicmotor.com

* Correspondence: wangqiwei1987@hotmail.com; Tel./Fax: +86-021-6958-9980

Abstract: Heavy-duty diesel engines equipped with turbochargers is an effective way to alleviate energy shortage and reduce gas emissions, but their compressor aerodynamic noise emissions have become an important issue that needs to be addressed urgently. Therefore, to study the aerodynamic noise emission characteristics of a compressor during the full operating range, experimental and numerical simulation methods were used to analyze the aerodynamic noise emissions. The results showed that aerodynamic noise's total sound pressure level (SPL) increased with increased speed under the test conditions. At low speeds, the total SPL of aerodynamic noise was affected by the mass flow of the compressor more obviously. The maximum difference of aerodynamic noise total SPL was 1.55 dB at 60,000 r/min under different mass flows. At the same speed, the compressor could achieve lower aerodynamic noise emissions by operating in the high-efficiency region (middle mass flows). In the compressor aerodynamic noises, the blade passing frequency (BPF) noise played a dominant role. The transient acoustic-vibration spectral characteristics and fluctuation pressure analysis indicated that BPF and its harmonic frequency noises were mainly caused by the unsteady fluctuation pressure. As the speed increased, the BPF noise contributed more to the total SPL of the aerodynamic noise, and its percentage was up to 75.35%. The novelty of this study was the analysis of the relationship between compressor aerodynamic noise and internal flow characteristics at full operating conditions. It provided a theoretical basis for reducing the heavy-duty diesel engine turbocharger compressor aerodynamic noise emissions.

Keywords: turbocharger compressor; aerodynamic noise; BPF noise; acoustic-vibration characteristics; unsteady fluctuation pressure; dynamic-static interferences

Citation: Huang, R.; Ni, J.; Wang, Q.; Shi, X.; Yin, Q. Experimental and Mechanism Study of Aerodynamic Noise Emission Characteristics from a Turbocharger Compressor of Heavy-Duty Diesel Engine Based on Full Operating Range. *Sustainability* **2023**, *15*, 11300. https://doi.org/10.3390/su151411300

Academic Editor: Luca Cioccolanti

Received: 26 June 2023
Revised: 11 July 2023
Accepted: 18 July 2023
Published: 20 July 2023

1. Introduction

Turbochargers can increase the specific power output of internal combustion engines and reduce gas emissions. Therefore, turbochargers are widely used in transportation [1–5]. However, in addition to the combustion and mechanical noise in the engine cylinder, the noise generated by the turbocharger becomes a non-negligible part of the engine noise source [6]. Due to the increase in the output power requirements of diesel engines, the turbocharger pressure ratio increases, leading to a rise in the compressor load and higher aerodynamic noise emissions [7,8]. In the existing literature, aerodynamic noise is considered the main noise source of turbochargers [9,10]. In recent years, excessive compressor aerodynamic noise emissions have become an important issue that needs to be addressed in heavy-duty diesel engine turbochargers.

The experimental method is one of the main technical approaches to studying the aerodynamic noise of a compressor. During the noise experiments, the sound spectrum is obtained by measuring the noise at the operating points of the compressor, thus visualizing the noise characteristics of the compressor [11]. Many fruitful works have been carried out by scholars. Li et al. [12] conducted experimental noise research on a turbocharger bench. Using an acoustic array, they measured the surface radiated noise of a gasoline engine turbocharger. The results showed that the compressor aerodynamic noise was the main source of turbocharger noise. Raitor et al. [13] studied the main noise sources of a centrifugal compressor and found that blade passing frequency (BPF), buzz-saw, and tip clearance noise were the main noise sources. Figurella et al. [14] showed that the discrete noise could be observed in the BPF of a compressor and its harmonic frequencies. Zhang et al. [15] explored the influence of differential tip clearance on the performance and noise of an axial compressor adopting the experiment method. The results indicated that the compressor's sound pressure level (SPL) was lowest at a relatively small tip clearance rather than zero. In addition, Galindo et al. [16] conducted experiments to explore the effect of inlet geometry on noise emissions of an automotive turbocharger compressor. A convergent-divergent nozzle could significantly improve surge margin and reduce noise emissions. In summary, there is still a lack of experimental research on heavy-duty diesel engine turbocharger compressors, and it is essential to carry out experimental studies on the aerodynamic noise of compressors in the full operating range to reduce the aerodynamic noises of compressors.

A review of the available literature showed that two methods are used for compressor noise measurements, but there are some problems when using these methods. The first method is based on radiated noise, measured using a microphone in an anechoic environment. However, the disadvantage is that it is difficult to distinguish between different noise components. Another method is to use pressure transducers to measure noise in the compressor duct, but this method relies on high-precision measurement equipment and has high test costs [17]. In addition, the experimental method is usually used to improve the acoustic performance of the compressor and cannot explore the relationship between the compressor's aerodynamic noise and flow characteristics. To remedy these shortcomings, scholars have used reliable numerical computational methods to conduct noise simulation studies on compressors.

Numerical simulation of compressor noise usually couples computational fluid dynamics (CFD) and computational aerodynamic acoustics (CAA) methods [18–20]. Dehner et al. [21] studied the noise emissions of a turbocharger centrifugal compressor from a spark-ignition engine using the experiments, CFD and modal decomposition methods. The results showed the whooshing noise primarily propagated along the duct in acoustic azimuthal modes. Galindo et al. [22] investigated the influence of four inlet geometries on compressor performances (noise emissions, compressor surge margin and efficiency). They found that a convergent nozzle could strongly reduce the intake orifice noise. In addition, Galindo et al. [23] explored the impact of tip clearance on noise generation and flow behavior of centrifugal compressors in near-surge conditions. The results showed that there were no significant changes in compressor acoustic signature when varying the tip clearance in near-surge conditions. Sundström et al. [24] predicted the flow field and characterized the acoustic near-field generation and propagation under a centrifugal compressor's stable and near-surge operating conditions. They found that an amplified broadband feature at two times the frequency of the rotating order of the shaft was captured under the near-surge condition.

Furthermore, Sundström et al. [25] conducted a numerical simulations based on the large eddy simulation method to explore vaneless diffuser rotating stall instability in a centrifugal compressor. Jyothishkumar et al. [26] characterized the flow structures and the associated instabilities near the stall point (before the surge). They found that there existed a flow-acoustics coupling at near-surge operating conditions. Broatch et al. [27] conducted experimental and numerical simulation investigations to analyze fluid phenomena related to whoosh noise under near-surge conditions. The results showed that a broadband noise in 1–3 kHz frequency band was detected in the experimental measurements during the simulated conditions. This whoosh noise was also captured by the numerical model.

Guo et al. [28] presented a numerical simulation of the stall flow phenomenon inside a turbocharger centrifugal compressor with a vaneless diffuser. They found a distinct stall frequency at the given compressor speed. In addition, Tomita et al. [29] studied the differences in phenomena of the two compressors with different structures were investigated with experimental and computational methods to reveal internal flow phenomena. However, analysis of the aerodynamic noise and interior flow characteristics of the heavy-duty diesel engine turbocharger compressors based on the full operating range is still lacking. Therefore, a compressor simulation study is needed to analyze the relationship between aerodynamic noise and flow characteristics.

To analyze the aerodynamic noise emission characteristics and mechanism of a heavy-duty diesel turbocharger compressor in the full operating range, the purpose of this study was to obtain the aerodynamic noise emissions law for the full operating range of a compressor for heavy-duty diesel engine using an experimental method and to analyze the relationship between different internal flow characteristics and aerodynamic noise characteristics adopting simulation method. The innovation was to refine the aerodynamic noise mechanism of the compressor for heavy-duty diesel engine turbochargers, propose that blade aerodynamic force and dynamic interference were the main discrete monophonic noise sources, and provide a theoretical basis for reducing the turbocharger compressor aerodynamic noise emissions of the heavy-duty diesel engine. The research framework of this study is shown in Figure 1. The remaining parts were organized as follows: Section 2 was the experimental facility and methods, which introduced the test device, measurement equipment, test conditions and data processing methods. Section 3 was the numerical simulation calculation of the compressor, introducing the model parameters setting and validating the model. Section 4 analyzed the test and simulation results, including the variation law of the aerodynamic noise emissions of the compressor in the full operating range (total SPL, BPF noise, acoustic-vibration characteristics, etc.), and explored the relationship between the aerodynamic noise emissions and unsteady fluctuation pressure, dynamic-static interference effects. In Section 5, the main research results of this study were summarized.

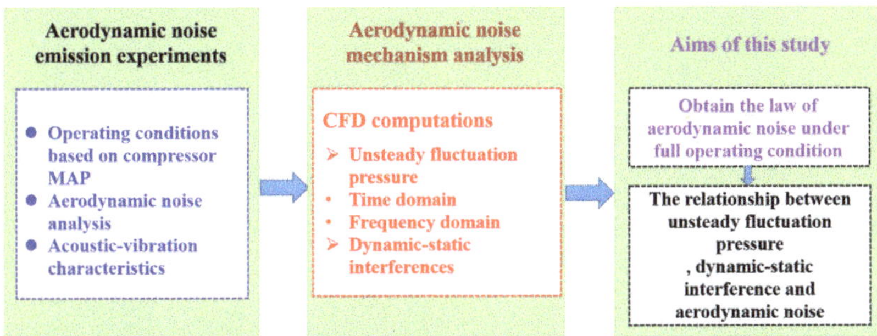

Figure 1. Investigation procedure.

2. Experimental Facility and Methods

2.1. Test System and Measurement Equipment

The compressor aerodynamic noise experiments were conducted on the turbocharger performance test bench, as shown in Figure 2. The compressor noise test system mainly included the turbocharger test bench, PCB-SN152495 type microphone, PCB-HT356B21 type vibration sensor, SIEMENS signal acquisition port and turbocharger test console. Among them, the turbocharger test bench had automatic data acquisition and processing functions, which could realize the precise control of turbocharger speed, compressor and turbine inlet and outlet parameters. The compressor inlet was equipped with pressure and temperature regulators, which could be automatically adjusted in real-time to ensure smooth temperature and pressure at the inlet. The turbine intake system was equipped with oil injection and ignition devices. The measuring range and accuracy of the aerodynamic noise test instruments are shown in Table 1.

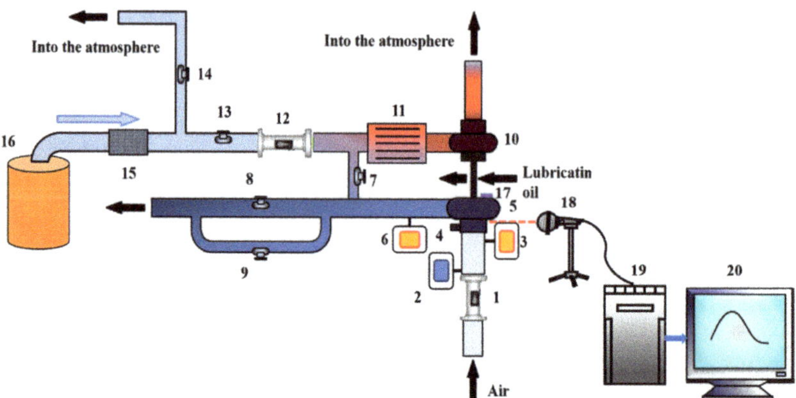

Figure 2. Schematic diagram of compressor aerodynamic noise test device. 1. Compressor inlet flowmeter. 2. Compressor inlet pressure sensor. 3. Compressor inlet temperature sensor. 4. Speed sensor. 5. Compressor. 6. Compressor outlet temperature sensor. 7. Automatic circulating valve. 8. Electric exhaust control valve. 9. Electric trimming valve. 10. Turbine. 11. Burner. 12. Turbine inlet flowmeter. 13. Turbine inlet control valve. 14. Air source vent valve. 15. Filter. 16. Air source. 17. Vibration sensors. 18. Microphone. 19. Signal acquisition port. 20. Computer.

Table 1. Measuring ranges, accuracies, and uncertainties of instruments.

Instruments	Parameters	Measuring Range	Accuracy	Uncertainty
Tachometer	Speed	0~400,000 r/min	0.1 r/min	0.01 r/min
Microphone	Noise	15~165 dB	0.1 dB	±0.02 dB
Vibration sensor	Vibration	±490 m/s^2 pk	1%	±0.2 m/s^2
Temperature sensor	Compressor inlet and outlet temperature	−200~400 °C	0.25 °C	±0.1 °C
Pressure sensor	Compressor inlet pressure	−175~35,000 Pa, −40~85 °C	0.05%	±0.01%
Pressure sensor	Compressor outlet pressure	0~700,000 Pa, −40~85 °C	0.05%	±0.01%
Temperature sensor	Turbine inlet and outlet temperature	−200~1372 °C	0.4%	±0.1 °C
Pressure sensor	Turbine inlet pressure	0~700,000 Pa, −40~85 °C	0.05%	±0.01%
Pressure sensor	Turbine outlet pressure	−175~35,000 Pa, −40~85 °C	0.05%	±0.01%

The aerodynamic noise test operating conditions consisted of the characteristic working points on the MAP diagram of the compressor, as shown in Figure 3. Among them, case 1 was defined as the near-choke region, case 2 as the high-efficiency region, and Case 3 as the near-surge region.

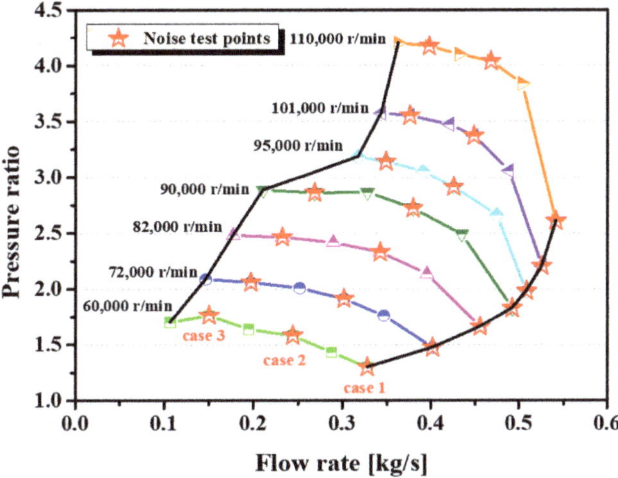

Figure 3. Aerodynamic noise test condition points.

2.2. Experimental Procedure and Data Processing

To ensure that the compressor inlet aerodynamic noise experiments were not affected by other noise sources, the compressor inlet piping was not installed with an inlet muffler. To avoid the effect of the ground and walls on the compressor aerodynamic noise measurement, the turbocharger axis was about 1.4 m from the ground, and the turbocharger case was greater than 1 m from the wall. Before the experiments, the noise of the turbocharger operating environment was measured and calibrated. In the experiments, the ambient noise in the test chamber was 42.5 dB, and the ambient temperature was 23.5 °C. The data collected during the test were processed by Simcenter Testlab 2021.1.

3. Numerical Simulation

3.1. Research Object

The research object of this study was a turbocharger compressor of a heavy-duty diesel engine. The compressor structure uses the splitter blade, and the diffuser uses a bladeless structure. The specific parameters are shown in Table 2. Chen et al. [30] proposed a novel pseudo-MAP method which was a contour map with the performance of only nine compressor characteristic operating points. Therefore, to make the simulation representative, three speeds of 60,000 r/min, 90,000 r/min and 110,000 r/min were chosen to represent the compressor's low, medium, and high speeds. In each speed line, three characteristic working points were selected to characterize the near-choke region (case 1), high-efficiency region (case 2) and near-surge region (case 3), and the total number of simulated working cases were 9 points, and the distribution of calculated working cases are shown in Figure 4.

Table 2. Compressor parameters.

Item	Value
Turbocharger compatibility	Heavy-duty diesel engine
Outlet diameter of impeller (mm)	94.4
Inlet diameter of impeller (mm)	66.46
Main blade number	7
Splitter blade number	7
Diffuser height (mm)	4.77
Design pressure ratio	4.5
Rated speed (r/min)	117,000
Outlet diameter of the diffuser (mm)	166.15
Inlet diameter of the diffuser (mm)	90
Type of cooling	Oil cooling

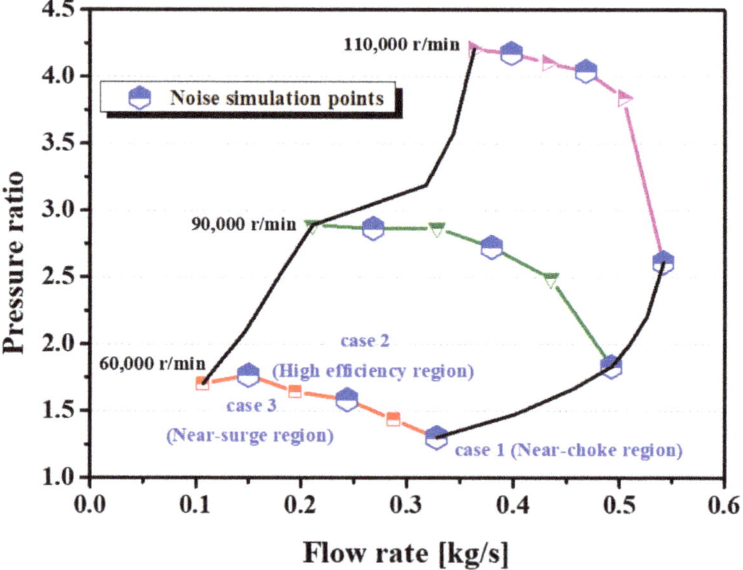

Figure 4. Aerodynamic noise simulation points.

3.2. Model Parameter Setting and Validation

To analyze the unsteady fluctuation pressure phenomenon of the compressor in detail, monitoring points were set in the inlet region, rotor region, diffuser region and volute region, respectively, as shown in Figure 5. In the compressor model setup, the impeller region was set as the rotating domain, and the rest were stationary domains. The intersection between the inlet section and the impeller rotating domain and between the impeller rotating domain and the diffuser domain was adopted as the transient rotor-stator intersection. Considering the complexity of the meshing of each fluid domain, structured meshes were used for the inlet extension, outlet extension and rotor region, and unstructured meshes were used for the rest of the stationary domains. To simulate the surface boundary layer of the blade and hub accurately, the y+ of the first grid layer near the wall was less than 1. The turbulence model adopted the Shear Stress Transport model, and the mathematical model adopted the Reynolds-averaged Navier–Stokes equation system. The model walls were all set as smooth, non-slip adiabatic walls [31]. The fluid was defined as an ideal gas, and the fluid viscosity was set as a function of temperature. The inlet of the compressor was set to the total temperature of 293.15 K and the total pressure of 101.325 kPa, and the outlet was set to the mass flow boundary.

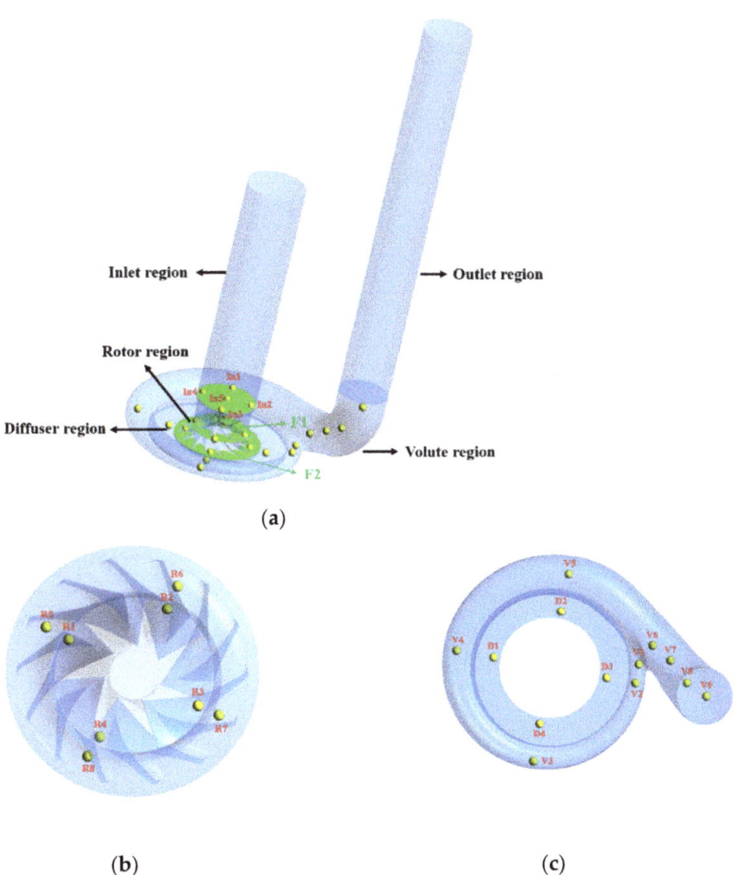

Figure 5. Compressor fluid domain and monitoring points distribution. (**a**) Compressor model (**b**) Rotor region (**c**) Diffuser and volute regions.

During the numerical simulation, the unsteady flow was calculated based on the steady flow simulation. The high resolution with second-order accuracy was selected as the option of the advection scheme. In addition, the second-order backward Euler with second-order accuracy was chosen as the option for the transient scheme.

Blade passing frequency (BPF) noise is the main component of compressor aerodynamic noise. The equation for calculating the blade passing frequency (BPF) is as follows:

$$f_B = \frac{nZ}{60} \tag{1}$$

where n is the compressor speed and Z is the number of blade sets.

The BPF corresponds to the harmonic frequency f_H is calculated by the following formula:

$$f_H = m f_B \ (m = 2, 3, 4, \ldots) \tag{2}$$

where, m is the number of orders.

In this study, the simulated compressor speeds included 60,000 r/min, 90,000 r/min and 110,000 r/min, and their corresponding BPFs were 7000 Hz, 10,500 Hz and 12,833 Hz, respectively. Focusing on the first fifth-order noise, the corresponding maximum frequencies were 35,000 Hz, 52,500 Hz and 64125 Hz, respectively. According to the Nyquist sampling law [17], the simulation time step is calculated as follows:

$$\Delta t \leq \frac{1}{2f_{max}} \tag{3}$$

where, f_{max} is the maximum frequency. Therefore, the time steps were known to be 1.42×10^{-5} s, 9.52×10^{-6} s and 7.79×10^{-6} s, respectively. Due to every 4° rotation of the impeller was a time step that satisfied the computational requirements, the time steps calculated according to the impeller rotation per 4° were about 1.11×10^{-5} s, 7.41×10^{-6} s and 6.06×10^{-6} s for the three speeds of 60,000 r/min, 90,000 r/min and 110,000 r/min, respectively.

In addition, in the transient simulations, the normalized root-mean-square residuals were used to determine convergence and control the termination of coefficient iterations. The sum of residuals to the sum of fluxes for a given variable in all cells must be reduced to less than 1×10^{-6} to ensure convergence of the computations.

Table 3 compares the various numerical parameters critical to modeling compressor aerodynamics and noise generation in this survey with other literature surveys.

Table 3. Comparison of the various numerical parameters critical to modeling compressor aerodynamics and noise generation in this survey with other literature surveys.

Study	Tip Diameter (mm)	Elements (Million)	Wheel Rotation (-)	Turbulence Method (-)	Boundary Conditions		Time Steps (°)
					Inlet	Outlet	
Sundström et al. [24]	88	9	Sliding	LES	Pressure	Mass flow	1
Fontenasi et al. [18]	-	9.6	Sliding	DES	Mass flow	Pressure	0.5
Broatch et al. [27]	48.6	9.6	Sliding	DES	Mass flow	Pressure	1
Karim et al. [9]	-	-	Sliding	LES	Pressure	Mass flow	-
Semlitsch et al. [32]	88	6	Sliding	LES	Mass flow	Pressure	5
Jyothishkumar et al. [26]	88	6	Sliding	LES	Mass flow	Pressure	5
Tomita et al. [29]	50	3.2	-	URANS (k-ε)	-	-	3.6/7.2
Guo et al. [28]	182.8	2.5	Sliding	URANS (k-ε)	Pressure	Mass flow	3
Dehner et al. [21]	-	5.5	Static	DES	Pressure	Mass flow	-
Galindo et al. [22]	-	10	Sliding	DES	Mass flow	Pressure	4
Galindo et al. [23]	-	9.5	Sliding	SST/DES	Mass flow	Pressure	1
Fardafshar et al. [33]	-	-	Static	SAS-SST	-	-	-
This investigation	-	4.9	Sliding	SST	Pressure	Mass flow	4

To improve the calculation accuracy, the mesh independence analysis was conducted in the study based on the Richardson extrapolation method [34]. The equations used by the extrapolation method are presented in Table 4.

Table 4. Equations for Richardson extrapolation method.

Factor	Equation
Representative mesh size h	$h = \left[\frac{1}{N}\sum_{i=1}^{N}(\Delta V_i)\right]^{1/3}$
Grid refinement factor r	$r = \frac{h_{coarse}}{h_{fine}}$
Apparent order p	$p = \frac{1}{ln(r_{21})}\lvert ln\lvert\varepsilon_{32}/\varepsilon_{21}\rvert + q(p)\rvert$
Approximate relative error e_a^{21}	$e_a^{21} = \left\lvert\frac{\phi_1-\phi_2}{\phi_1}\right\rvert$
Extrapolated relative error e_{ext}^{21}	$e_{ext}^{21} = \left\lvert\frac{\phi_{ext}^{12}-\phi_1}{\phi_{ext}^{12}}\right\rvert$
Grid convergence index GCI_{fine}^{21}	$GCI_{fine}^{21} = \frac{1.25e_a^{21}}{r_{21}^p-1}$

Three meshes were used for the independence mesh study: Mesh$_1$ with $N_1 = 6{,}730{,}132$ elements, Mesh$_2$ with $N_2 = 4{,}932{,}456$ elements, and Mesh$_3$ with $N_3 = 3{,}102{,}894$ elements. The design point was selected as the simulation corresponded to a mass flow of 0.2439 kg/s, a speed of 60,000 r/min, an inlet pressure of 101.325 kPa and an inlet temperature of 293.15 K. The discretization errors are listed in Table 5, which contains all the important parameters obtained from the mesh independence study. As can be seen in Table 5, the refinement factors for all grids were greater than 1.3. The variables used to determine grid convergence were the pressure ratio and efficiency, monitored in each simulation.

Table 5. Discretization errors of three meshes.

	ϕ = **Pressure Ratio**	ϕ = **Efficiency**
N_1, N_2, N_3	6,730,132, 4,932,456, 3,102,894	6,730,132, 4,932,456, 3,102,894
r_{31}	2.4192	2.4192
r_{21}	1.8292	1.8292
r_{32}	1.3226	1.3226
ϕ_1	1.5864	0.8146
ϕ_2	1.5841	0.8120
ϕ_3	1.5736	0.8017
p	0.7932	2.2437
ϕ_{ext}^{21}	1.5901	0.8155
e_a^{21}	0.145%	0.3192%
e_{ext}^{21}	0.2327%	0.1101%
GCI_{fine}^{21}	0.295%	0.1387%

Figure 6 shows the Extrapolation results for pressure ratio and efficiency to the number of elements. As can be seen from the figure, there was some error in the calculation results of the N_3 mesh, but the N_2 mesh and N_1 mesh were the same. In addition, the calculation time used in the N_2 mesh was reduced by 33.3% compared to the calculation time implemented for the N_1 mesh. Therefore, Mesh$_2$ comprised N_2 = 4,932,456 elements was selected for the numerical simulation study.

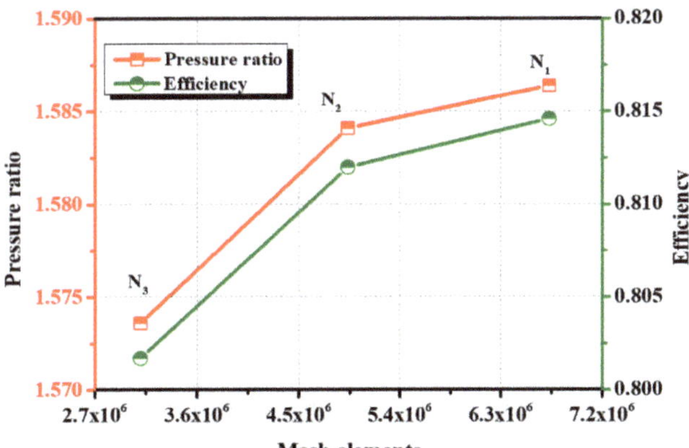

Figure 6. Extrapolation results in pressure ratio and efficiency to several elements at a speed of 60,000 r/min.

Figure 7 compares the experimental and simulated values of the pressure ratio and efficiency of the compressor. At low speeds, the experimental values matched well with the simulated values. For the pressure ratio and efficiency, the maximum errors were 4.89% and 3.92%, respectively, occurring at the 110,000 r/min speed line. This difference was attributed to the simplification of geometry's secondary features, the simulation's heat transfer parameter settings, and manufacturing errors [35–37].

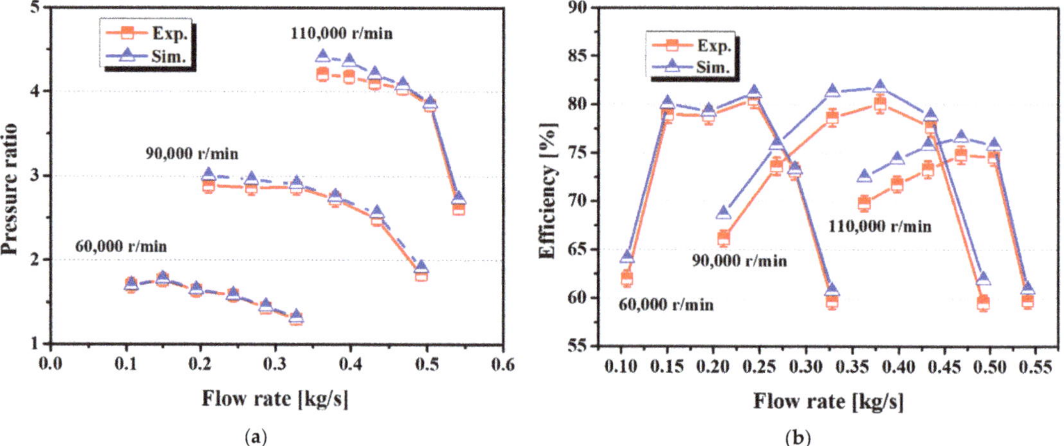

Figure 7. Compressor model test verification. (**a**) Pressure ratio (**b**) Efficiency.

4. Result and Discussions

4.1. Experimental Analysis of Aerodynamic Noise Emission Characteristics

The aerodynamic noise components are shown in Figure 8. In this study, the discrete noise components in the aerodynamic noise of the compressor were investigated, including discrete monotone and multi-monotone noise.

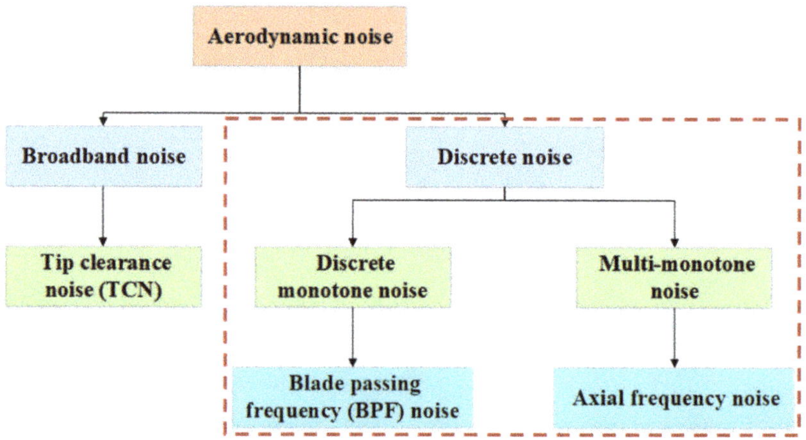

Figure 8. Aerodynamic noise emissions composition.

Figure 9 shows the aerodynamic noise emissions spectrum of the compressor under different operating conditions. The sound pressure level (SPL) of aerodynamic noise decreased with increasing frequency at the same speed. This was consistent with the findings of Zuo et al. [38].

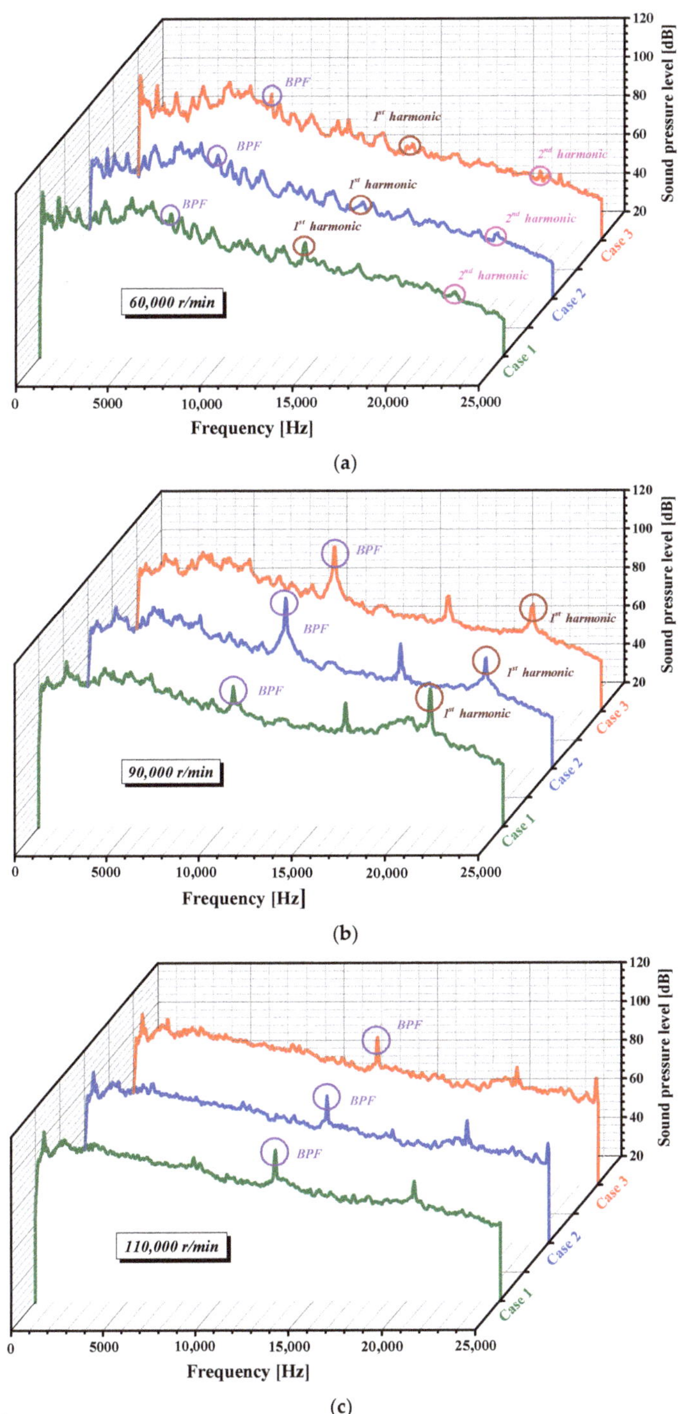

Figure 9. Aerodynamic noise emissions spectral characteristics of the compressor under different operating conditions. (**a**) 60,000 r/min (**b**) 90,000 r/min (**c**) 110,000 r/min.

For the same case, with the compressor speed increased, significant peaks were observed at the BPF and its harmonic frequency (Figure 9b,c). This was because the mutual disturbance frequency between the blade and the airflow increased as the speed increased, increasing the noise at the BPF and its harmonic frequency. In addition, among the noises at the harmonic frequencies corresponding to the BPF, the first-order harmonic noise (1st harmonic) was the largest. As the speed increased, the noise at the harmonic frequency corresponding to the BPF moved toward increasing frequency. Among them, no noise at the harmonic frequency corresponding to BPF was observed in the frequency range of 0 to 25,000 Hz at 110,000 r/min (as shown in Figure 9c). This was because the increase in speed could increase the BPF of the compressor, which led to an increase in its corresponding harmonic frequency.

Figure 10 shows the cloud plot of the aerodynamic noise emissions for the compressor at the full operating range. In the same case, the aerodynamic noise emissions decreased with increased frequency. Among the three cases, the SPL distribution of the aerodynamic noise in the high-efficiency region (case 2) was the smallest overall at the same speed. Specifically, compared with the near-choke region (case 1), the distribution of the SPL for the aerodynamic noise in the range of 10,000 to 20,000 Hz in case 2 was smaller. Compared with the near-surge region (case 3), the distribution of the SPL for the aerodynamic noise in the range of 0 to 5000 Hz in case 2 was also smaller. The main reason was that the airflow deteriorated when the compressor was operated in the near-choke region compared to the high-efficiency region. While the rotational stall phenomenon occurred when the compressor was operated in the near-surge region [39,40], and the airflow turbulence intensity increased, which increased the compressor's aerodynamic noise emissions. Therefore, at the same speed, the compressor operating in the high-efficiency region could achieve lower aerodynamic noise emissions.

To quantitatively analyze the aerodynamic noise of the compressor, the total SPL of the aerodynamic noise and the BPF noise are calculated as shown in Equations (4) and (5) [41]:

$$L_{total} = 10\log\left(\sum_{i=1}^{n} 10^{\frac{L_i}{10}}\right) \tag{4}$$

$$L_{BPF} = 10\log\left(\sum_{i=1}^{m} 10^{\frac{L_m}{10}}\right) \tag{5}$$

where, L_{total}, L_{BPF}, L_i and L_m are the total SPL, SPL of BPF, SPL at fixed frequency point and SPL at BPF and its harmonics, respectively. m and n are the number of BPF harmonic and frequency points.

The total SPL of aerodynamic and BPF noise of the compressor at the full operating range is shown in Figure 11. For the same case, the total SPL of aerodynamic and BPF noise increased as the speed increased. Among them, compared with 60,000 r/min, the total SPLs of aerodynamic noise and BPF noises for case 1, case 2 and case 3 at 110,000 r/min increased by 4.32%, 5.18%, 4.77% and 10.62%, 10.27% and 9.81%, respectively. This was mainly due to the increase in compressor speed, the frequency of mutual disturbance between impeller blades and airflow per unit time increased, and the amplitude of fluctuation pressure increased, which increased aerodynamic noise emissions. At the same speed, the total SPL of each case from high to low was near-choke region (case 1) > near-surge region (case 3) > high-efficiency region (case 2). The reasons are shown above.

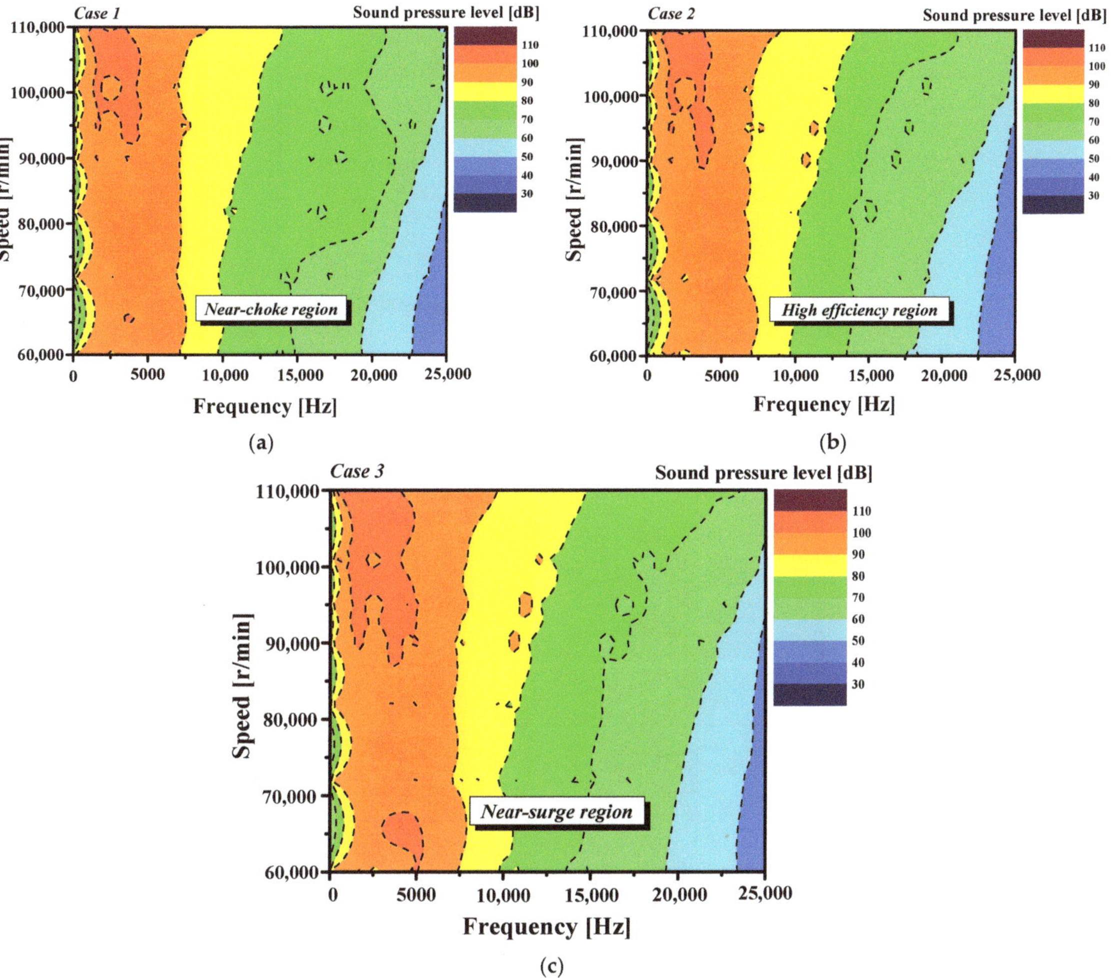

Figure 10. Aerodynamic noise emissions contour clouds of compressors under different operating conditions. (**a**) Case 1 (**b**) Case 2 (**c**) Case 3.

It is further observed that the maximum difference in the total SPL of aerodynamic noise between the three cases was 1.55 dB at 60,000 r/min, while the maximum difference in the total SPL was 0.61 dB at 110,000 r/min. This was because, at low speeds, the change of air inlet volume brought by the change of compressor mass flow occupied the main position in the compressor aerodynamic noise generation. At the same disturbance frequency, the interference effect between the blade and airflow was enhanced. As the speed increased, the effect of compressor mass flow on aerodynamic noise gradually decreased. Therefore, at low speeds, the total SPL of compressor aerodynamic noise was more obviously influenced by the mass flow.

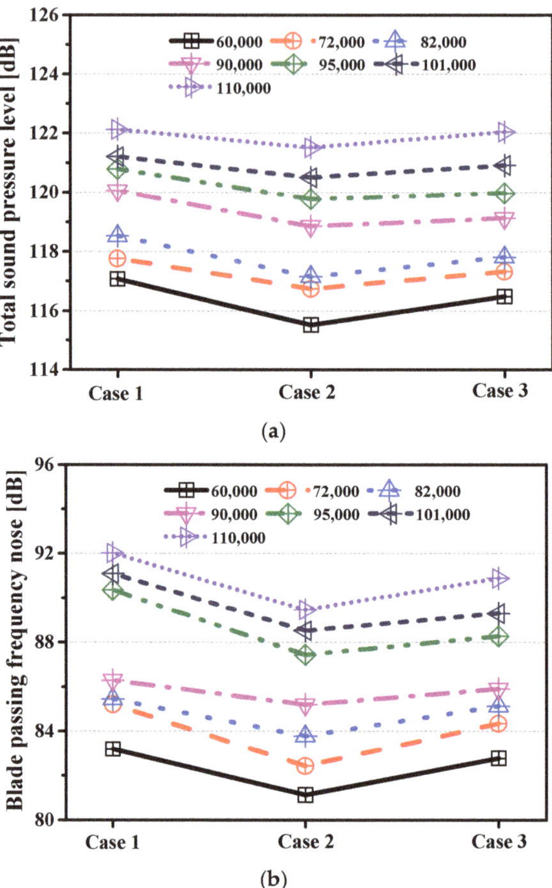

Figure 11. Total sound pressure level and blade passing frequency noise of the compressor under full operating range. (**a**) Total sound pressure level (**b**) BPF nose.

To further analyze the contribution of BPF noise to the total SPL of aerodynamic noise, the proportion of the compressor BPF noise to the total SPL of aerodynamic noise under the full operating range is shown in Figure 12. From the figure, it can be seen that the percentage of BPF noise exceeded 70% in all test conditions. This indicated that the BPF noise was the main noise component in the aerodynamic noise on the inlet side of the compressor. With the speed increased, the percentage of BPF noise increased. Among them, at the speed of 110,000 r/min, the largest proportion of BPF noise was 75.35% for the above reasons. Therefore, in the aerodynamic noise, the compressor speed had a high contribution to the BPF noise.

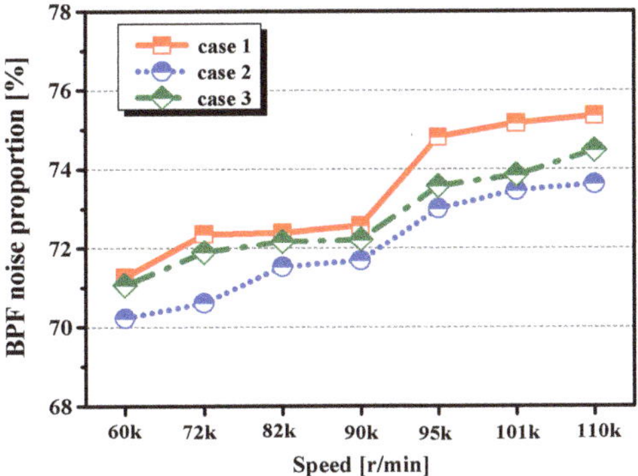

Figure 12. Blade passing frequency noise proportion.

The above analysis shows that the compressor operating in the high-efficiency region (case 2) could achieve lower aerodynamic noise emissions. To further analyze the aerodynamic noise emissions law in the high-efficiency region of the compressor, the transient acoustic-vibration spectrum distribution in the high-efficiency region at different speeds is shown in Figure 13. It can be seen that obvious peaks of aerodynamic noise and vibration acceleration were observed at the BPF of the compressor and its harmonic frequency at different speeds, and the peaks were more obvious as the speed increased.

Further observations revealed that more pronounced noise peaks were observed in the 1000 to 5000 Hz range, while no significant peaks were found in the vibration acceleration spectrum corresponding to the frequencies. This was because the noise corresponding to the frequencies below the BPF was caused by the secondary flow through the gap between the compressor blade tip and the clearance of the casing [23]. Therefore, it can be inferred that the noise and vibration at the BPF and its harmonic frequencies were caused by the fluctuating pressure of the compressor.

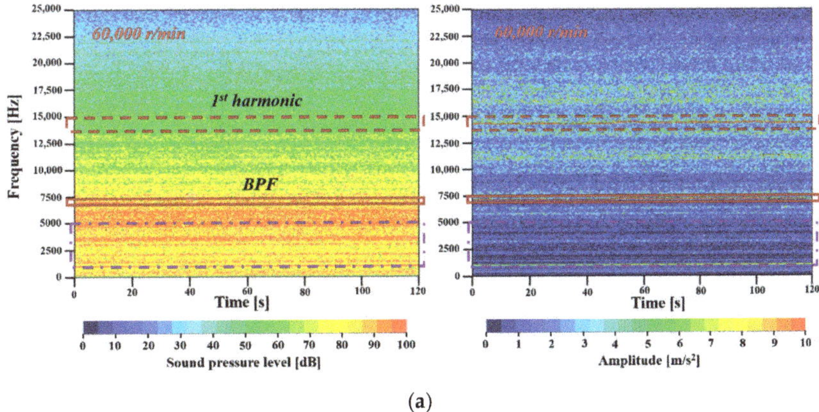

(a)

Figure 13. *Cont.*

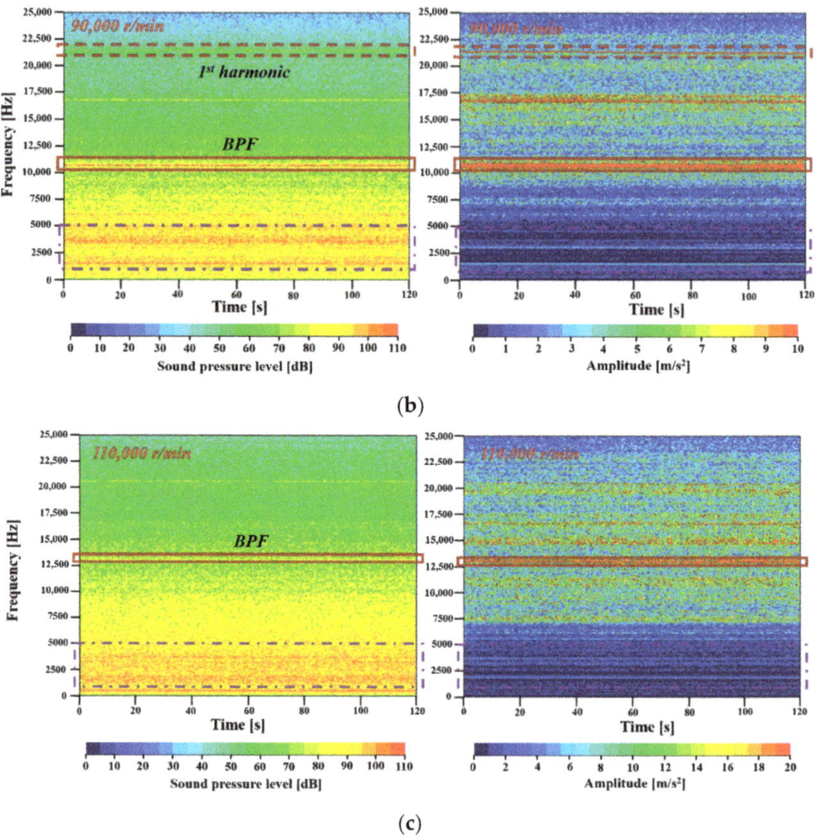

Figure 13. Transient acoustic-vibration characteristics of the compressor under high-efficiency regions. (**a**) 60,000 r/min (**b**) 90,000 r/min (**c**) 110,000 r/min.

4.2. Numerical Analysis of Compressor Aerodynamic Noise Emissions Mechanism

4.2.1. Unsteady Fluctuation Pressure

In the aerodynamic noise prediction of the compressor, the source information is the unsteady fluctuation pressure at the source surface. Therefore, it is essential to study the relationship between the unsteady fluctuation pressure and the aerodynamic noise to analyze the aerodynamic noise mechanism of the compressor. The analysis process of this section is shown in Figure 14.

During the change of fluctuation pressure at the monitoring points with a total time length of 0.005 s, the compressor's rotations were 5, 7.5 and 9.1 revolutions for speeds of 60,000 r/min, 90,000 r/min and 110,000 r/min, respectively. During the fluctuation pressure in the period of one revolution for the compressor, there were seven pressure fluctuation peaks consistent with the number of blade groups of the compressor.

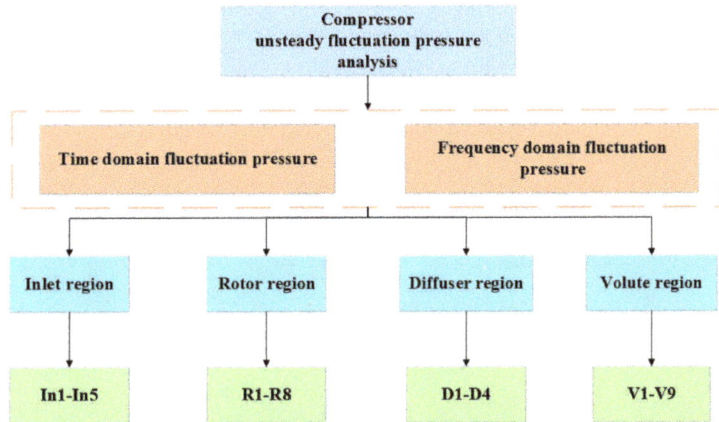

Figure 14. Fluctuation pressure analysis process of the compressor.

The time domain fluctuation pressure distribution at the inlet region of the compressor in the high-efficiency region (case 2) is shown in Figure 15. The figure shows that for the same monitoring point, the fluctuation pressure became dramatic as the speed increased. This was mainly due to the increased frequency of the impeller blades cutting the airflow mass per unit of time, and the fluctuation pressure was more obvious. At the same speed, the closer to the wall position, the greater the amplitude of fluctuation pressure at the monitoring point (In1 and In3 were the near-wall points, and In5 was the center point). This was mainly due to the interference between the high-speed rotating impeller and the incoming airflow, which created a steady periodic pressure fluctuation. The pulsation was stronger at the tip of the impeller while gradually decreasing outward from the leading edge of the impeller. Therefore, the fluctuation pressure in the near-wall area was more pronounced than at the center of the compressor inlet.

It was further observed that at low compressor speed (60,000 r/min), certain fluctuations between cycles at the same monitoring point for each compressor rotation (as shown in Figure 15a). As the speed increased, the amplitude of fluctuations between cycles at the same monitoring point decreased (Figure 15b,c). This was because, in the high-efficiency region, as the compressor speed increased, the mass flow increased. In addition, during each revolution of the compressor, the collision area between the blades and the airflow increased, and the interference time between each group of blades and the airflow became shorter, which resulted in weaker fluctuations between the cycles.

In the frequency domain analysis of aerodynamic noise, to make the unsteady fluctuation pressure characteristics of the compressor more obvious, the pressure coefficient C_P was for characterization, and its calculation formula is as follows:

$$C_P = \frac{2\left(P - P_j\right)}{\rho v^2} \tag{6}$$

$$V = \frac{\pi n D}{60} \tag{7}$$

where P is the pressure at the monitoring point at a certain moment (Pa), P_j is the average pressure at the monitoring point in the time range (Pa), ρ is the air density. Its value is 1.29 kg/m^3, V is the impeller outer edge exit circumferential velocity (m/s), n is the compressor speed (r/min), and D is the impeller exit diameter (m).

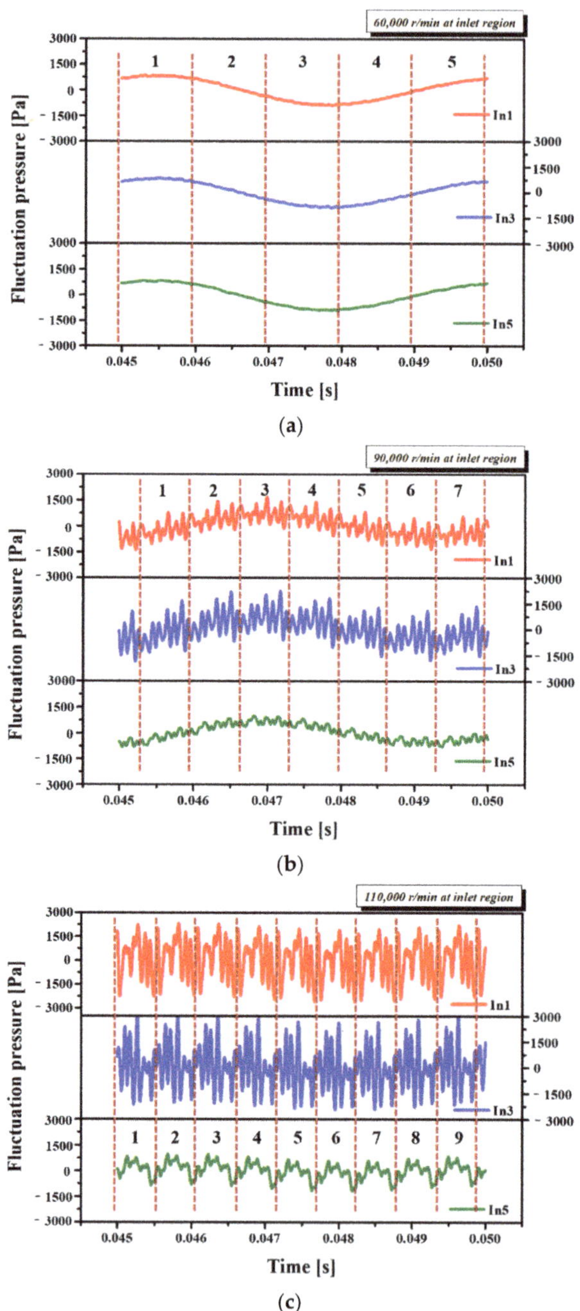

Figure 15. Time domain fluctuation pressure distributions of the compressor inlet region under different speeds. (**a**) 60,000 r/min (**b**) 90,000 r/min (**c**) 110,000 r/min.

The frequency domain fluctuation pressure distribution in the inlet region obtained by Fast Fourier Transform (FFT) of the fluctuation pressure changes at each monitoring point of the compressor inlet within 0.005 s is shown in Figure 16. It can be seen that for the same monitoring point, the peak of fluctuation pressure at each frequency increased gradually with an increase in the compressor speed, which indicated that the SPL of the aerodynamic noise on the inlet side of the compressor increased with an increase in the speed. In addition, at medium and high speeds, a more obvious peak of fluctuation pressure was observed at the BPF of each monitoring point, indicating that the fluctuation pressure formed by the interference between the impeller blades and the incoming airflow had a certain contribution to the BPF noise on the inlet side of the compressor.

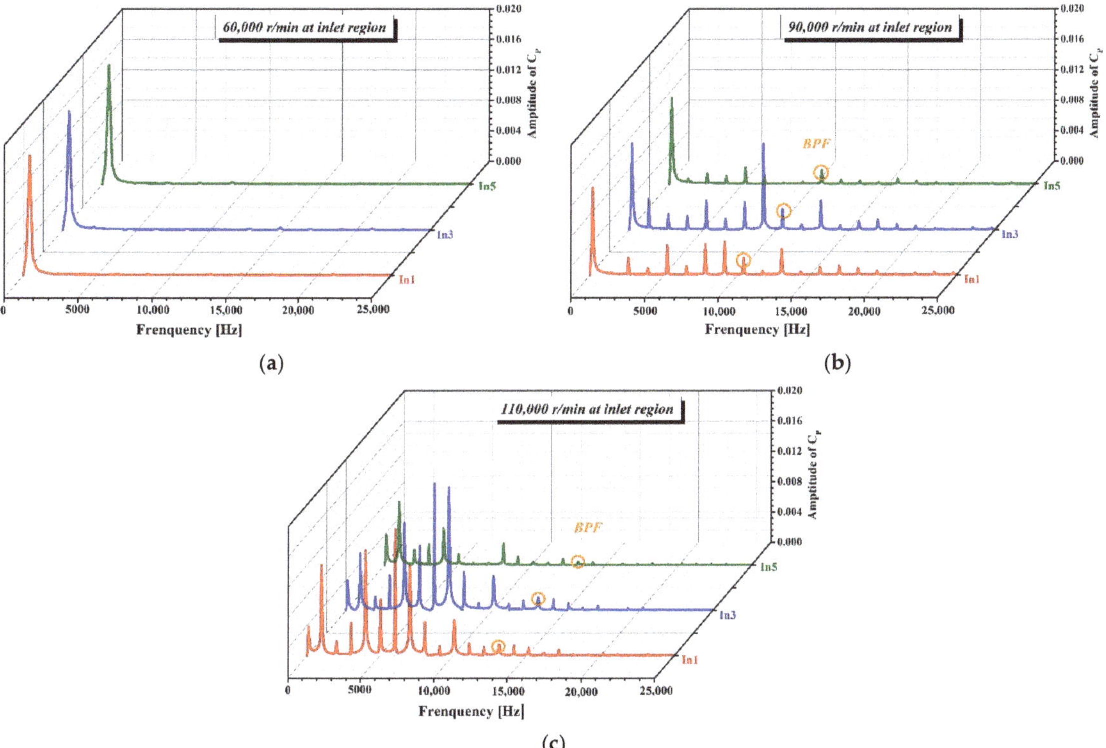

Figure 16. Frequency domain fluctuation pressure distributions of the compressor inlet region under different speeds. (**a**) 60,000 r/min (**b**) 90,000 r/min (**c**) 110,000 r/min.

The time domain fluctuation pressure distribution of the rotor region in the high-efficiency region is shown in Figure 17. The figure shows that the fluctuation pressure amplitude at R5 was higher than that of R1 at the same speed, which was caused by the higher airflow pressure at the location of R5. In addition, the fluctuation pressure amplitude increased as the speed increased for the same monitoring point. This was due to the increase in the frequency of blade and inlet airflow disturbances caused by the increase in speed, increasing the fluctuation pressure amplitude.

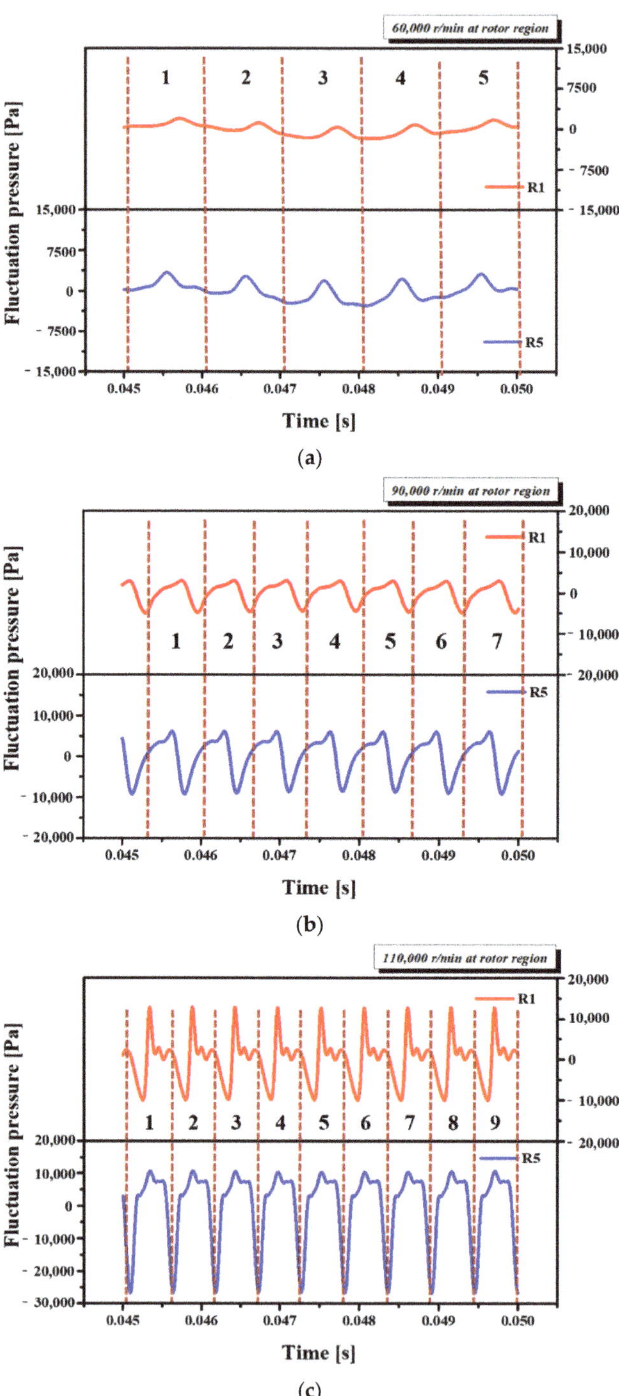

Figure 17. Time domain fluctuation pressure distributions of the compressor rotor region under different speeds. (**a**) 60,000 r/min (**b**) 90,000 r/min (**c**) 110,000 r/min.

Figure 18 shows the compressor rotor region's frequency domain fluctuation pressure distribution in the high-efficiency region. The figure shows that the peak of fluctuation pressure at each monitoring point mainly appeared at the compressor shaft frequency and its harmonic frequency at different speeds. This was because of the high blade tip speed of the compressor and the influence of aerodynamic force, which resulted in the peak of multi-monotone noise being more prominent. At the same speed, compared with the R1, the peak of fluctuation pressure at the harmonic frequency corresponding to the axial frequency of the R5 was increased. This was because the location of R5 was influenced by both the main blades and the splitter blades of the compressor.

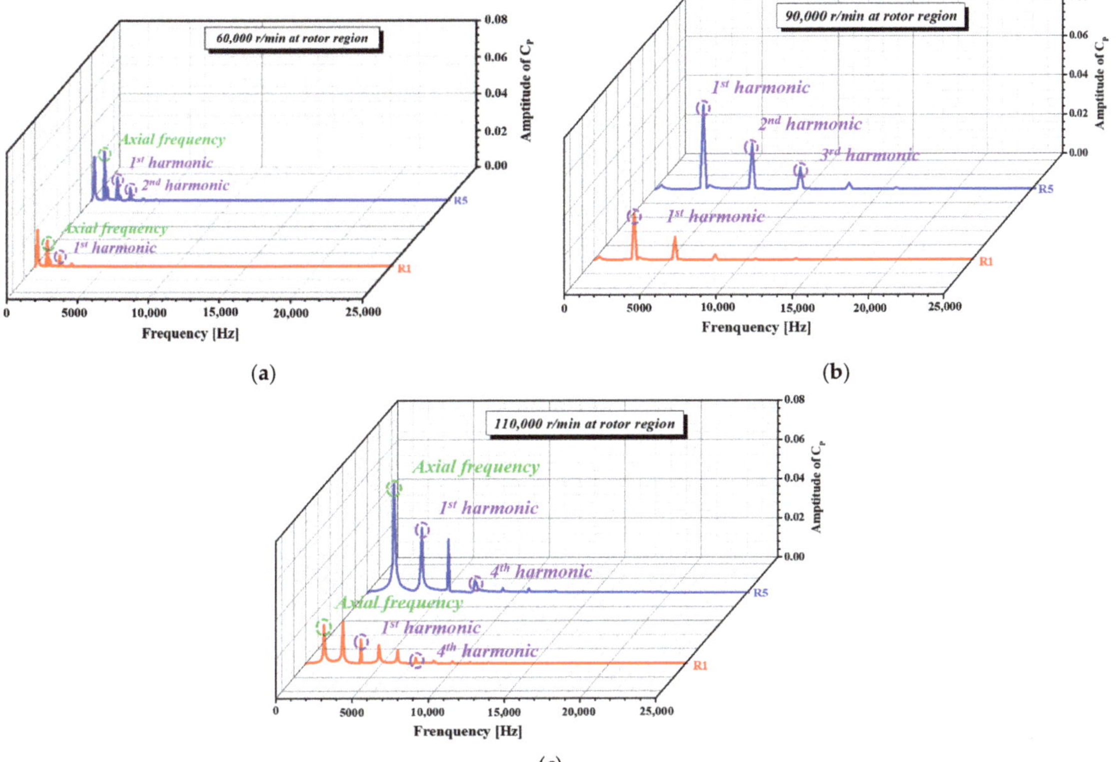

Figure 18. Frequency domain fluctuation pressure distributions of the compressor rotor region under different speeds. (**a**) 60,000 r/min (**b**) 90,000 r/min (**c**) 110,000 r/min.

Figure 19 shows the pressure distribution of the cross-section where the R1 and R5 were located. It can be seen from the figure that the main blade mainly influenced section F1 and the pressure difference between the pressure surface, and the suction surface of each blade was not obvious, resulting in a small peak of fluctuation pressure at R1. Section F2 was influenced by both the main blade and splitter blade and the pressure difference between the pressure surface and suction surface of each blade increased, which increased the peak of fluctuation pressure for R5.

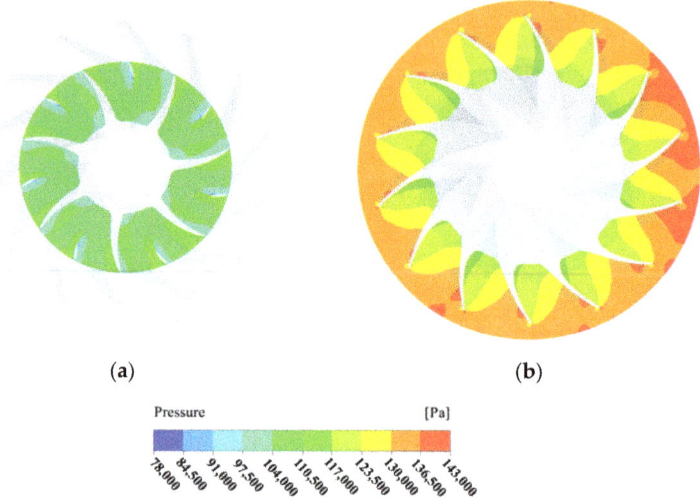

(a)   (b)

Figure 19. Pressure distribution in the compressor rotor region at 60,000 r/min of the high-efficiency region. (**a**) F1 (**b**) F2.

Figure 20 shows the time domain fluctuation pressure distribution of the compressor diffuser region in the high-efficiency region. The figure shows that for the same monitoring point, the fluctuation pressure amplitude increased with the compressor speed for the reasons shown above. In addition, seven pressure fluctuation peaks were observed during the fluctuation pressure variation of one compressor rotation, which was consistent with the number of compressor blade sets being seven. For the same monitoring point, there were certain fluctuations between each compressor cycle at different speeds.

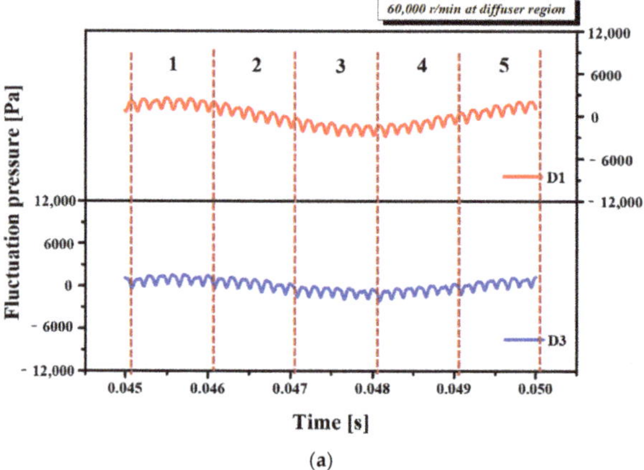

Figure 20. *Cont.*

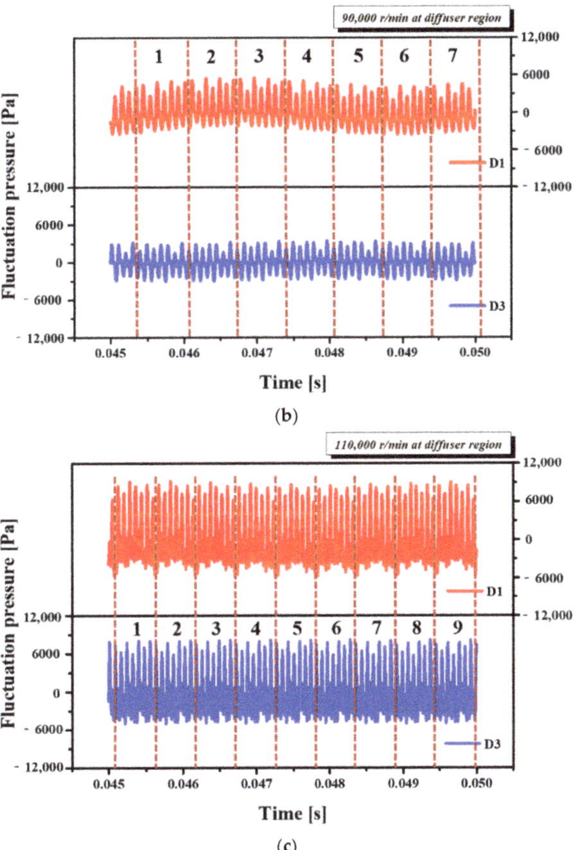

Figure 20. Time domain fluctuation pressure distributions of the compressor diffuser region under different speeds. (**a**) 60,000 r/min (**b**) 90,000 r/min (**c**) 110,000 r/min.

The frequency domain fluctuation pressure distribution in the diffuser region of the compressor is shown in Figure 21. The figure shows that for the same monitoring point, the main peaks of fluctuation pressure at each monitoring point appeared at the BPF and its first-order harmonic frequency as the speed increased. This was because the fluctuation pressure in the diffuser flow path was mainly caused by the dynamic-static interference between the impeller blades and the diffuser. Therefore, the fluctuation pressure of the compressor diffuser contributed more to the discrete monophonic noises (noise at BPF and its harmonic frequency).

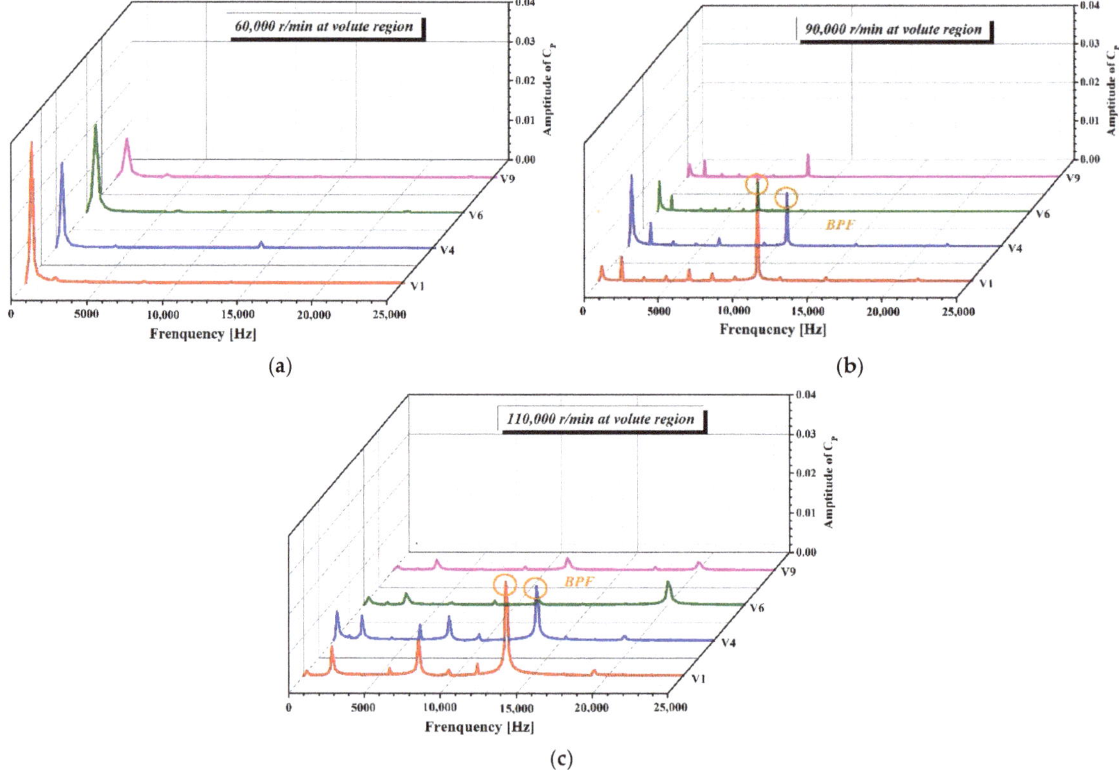

Figure 23. Frequency domain fluctuation pressure distributions of the compressor volute region under different speeds. (**a**) 60,000 r/min (**b**) 90,000 r/min (**c**) 110,000 r/min.

4.2.2. Dynamic-Static Interferences

From the above analysis, it was clear that unsteady fluctuation pressure was the main reason for discrete monophonic noise at the BPF and its harmonic frequencies. Among them, the unsteady fluctuation pressure was mainly caused by the dynamic-static interference effects, including the turbulent interference between the blades and the incoming airflow and the periodic cutting vortex mass of the impeller blades [42].

Figure 24 shows the static entropy and turbulent kinetic energy (TKE) distribution of the impeller and diffuser in the high-efficiency region of the compressor. From the figure, it can be seen that there was obvious static entropy and TKE changes at the blade's leading edge, mainly caused by the mutual interference between the impeller blade and the incoming airflow. In addition, there were also static entropy and TKE changes at the trailing edge of the blade, which reflected the dynamic-static interference between the impeller and the pressure spreader. Comparing these two areas of dynamic-static interference, it was found that there was no significant difference between them, indicating that the mutual interference between the impeller and the inlet airflow and the dynamic-static interference between the impeller and the diffuser had a high contribution to the noise at the BPF and its harmonic frequency.

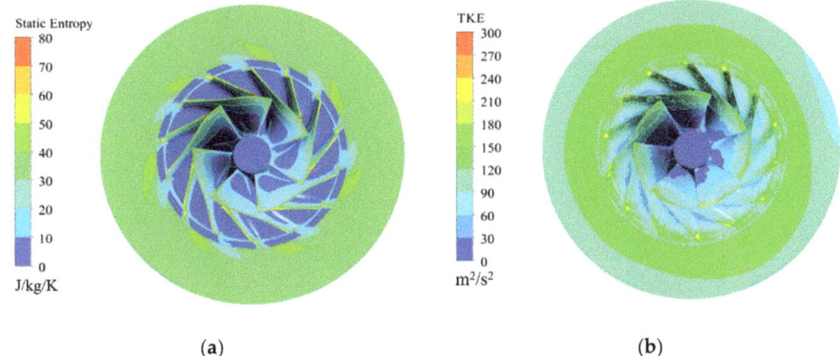

(a)   (b)

Figure 24. Static entropy and TKE distribution of the compressor at 60,000 r/min of the high-efficiency region. (**a**) Static entropy (**b**) TKE.

Figure 25 shows the distribution of static entropy and TKE for the impeller blade expansion degree of 90% in the high-efficiency region of the compressor. As can be seen from the figure, the large values of static entropy and TKE at the trailing edge of the blade indicated that there was an obvious interference effect between the impeller and the diffuser, and the noise induced by it at the BPF and its harmonic frequency was more prominent.

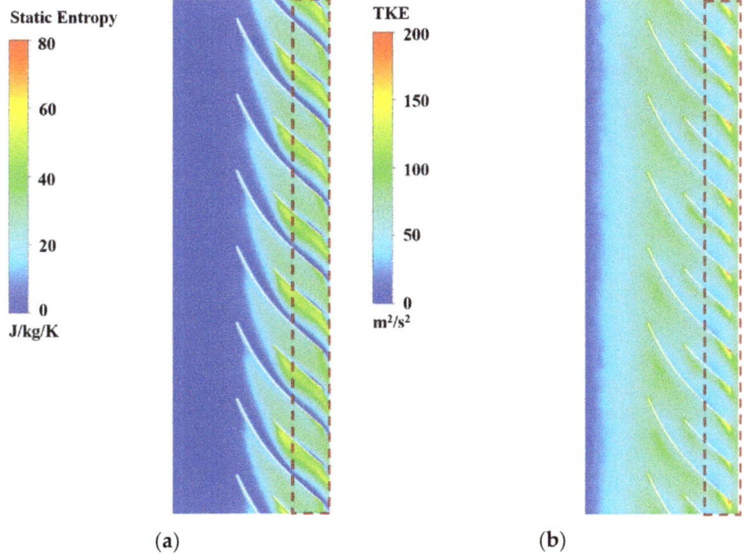

(a)   (b)

Figure 25. Blade-to-blade view of Static entropy and TKE distribution for the compressor at 60,000 r/min of the high-efficiency region. (**a**) Static entropy (**b**) TKE.

4.3. Comparative Analysis of the Results and the Existing Studies

To further emphasize the novelty of this study, a detailed comparison was made between this study and other literature, as shown in Table 6.

Table 6. Comparison of the investigation and other literature survey.

Study	Research Object	Research Methods	Main Conclusions
Li et al. [12]	Turbocharger for small vehicles	Experiment	The compressor was the main noise source of the turbocharger.
Raitor et al. [13]	Two centrifugal compressors of SRV2 and SRV4	Experiment	(1) At the design speed, blade tone and buzz-saw noise were the main contributors. (2) On the inlet, rotor-alone noise was the main source.
Figurella et al. [14]	An automotive centrifugal compressor	Experiment	Near-choke condition, discrete tones, including rotor-order frequency and its harmonics (blade-pass), were observed.
Zhang et al. [15]	Single-stage compressor with low speed and high load	Experiment	The sound pressure level of the compressor was lowest at a relatively small tip clearance rather than zero.
Galindo et al. [16]	A small automotive turbocharger compressor	Experiment	The effects of several inlet geometries on compressor performance were investigated.
Dehner et al. [21]	A turbocharger centrifugal compressor from a spark-ignition internal combustion engine	Numerical simulation, experiments and modal decomposition	(1) Revealed the presence of rotating instabilities that may interact with the rotor blades to generate noise. (2) The whooshing noise primarily propagated along the duct as acoustic azimuthal modes.
Galindo et al. [22]	An automotive centrifugal compressor	Numerical simulation	The use of a convergent-divergent nozzle could strongly reduce the intake orifice noise.
Galindo et al. [23]	A 49 mm exducer diameter centrifugal compressor	Numerical simulation	In near-surge conditions, there were no significant changes in the compressor acoustic signature when varying the tip clearance.
Sundström et al. [24]	An automotive centrifugal compressor	Numerical simulation	For the near-surge condition, an amplified broadband feature at two times the frequency of the rotating order of the shaft (possible whoosh noise) was captured. In addition, an amplified feature of around 50% of the rotating order was captured.
Jyothishkumar et al. [26]	A turbocharger centrifugal compressor from a heavy truck engine	Numerical simulation	There existed a flow-acoustics coupling at near-surge operating conditions.
Broatch et al. [27]	A turbocharger centrifugal compressor from a diesel engine	In-duct experimental measurements and numerical simulation	(1) A suitable comparison methodology was developed, relying on pressure decomposition. (2) Whoosh, noise at the outlet duct was detected in experimental and numerical spectra.
Guo et al. [28]	A turbocharger centrifugal compressor	Numerical simulation	A distinct stall frequency at the given compressor speed.
This investigation	A turbocharger centrifugal compressor from a heavy-duty diesel engine	Based on radiated noise experimental measurements and numerical simulation	(1) The aerodynamic noise characteristics of the centrifugal compressor in the full operating range based on the radiation noise experimental measurements were given. (2) For the centrifugal compressor from a heavy-duty diesel engine, the blade passing frequency and its harmonic frequency noise were mainly caused by the unsteady fluctuation pressure. (3) The impeller blades interfere with the incoming airflow, and the dynamic-static interference between the impeller and the diffuser significantly contributed to the BPF and its harmonic frequency noise.

As seen from Table 6, the existing literature focused on centrifugal compressors for small automotive turbochargers and the research methods used in the studies, including experiment, numerical simulation, and a combination of experiment and numerical simulation. The findings included the analysis of aerodynamic noise component composition, aerodynamic noise characteristics of centrifugal compressors under specific operating conditions, and flow characterization. However, there were fewer studies on the aerodynamic noise characterization of centrifugal compressors for heavy-duty diesel engine turbochargers under the full operating range. In addition, there was a lack of analysis based on the linkage between aerodynamic noise characteristics and internal flow characteristics of centrifugal compressors under specific operating conditions. Therefore, the innovation of the study was to refine the aerodynamic noise mechanism of the compressor for heavy-duty diesel engine turbochargers, propose that blade aerodynamic force and dynamic interference were the main discrete monophonic noise sources, and provide a theoretical basis for reducing the heavy-duty diesel engine turbocharger compressor aerodynamic noise emissions.

5. Conclusions

This study conducted an experimental and simulation investigation of the aerodynamic noise emission of a heavy-duty diesel engine turbocharger compressor in the full operating range. To analyze the aerodynamic noise mechanism, the relationship between the unsteady fluctuation pressure, dynamic-static interference and aerodynamic noise was studied based on numerical simulation methods. The main conclusions were as follows:

1. Under the test conditions, the sound pressure level (SPL) of the aerodynamic noise for the compressor of a heavy-duty diesel engine increased with an increase in speed. At the same speed, the compressor operating in the high-efficiency region (middle mass flows) could achieve lower aerodynamic noise emissions.
2. At low speeds, the total SPL of aerodynamic noise was more obviously affected by the mass flow of the compressor. At 60,000 r/min, the maximum difference in the total SPL of aerodynamic noise was 1.55 dB at different mass flows.
3. Compared with 60,000 r/min, the total SPLs of aerodynamic noise and BPF noises for the near-choke region, high-efficiency region and near-surge region at 110,000 r/min increased by 4.32%, 5.18%, 4.77% and 10.62%, 10.27% and 9.81%, respectively.
4. In the compressor aerodynamic noise, the blade passing frequency (BPF) noise occupies a dominant position. As the engine and compressor speeds increased, the BPF noise contributed more to the total SPL of the aerodynamic noise, and its proportion was up to 75.35%.
5. From the analysis of the transient acoustic-vibration spectrum and fluctuation pressure, it could be seen that BPF and its harmonic frequency noise were mainly caused by the unsteady fluctuation pressure of the compressor.
6. In the compressor, the impeller blades interfere with the incoming airflow, and the dynamic-static interference between the impeller and the diffuser greatly contributes to the BPF and its harmonic frequency noise.

Author Contributions: Methodology, Q.W.; Formal analysis, R.H.; Investigation, R.H. and Q.Y.; Resources, J.N.; Data curation, Q.W.; Writing—original draft, R.H.; Writing—review & editing, R.H., X.S. and Q.Y.; Visualization, X.S.; Project administration, J.N.; Funding acquisition, J.N. and Q.W. All authors have read and agreed to the published version of the manuscript.

Funding: This research was funded by National Natural Science Foundation of China Youth Science Foundation project (grant number 22102116) and the State Key Laboratory of Internal Combustion Engine Reliability Open Subject Foundation of China (grant number skler-202114).

Institutional Review Board Statement: Not applicable.

Informed Consent Statement: Not applicable.

Data Availability Statement: The data presented in this study are available in the main text of the article.

Acknowledgments: The team of authors acknowledges anonymous reviewers for their feedback, which certainly improved the clarity and quality of this paper.

Conflicts of Interest: The authors declare that they have no known competing financial interests or personal relationships that could have appeared to influence the work reported in this paper.

References

1. Bao, J.; Wang, H.; Wang, R.; Wang, Q.; Di, L.; Shi, C. Comparative experimental study on macroscopic spray characteristics of various oxygenated diesel fuels. *Energy Sci. Eng.* **2023**, *11*, 1579–1588. [CrossRef]
2. Huang, R.; Ni, J.; Cheng, Z.; Wang, Q.; Shi, X.; Yao, X. Assessing the effects of ethanol additive and driving behaviors on fuel economy, particle number, and gaseous emissions of a GDI vehicle under real driving conditions. *Fuel* **2021**, *306*, 121642. [CrossRef]
3. Mądziel, M. Vehicle Emission Models and Traffic Simulators: A Review. *Energies* **2023**, *16*, 3941. [CrossRef]
4. Bao, J.; Qu, P.; Wang, H.; Zhou, C.; Zhang, L.; Shi, C. Implementation of various bowl designs in an HPDI natural gas engine focused on performance and pollutant emissions. *Chemosphere* **2022**, *303*, 135275. [CrossRef] [PubMed]
5. Shi, C.; Chai, S.; Wang, H.; Ji, C.; Ge, Y.; Di, L. An insight into direct water injection applied on the hydrogen-enriched rotary engine. *Fuel* **2023**, *339*, 127352. [CrossRef]
6. Broatch, A.; Ruiz, S.; García-Tíscar, J.; Roig, F. On the influence of inlet elbow radius on recirculating backflow, whoosh noise and efficiency in turbocharger compressors. *Exp. Therm. Fluid Sci.* **2018**, *96*, 224–233. [CrossRef]
7. Abom, M.; Kabral, R. *Turbocharger Noise—Generation and Control*; SAE Technical Paper 2014-36-0802; SAE International: Warrendale, PA, USA, 2014.
8. Sharma, S.; Broatch, A.; García-Tíscar, J.; Allport, J.; Nickson, A. Acoustic characterisation of a small high-speed centrifugal compressor with casing treatment: An experimental study. *Aerosp. Sci. Technol.* **2019**, *95*, 105518. [CrossRef]
9. Karim, A.; Miazgowicz, K.; Lizotte, B.; Zouani, A. *Computational Aero-Acoustics Simulation of Compressor Whoosh Noise in Automotive Turbochargers*; SAE Technical Paper 2013-01-1880; SAE International: Warrendale, PA, USA, 2013. [CrossRef]
10. Wolfram, D.; Carolus, T.H. Experimental and numerical investigation of the unsteady flow field and tone generation in an isolated centrifugal fan impeller. *J. Sound Vib.* **2010**, *329*, 4380–4397. [CrossRef]
11. Torregrosa, A.J.; Broatch, A.; Navarro, R.; García-Tíscar, J. Acoustic characterization of automotive turbocompressors. *Int. J. Engine Res.* **2015**, *16*, 31–37. [CrossRef]
12. Li, H.; Ning, M.; Hou, L.; Zhou, T. Experimental Study on the Noise Identification of the Turbocharger. *Lect. Notes Electr. Eng.* **2011**, *121*, 1–7. [CrossRef]
13. Raitor, T.; Neise, W. Sound generation in centrifugal compressors. *J. Sound Vib.* **2008**, *314*, 738–756. [CrossRef]
14. Figurella, N.; Dehner, R.; Selamet, A.; Tallio, K.; Miazgowicz, K.; Wade, R. Noise at the mid to high flow range of a turbocharger compressor. *Noise Control. Eng. J.* **2014**, *62*, 306–312. [CrossRef]
15. Zhang, M.; Dong, X.; Li, J.; Sun, D.; Sun, X. Effect of differential tip clearance on the performance and noise of an axial compressor. *Aerosp. Sci. Technol.* **2023**, *132*, 109070. [CrossRef]
16. Galindo, J.; Tiseira, A.; Navarro, R.; Tarí, D.; Meano, C.M. Effect of the inlet geometry on performance, surge margin and noise emission of an automotive turbocharger compressor. *Appl. Therm. Eng.* **2017**, *110*, 875–882. [CrossRef]
17. Zhong, C.; Hu, L.; Gong, J.; Wu, C.; Wang, S.; Zhu, X. Effects analysis on aerodynamic noise reduction of centrifugal compressor used for gasoline engine. *Appl. Acoust.* **2021**, *180*, 108104. [CrossRef]
18. Fontanesi, S.; Paltrinieri, S.; Cantore, G. CFD analysis of the acoustic behavior of a centrifugal compressor for high performance engine application. *Energy Procedia* **2014**, *45*, 759–768. [CrossRef]
19. Sun, H.; Lee, S. Numerical prediction of centrifugal compressor noise. *J. Sound Vib.* **2004**, *269*, 421–430. [CrossRef]
20. Zhang, J.; Chu, W.; Zhang, H.; Wu, Y.; Dong, X. Numerical and experimental investigations of the unsteady aerodynamics and aero-acoustics characteristics of a backward curved blade centrifugal fan. *Appl. Acoust.* **2016**, *110*, 256–267. [CrossRef]
21. Dehner, R.; Sriganesh, P.; Selamet, A.; Miazgowicz, K. Generation Mechanism of Broadband Whoosh Noise in an Automotive Turbocharger Centrifugal Compressor. *J. Turbomach.* **2021**, *143*, 121003. [CrossRef]
22. Galindo, J.; Gil, A.; Navarro, R.; Tarí, D. Analysis of the impact of the geometry on the performance of an automotive centrifugal compressor using CFD simulations. *Appl. Therm. Eng.* **2018**, *148*, 1324–1333. [CrossRef]
23. Galindo, J.; Tiseira, A.; Navarro, R.; López, M. Influence of tip clearance on flow behavior and noise generation of centrifugal compressors in near-surge conditions. *Int. J. Heat Fluid Flow* **2015**, *52*, 129–139. [CrossRef]
24. Sundström, E.; Semlitsch, B.; Mihaescu, M. Centrifugal compressor: The sound of surge. In Proceedings of the 21st AIAA/CEAS Aeroacoustics Conference, Dallas, TX, USA, 22–26 June 2015.
25. Sundström, E.; Mihăescu, M.; Giachi, M.; Belardini, E.; Michelassi, V. Analysis of Vaneless Diffuser Stall Instability in a Centrifugal Compressor. *Int. J. Turbomach. Propuls. Power* **2017**, *2*, 19. [CrossRef]
26. Jyothishkumar, V.; Mihaescu, M.; Semlitsch, B.; Fuchs, L. Numerical flow analysis in centrifugal compressor near surge condition. In Proceedings of the 43rd AIAA Fluid Dynamics Conference, San Diego, CA, USA, 24–27 June 2013. [CrossRef]
27. Broatch, A.; Galindo, J.; Navarro, R.; García-Tíscar, J. Methodology for experimental validation of a CFD model for predicting noise generation in centrifugal compressors. *Int. J. Heat Fluid Flow* **2014**, *50*, 134–144. [CrossRef]
28. Guo, Q.; Chen, H.; Zhu, X.C.; Du, Z.H.; Zhao, Y. Numerical simulations of stall inside a centrifugal compressor. Proceedings of the Institution of Mechanical Engineers. *J. Power Energy* **2007**, *221*, 683–693. [CrossRef]
29. Tomita, I.; Ibaraki, S.; Furukawa, M.; Yamada, K. The Effect of Tip Leakage Vortex for Operating Range Enhancement of Centrifugal Compressor. *J. Turbomach.* **2013**, *135*, 051020. [CrossRef]
30. Chen, Q.; Ni, J.; Wang, Q.; Shi, X. Match-based pseudo-MAP full-operation-range optimization method for a turbocharger compressor. *Struct. Multidiscip. Optim.* **2019**, *60*, 1139–1153. [CrossRef]

31. Fan, H.; Ni, J.; Shi, X.; Qu, D.; Zheng, Y.; Zheng, Y. Simulation of a Combined Nozzled and Nozzleless Twin-Entry Turbine for Improved Efficiency. *J. Eng. Gas Turbines Power* **2019**, *141*, 051019. [CrossRef]
32. Semlitsch, B.; JyothishKumar, V.; Mihaescu, M.; Fuchs, L.; Gutmark, E.; Gancedo, M. *Numerical Flow Analysis of a Centrifugal Compressor with Ported and without Ported Shroud*; SAE Technical Paper 2014-01-1655; SAE International: Warrendale, PA, USA, 2014. [CrossRef]
33. Fardafshar, N.; Koutsovasilis, P. Ported Shroud Influence on the Aero-Acoustic Properties of Automotive Turbochargers: Quantification by Means of Simulation and Measurement. DAGA 2018—44. Jahrestagung für Akustik. 2018. Available online: https://www.google.com.hk/url?sa=t&source=web&cd=&cad=rja&uact=8&ved=2ahUKEwjogp3tmZyAAxVvg1YBHQ0kB6 0QFnoECAkQAQ&url=https%3A%2F%2Fpub.dega-akustik.de%2FDAGA_2018%2Fdata%2Farticles%2F000021.pdf&usg= AOvVaw1AnSuOQ2osRHo3dyat0V3z&opi=89978449 (accessed on 25 June 2023).
34. Celik, I.B.; Ghia, U.; Roache, P.J.; Freitas, C.J. Procedure for Estimation and Reporting of Uncertainty Due to Discretization in CFD Applications. *J. Fluids Eng.* **2008**, *130*, 078001.
35. Huang, R.; Ni, J.; Fan, H.; Shi, X.; Wang, Q. Investigating a New Method-Based Internal Joint Operation Law for Optimizing the Performance of a Turbocharger Compressor. *Sustainability* **2023**, *15*, 990. [CrossRef]
36. Yin, S.; Ni, J.; Fan, H.; Shi, X.; Huang, R. A Study of Evaluation Method for Turbocharger Turbine Based on Joint Operation Curve. *Sustainability* **2022**, *14*, 9952. [CrossRef]
37. Shi, C.; Chai, S.; Di, L.; Ji, C.; Ge, Y.; Wang, H. Combined experimental-numerical analysis of hydrogen as a combustion enhancer applied to wankel engine. *Energy* **2023**, *263*, 125896. [CrossRef]
38. Zuo, S.G.; He, H.J.; Wu, X.D.; Wei, K.J. Numerical Analysis on Effects of Inlet Bent Pipe's Position on a Centrifugal Compressor's Aerodynamic Noise. *Appl. Mech. Mater.* **2014**, *633–634*, 1196–1201. [CrossRef]
39. Zamiri, A.; Choi, M.; Chung, J.T. Effect of blade squealer tips on aerodynamic performance and stall margin in a transonic centrifugal compressor with vaned diffuser. *Aerosp. Sci. Technol.* **2022**, *123*, 107504. [CrossRef]
40. Guan, D.; Sun, D.; Xu, R.; Bishop, D.; Sun, X.; Ni, S.; Du, J.; Zhao, D. Experimental investigation on axial compressor stall phenomena using aeroacoustics measurements via empirical mode and proper orthogonal decomposition methods. *Aerosp. Sci. Technol.* **2021**, *112*, 106655. [CrossRef]
41. Liu, C.; Cao, Y.; Zhang, W.; Ming, P.; Liu, Y. Numerical and experimental investigations of centrifugal compressor BPF noise. *Appl. Acoust.* **2019**, *150*, 290–301. [CrossRef]
42. Xu, Q.; Wu, J.; Wu, L.; Yang, H.; Wang, Z.; Wei, Y. Pressure and velocity fluctuations characteristics of the tip clearance flow in an axial compressor stage at the near-stall condition. *Aerosp. Sci. Technol.* **2022**, *129*, 107796. [CrossRef]

Article

Research on Combustion and Emission Characteristics of a N-Butanol Combined Injection SI Engine

Weiwei Shang [1,2], Xiumin Yu [1], Kehao Miao [1], Zezhou Guo [1,*], Huiying Liu [2] and Xiaoxue Xing [2]

[1] College of Automotive Engineering, Jilin University, Changchun 130022, China; shangww15@mails.jlu.edu.cn (W.S.); yuxm@jlu.edu.cn (X.Y.); miaokh1521@jlu.edu.cn (K.M.)
[2] Electronic Information Engineering College, Changchun University, Changchun 130012, China; huiying20@mails.jlu.edu.cn (H.L.); xingxx@ccu.edu.cn (X.X.)
* Correspondence: guozz@jlu.edu.cn

Abstract: Using n-butanol as an alternative fuel can effectively alleviate the increasingly prominent problems of fossil resource depletion and environmental pollution. Combined injection technology can effectively improve engine combustion and emission characteristics while applying combined injection technology to n-butanol engines has not been studied yet. Therefore, this study adopted butanol port injection plus butanol direct injection mode. The engine test bench studied the combustion and emission performance under different direct injection ratios (NDIr) and excess air ratios (λ). Results show that with increasing NDIr, the engine torque (Ttq), peak in-cylinder pressure (Pmax), peak in-cylinder temperature (Tmax), and the maximum rate of heat release (dQmax), all rise first and then drop, reaching the maximum value at NDIr = 20%. The θ0-90 and COV_{IMEP} decrease first and then increase as NDIr increases. NDIr = 20% is considered the best injection ratio to obtain the optimal combustion performance. NDIr has little affected on CO emission, and the NDIr corresponding to the lowest HC emissions are concentrated at 40% to 60%, especially at lean burn conditions. NOx emissions increase with increasing NDIr, especially at N20DI, but not by much at NDIr of 40–80%. With the increase in NDIr, the number of nucleation mode particles, accumulation mode particles, and total particle decrease first and then increase. Therefore, the n-butanol combined injection mode with the appropriate NDIr can effectively optimize SI engines' combustion and emission performance.

Keywords: n-butanol; combined injection mode; combustion; gaseous emissions; particle number

Citation: Shang, W.; Yu, X.; Miao, K.; Guo, Z.; Liu, H.; Xing, X. Research on Combustion and Emission Characteristics of a N-Butanol Combined Injection SI Engine. *Sustainability* **2023**, *15*, 9696. https://doi.org/10.3390/su15129696

Academic Editors: Jinxin Yang, Jianbing Gao, Cheng Shi and Peng Zhang

Received: 25 May 2023
Revised: 13 June 2023
Accepted: 14 June 2023
Published: 16 June 2023

1. Introduction

In recent years, with the rapid development of automotive industries, the depletion of fossil sources and the aggravation of environmental pollution are becoming increasingly prominent [1,2]. Thus the research on green, alternative, and renewable fuels has attracted the attention of many scholars [3,4]. At the same time, advanced technical methods in the field of internal combustion engines are also very important in improving engine performance [5]. Therefore, the effective combination of alternative fuel characteristics and fuel injection modes will play a crucial role in the engine's performance, which is worth investigating.

The alternative fuel of internal combustion engines can be divided into two categories according to the different physical states of fuel: gas and liquid. Among the alternative fuels studied and used in the internal combustion engine at present, liquid alternative fuel mainly includes biomass fuel, alcohol fuel (ethanol, methanol, butanol), gas alternative fuel mainly includes liquefied petroleum gas (LPG), compressed natural gas (CNG), dimethyl ether hydrogen and so on [6]. Among all alternative fuels, alcohol fuel, as a renewable oxygen-containing biofuel, has been widely concerned by scholars at home and abroad because of its wide range of sources and clean combustion. Alcohols are considered one of the most promising alternative fuels for ignition engines, and their application in automobiles has been extensively studied [7]. Table 1 shows the properties of typical alcohol fuels [8]. Compared with methanol and ethanol fuel, butanol fuel has the advantages of higher caloric value and low latent heat of evaporation [9]. Therefore, butanol as a new alternative energy for engines has attracted more and more attention in recent years [10,11].

Table 1. Properties of alcohol fuels.

Fuel Properties	Methanol	Ethanol	N-butanol
Molecular formula	CH_3OH	C_2H_5OH	C_4H_9OH
Viscosity (Pa·s) at 20 °C	0.61	0.789	0.808
Research octane number	109–136	108–129	96–98
Laminar flame speed (cm/s)	52	48	48
Latent heat of evaporation (kJ/kg)	1103	840	582
Lower caloric value (MJ/kg)	19.7	26.8	33.1
Flammability limits (% vol.)	6.0–36.5	4.3–19	1.4–11.2
Stoichiometric air–fuel ratio	6.49	9.02	11.21

There are four isomers of butanol, and the different ways the carbon chain and hydroxyl group are connected the lead to the different uses of the isomers. N-butanol has been widely used as a promising alternative fuel for gasoline because of its better combustion performance and the similar physical and chemical properties to gasoline [12]. In addition, n-butanol has many advantages compared with Methanol and Ethanol, such as higher heating value and viscosity and lower volatility and vaporization heat. N-butanol can be obtained not only from coal but also by biological methods [13]. The raw materials are from various sources, including wheat, corn, and other crops, as well as agricultural waste straw [14]. According to Butyl Fuel Company's research data, 270 mL n-butanol can be produced by microbial fermentation with 1 L corn as raw material, and the cost is only 0.317 US dollars/liter. Additionally, with the continuous development of n-butanol production technology, the production cost is expected to be further reduced [15], which provides strong support for using n-butanol as an alternative fuel for engines [16].

To achieve the goal of high-efficiency and low emission of engines, new technologies of engines have been introduced in recent years, such as the combined injection technique [17]. The common injection mode of internal combustion engines includes port fuel injection (PFI) and direct injection (DI) [18]. Compared to port fuel injection, GDI enables precise fuel injection control in all operating conditions, reducing fuel consumption [19]. DI can form a localized concentration area near the spark plug and form a layered mixture in the whole cylinder to promote a better combustion effect and improve combustion thermal efficiency [20]. However, DI also has problems such as higher demand for fuel supply pressure, HC, and more emission of particulate matter [21]. PFI allows fuel and air to mix better in the inlet [22]. Therefore, the two injection methods have their own advantages and disadvantages. So, combining two injection modes, that is, the realization of compound injection in one engine through two injection systems, is of great significance to optimizing engine performance.

The research on butanol as engine fuel mainly focuses on n-butanol blends with other fuels and the pure n-butanol engine. Concerning engines of n-butanol blends with other fuels, J. Yang et al. [23] explored the combustion and emission characteristics of the hydrogen/n-butanol and hydrogen/gasoline rotary engines. They found that hydrogen addition improved the combustion and emission characteristics of gasoline and n-butanol rotary engines. Additionally, for both n-butanol and gasoline rotary engines, adding hydrogen can improve the brake thermal efficiency, shorten the development and propagation periods, reduce the coefficient of variation, and reduce the emission of HC and CO. However, NOx emission increased slightly after blending hydrogen. V. Thangavel et al. [9] studied the performance of a single-cylinder SI engine with both ethanol and gasoline injected into the inlet and performed a comparative analysis with n-butanol gasoline operation with the combined injection strategy in the intake port of a single-cylinder SI engine.

They found that at high torque conditions, the benefits of mixing ethanol with gasoline are significant. Compared to ethanol, the amount of n-butanol must be increased with increasing torque for better performance and low emissions. R. a. Ravikumar et al. [24] studied the combustion and emission characteristics of a twin spark ignition engine using n-butanol and gasoline dual fuel. They found that blending B35 resulted in lower carbon monoxide emissions, lower unburned hydrocarbon emissions, and lower nitrogen oxide emissions than pure gasoline. E. Agbro et al. [25] investigated the impact of n-butanol addition on the combustion performance and knocked properties under supercharged spark-ignition engine conditions; their results indicated that blending n-butanol can improve the anti-knock properties of gasoline and improve engine efficiency by using a higher compression ratio. Z. Guo et al. [19] tested gasoline/n-butanol blends with n-butanol volumetric ratios of 0%, 20%, 40%, 60%, 80%, and 100% in a DI SI engine. Results showed that a 20% butanol blending ratio could effectively reduce particle and gaseous emissions. M. Saraswat et al. [26] investigated the effect of different volume basis of butanol-gasoline and butanol/diesel mixture on IC engines. The results show that blending butanol can improve the power, torque, brake specific energy consumption, Hydrocarbons, Carbon-mono-oxides and NO emissions, but the NOx and CO_2 emissions are higher than those of gasoline and diesel. F. Meng et al. [27] investigated the combustion and emissions performance of a combined injection hydrogen/n-butanol dual-fuel engine with hydrogen addition fractions (0%, 2.5%, and 5%). The obtained results demonstrated the power and fuel economy performance of n-butanol engines are improved after adding hydrogen, and the HC and CO emissions drop while the NOx emissions sharply rise. T. Su et al. [28] researched the performance of a hydrogen/n-butanol dual fuel rotary engine with a n-butanol and hydrogen port-injection system. The test results indicated that blending hydrogen increased the brake thermal efficiency and in-cylinder temperature and reduced CO and CO_2 emissions effectively. Compared to n-butanol blends with other fuels, the pure n-butanol as an SI engine fuel is less researched. S. S. Merola et al. [29] studied the influence of injection timing on a wall-guided direct injection SI engine fueled with n-butanol. They found that late injection timing reduced soot but resulted in higher HC emissions and poorer performance than the optimum point. Additionally, they proved that fuel impingement on the piston crown is the main influencing factor for soot formation. N. S. a. Sandhu, X. a. Yu, and S. a. Leblanc et al. [30] analyzed the combustion characteristics of neat n-butanol under spark ignition operation using a single-cylinder SI engine. They found that n-butanol has similar physicochemical property and fuel characteristics to that of gasoline and has lower NOx, unburnt HC emission and CO_2 emissions. The above research indicates that pure butanol is feasible to replace traditional fuel as engine fuel. However, the performance of pure butanol engines needs to be further optimized.

Therefore, in the context of the world energy crisis, environmental pollution, and increasingly stringent emission regulations, it is of vital significance to adopt emerging technologies to improve the combustion performance of pure butanol engines and provide better theoretical and experimental support for the practical application of pure butanol engines [31]. Although n-butanol as an SI engine fuel has been partially investigated, most of them focus on n-butanol blended with other fuels. The performance of traditional pure butanol engines is unsatisfactory. Therefore, we proposed a new combustion idea as follows: the combined injection mode of partial butanol direct injection in the cylinder and partial butanol port injection, which has no research on SI engines fueled with n-butanol. We studied the influence of the combined injection mode on the combustion and emission characteristics of the butanol engine and explored the optimization potential of advanced technology on butanol engine performance. In this experiment, the combined injection mode adopts part of the n-butanol port injection plus part of the n-butanol direct injection. By engine test bench, the variations in combustion characteristics, we examined gaseous and particle emissions of SI engines due to changes in the direct injection ratio (DIr) and λ have been studied in this paper.

Although n-butanol as an SI engine fuel has been partially investigated, most focus on n-butanol blended with other fuels. The performance of traditional pure butanol engines is not satisfactory. Therefore, the effects of compound injection on combustion and emission characteristics of pure butanol engines were experimentally studied in this paper. The main contributions of this paper are as follows: (1) we proposed a new injection mode that is the combined injection mode of partial butanol direct injection in the cylinder and partial butanol port injection, which has no research on SI engines fueled with n-butanol. (2) The effects of direct injection ratio (DIr) and λ on combustion characteristics, gas emission, and particle emission of SI n-butanol engine under combined injection mode were investigated. (3) The potential of the combined injection technique in optimizing n-butanol engine performance was explored. Therefore, the purpose of this study is to use combined injection technology to solve the problem that the low saturated steam pressure of butanol may form wall flow when pure port butanol injection, while pure butanol direct injection may cause wall oil film due to the high viscosity of butanol. Coupling the two injection modes, butanol combined injection may improve their evaporation, mixing, and combustion, optimizing the butanol engine's performance.

2. Experimental Setup and Procedure

2.1. Engine and Instrument

In this study, the experiments were performed on a four-cylinder four-stroke water-cooled combined injection SI engine, a dual injection engine with port and direct injection. The main technical parameters of the original engine are listed in Table 2. Since the original direct injection system was designed for gasoline fuel, the fuel injected in the cylinder directly in this paper is butanol. Therefore, the high-pressure oil pump with direct injection in the original cylinder was not used in this test, but the high-pressure nitrogen cylinder pressurized method was used, and the direct injection pressure was 5 MPa. The original engine's oil pump still supplies the low-pressure port injection with an injection pressure of 0.3 MPa. The dual injection systems of the engine included partial n-butanol direct injection and partial n-butanol port injection. The experiment system structure image of the test engine is shown in Figure 1. Figure 1 shows that engine control parameters such as injection timing and injection durations for n-butanol and the throttle opening were all accurately controlled by the electronic control unit (ECU). The direct injection ratio is controlled precisely by adjusting the injection pulse width of the injector in real time.

Table 2. Main parameters of Engine.

Engine Parameter	Parameter Values
Engine Type	four cylinders; combined injection; naturally aspirated; water cooled; spark-ignition
Compression ratio	9.6
Bore/mm	82.5
Stroke/mm	92.8
Displaced volume/mL	1984
Maximum power/kW	132 (5000–6000 rpm)
Maximum torque/N·m	320 (1800–5000 rpm)

The specific information on the experimental instruments is listed in Table 3. The test engine is coupled to a CW160-type eddy current dynamometer to maintain a constant speed and measure torque. The values of each test point in the actual measurement were tested three times to take the average value. The n-butanol fuel flow was acquired by DMF-1-1AB and Ono Sokki DF-2420 flow meters. The λ was measured by an ETAS Lambda Meter 4 broadband oxygen sensor, and crank angle signals were collected with a Kistler-2614B4 crank angle encoder. CO, HC and NO_x exhaust gas were measured by an AVL DICOM 4000 five component tail gas λ analyzer. The in-cylinder pressure was obtained by an AVL GU13Z-24 pressure transducer mounted in cylinder 2. The collected cylinder pressure and crank angle signals are fed into the Dewesoft SRIUSi combustion analyzer for analyzing and calculating the combustion data. The combustion analyzer collects 200 cycle values after the test is stabilized for every test point to take the average value. The gaseous emissions were measured by an AVL-DICOM 4000 five-component tail gas analyzer. DMS500 fast particle analyzer was used to measure particle emissions.

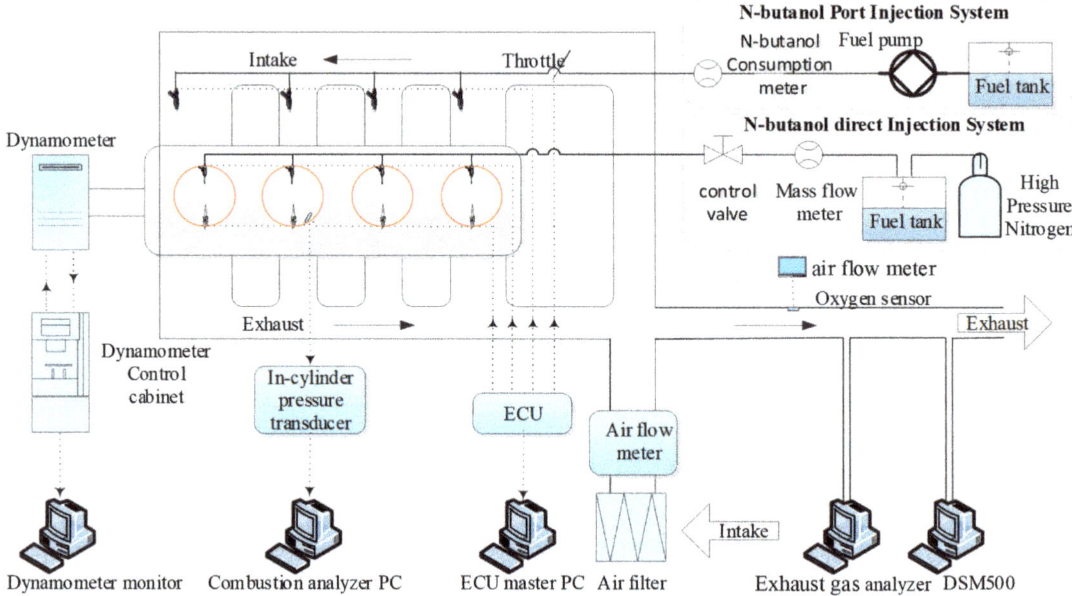

Figure 1. A schematic diagram of the test engine.

Table 3. The main test equipment of the experiment.

Parameter	Manufacturer	Range	Precision	Production Type
Speed	Luoyang Nanfeng Electromechanic Equipment Manufacturing Co., Ltd. (Luoyang, China)	0–6000 rpm	$\leq\pm1$ rpm	CW160
Crank angle	Kistler Instrument China Ltd. (China)	0–720°	$\leq\pm0.5°$	Kistler-2614B4
Excess air ratio	ETAS Engineering TOOLS (Germany)	0.700–32.767	$\leq\pm1.5\%$	LAMBDA LA4
Torque	Luoyang Nanfeng Electromechanic Equipment Manufacturing Co., Ltd. (Luoyang, China)	0–600 N m	$\leq\pm0.28$ N·m	CW160
Cylinder pressure	DEWETRON GmbH. (Austria)	0–20 MPa	$\leq\pm0.3\%$	AVL-GU13Z-24
n-butanol mass flow rate	Ono Sokki DF-2420 (Japan)	0.2~82 kg/h	$\leq\pm$ g/s	DF-2420
Carbon monoxide (CO)	AVL List GmbH (Austria)	0–10% vol	$\leq\pm0.01\%$ vol	AVL DICOM 4000
Hydrocarbon (HC)	AVL List GmbH (Austria)	0–20,000 ppm vol	$\leq\pm1$ ppm	AVL DICOM 4000
Nitrogen oxides (NOx)	AVL List GmbH (Austria)	0–5000 ppm vol	$\leq\pm1$ ppm	AVL DICOM 4000
Particle number concentration	British combustion (England)	0–1011 dN/dlogDp/cm^3	$\leq\pm1.4\times10^4$ dN/dlogDp/cm^3	DMS500

2.2. Injection Modes and Definitions

The fuel used in the experiment was 99.7% purity n-butanol. For ease of expression and understanding, the naming of different injection modes of n-butanol tested is reported in Table 4.

Table 4. The naming of different injection modes.

The Naming of Different Injection Modes	NPI	N20DI	N40DI	N60DI	N80DI	NDI
N-butanol direct injection ratio (NDIr)	0%	20%	40%	60%	80%	100%

2.3. Experimental Method

The experiment was mainly focused on the combustion and emission characteristics of an n-butanol SI engine adopting different injection modes under lean burn conditions. The experiments were conducted at n-butanol direct injection ratios 0%, 20%, 40%, 60%, 80%, 100%, and five excess air ratios 0.9, 1, 1.1, 1.2, 1.3. The engine was run at a constant speed of 1500 rpm, and the throttle opening was kept at 10%, a typical urban condition. The engine coolant temperature was kept at (85 ± 5) °C. The injection timings of the cylinder direct injector and port fuel injector were 180° and 300 °CA BTDC. The direct injection pressure was set at 5 MPa, and the injection pressures of the port fuel injector were 0.3 MPa. The injection duration is adjusted constantly according to different NDIr. To highlight the influence of NDIr, the ignition advance angles were all fixed at 15 °CA unless otherwise specified.

3. Results and Discussion

Fuel injection modes, NDIr, and λ have an important influence on the combustion and emission characteristics of the engine. This paper discusses the influence of direct injection mode on butanol engine performance from three aspects: combustion characteristics, gas emission characteristics, and particle emission characteristics.

3.1. Combustion Characteristics

Figure 2 shows Ttq with ignition time at different NDIr and λ. The error bar indicates standard deviations for each experimental data. The value of λ is set to 0.9, 1, and 1.2, which, respectively, represent the three conditions of rich burn, stoichiometry, and lean born. It can be seen that Ttq increases first and then decreases with the advance of ignition time at different NDIr and λ. This can be explained by the following reasons. With the increase in the ignition advance angle, more fuel burns before the TDC leads to a rise in the compression negative work. When the ignition time is too late, post-combustion occurs, resulting in a decrease in torque. We also can see that the torque of N20DI is nearly the highest among different NDIr. The specific reasons will be analyzed in detail in the next section. In addition, Figure 2a–c show that under different λ, the maximum torque corresponding to the ignition time is different. When the values of λ are 0.9 and 1, the MBT is 15 °CA BTDC. While MBT is 20 °CA BTDC at λ = 1.2. This is mainly caused by the slow combustion speed under lean burn conditions. The slow combustion speed makes the whole combustion duration long, and the ignition time needs to be advanced to obtain the appropriate combustion phase.

Figure 3 displays the Ttq with NDIr at different values of λ. It can be seen that Ttq rises first and then drops with the increase in NDIr, reaching the maximum value at NDIr = 20%. The Ttq values of N20DI increase by about 3.3%, 0.4%, 9.1%, 4.1%, and 2.6% for λ values of 0.9, 1.0, 1.1, and 1.2, respectively. This can be attributed to the following reasons. On the one hand, when n-butanol is directly injected into the cylinder, it will cause a localized over-dense layered mixture near the spark plug, promote combustion in the cylinder and increase Ttq [32]. On the other hand, the n-butanol injected directly into the cylinder evaporates and absorbs heat, causing a drop in the temperature of the cylinder. Additionally, with the increase of NDIr, the local rich-fuel region increases, resulting in the formation of an inhomogeneous mixture and, finally, mass-spread combustion [21]. When NDIr is lower (20%), all these beneficial factors predominate and encourage the Ttq to rise. Therefore, Ttq rises first and then drops as NDIr increases. Moreover, compared with the NDI, the NPI injection mode allows more time to obtain a better air-fuel mix, resulting in a more complete combustion of the in-cylinder mixture and a higher Ttq. In addition, Ttq decreases gradually with the increase in λ, which could be attributed to the rise in λ making the mixture of n-butanol and air lean so that cycle fuel feeding decreases, resulting in the decrease in Ttq.

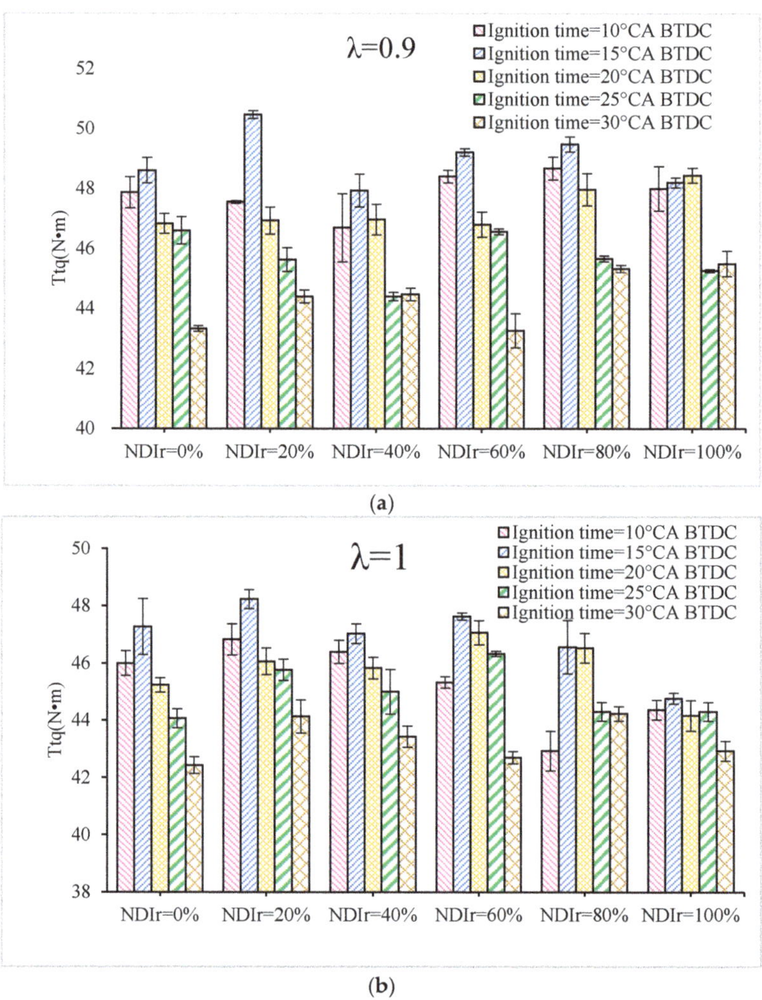

Figure 2. *Cont.*

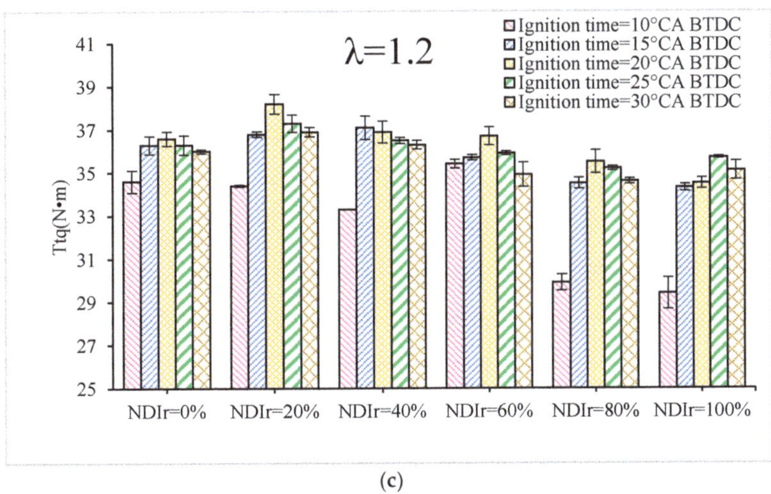

(c)

Figure 2. Ttq with ignition time at different NDIr under λ = 0.9, 1, 1.2. (**a**) λ = 0.9; (**b**) λ = 1; (**c**) λ = 1.2.

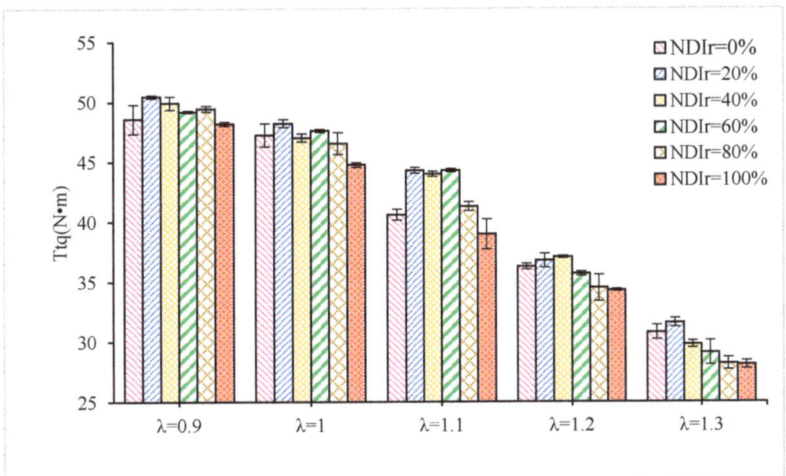

Figure 3. Ttq with NDIr at different values of λ.

Figure 4 displays the Pmax with NDIr at different values of λ. From Figure 4, we can see that Pmax shows similar trends as Ttq. Pmax goes up slightly and then down with the rising NDIr values, reaching the maximum value at NDIr = 20% and minimum value at NDIr = 100%. A small amount of butanol directly injected into the cylinder can result in a localized concentrated stratified mixture near the spark plug, which improves the combustion speed and promotes the combustion more completely. Thus Pmax is increased. As NDIr increases further, most of the fuel is injected directly into the cylinder resulting in a shorter time for the fuel to mix with the air. Thus the heterogeneous mixture weakens the advantage of the stratified mixture in the ignition, which results in incomplete combustion and decreases the Pmax. Therefore, Pmax increases first and then gradually decreases. In addition, Pmax decreases gradually with the increase in λ, which could be attributed to the rise in λ making the mixture of n-butanol and air lean so that cycle fuel feeding decreases, resulting in the decrease in Pmax.

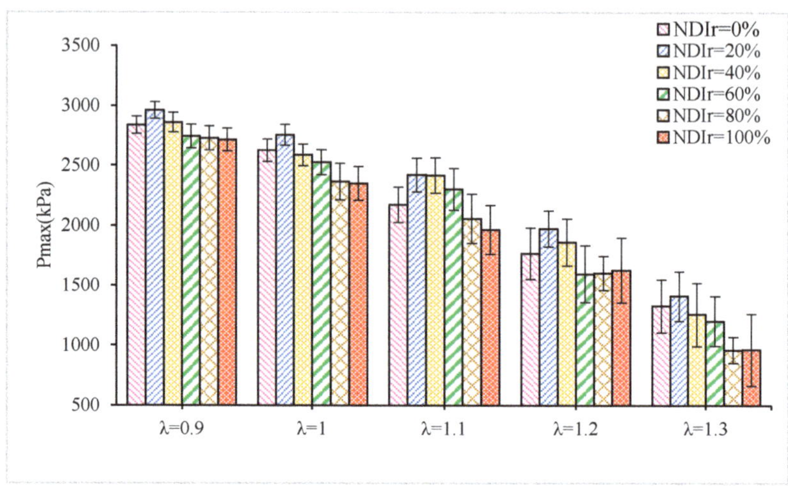

Figure 4. Pmax with NDIr at different values of λ.

From Figure 5, it can see that Tmax with NDIr at different λ. It can be seen that the Tmax shows a slight increase and then a gradual decrease with the increasing DIr, reaching the maximum value at NDIr = 20%. The results may be explained in two ways. Firstly, a dual injection could effectively combine the benefit of NDI and NPI, a small amount of n-butanol is injected directly into the cylinder, causing a localized concentration of the layered mixture near the spark plug, which promotes the combustion in the cylinder and leads to an increase in Tmax [13]. On the other hand, it can be attributed to is the properties of butanol itself. Because of the higher latent heat of vaporization, the n-butanol injected directly into the cylinder evaporates and absorbs heat, resulting in a decrease in Tmax [30]. The latter predominates when NDIr exceeds 20%, so the Tmax increases first and then decreases as the NDIr increases.

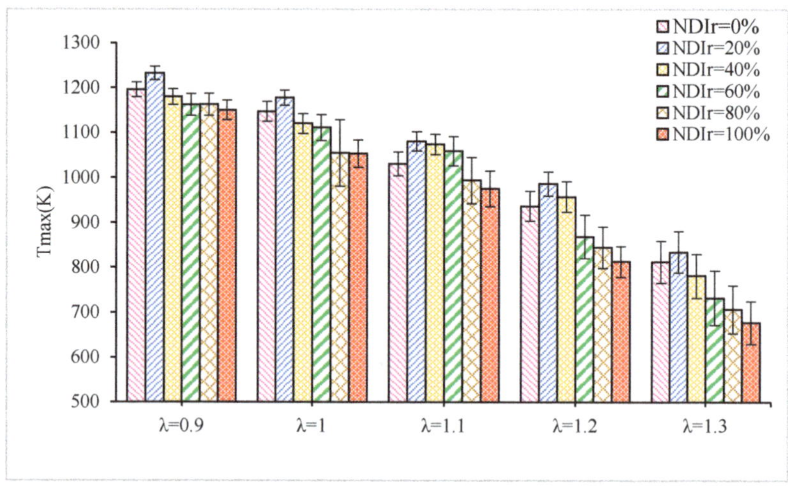

Figure 5. Tmax with NDIr at different values of λ.

Figures 6 and 7 show the maximum heat release rate (dQmax) and the dQmax (AdQmax) position with NDIr under different λ. It is shown that the dQmax has the same tendency as the Tmax. Additionally, the AdQmax decreases initially and then increases. It is clear that dual injection with 20% and 40% NDIr results in a higher dQmax. Meanwhile, the peak phase moves forward correspondingly, which enables combustion to release more energy around the TDC. This is mainly because the locally stratified mixture formed near the spark plug due to the small amount of n-butanol injected directly into the cylinder can ensure the ignition's reliability, creating a more stable flame core and promoting the early propagation of flame, thus promoting the exothermic process of the mixture.

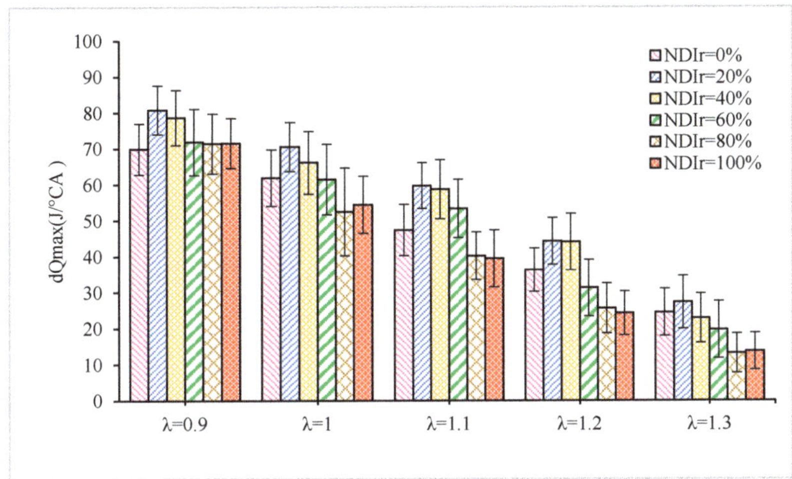

Figure 6. dQmax with NDIr at different values of λ.

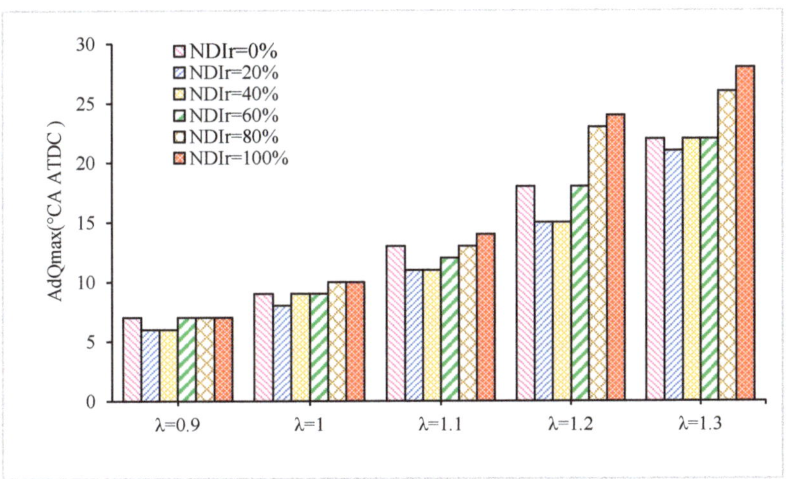

Figure 7. AdQmax with NDIr at different values of λ.

In this paper, the θ0-90 is defined as the crank angles for which 0–90% of the fuel mass has been burned. Figure 8 shows that the θ0-90 decreases first and then increases as NDIr increases. Figure 8 shows that compared to the single injection mode, the dual injection mode with NDIr values of 20% and 40% has obvious advantages in terms of θ0-90, especially under the larger λ. Compared to the results obtained with NPI and NDI, at λ = 0.9, 1, 1.1, 1.2, 1.3, the θ0-90 of N20DI decreases by 2 °CA, 2 °CA, 3 °CA, 4 °CA, 5 °CA, and, 2 °CA, 4 °CA, 7 °CA, 15 °CA, and 14 °CA, respectively. Combined with the previous analysis, dual injection mode with a relatively small NDIr can form a good layered mixture, which is beneficial to the ignition and the formation of a stable flame nucleus in the initial stage. Therefore, the smaller NDIr greatly increases the combustion speed. In addition, when the NDIr is constant, θ0-90 extends with the increase in λ. The increase in λ means the mixture gets leaner and the fuel burning rate slower, thus extending θ0-90.

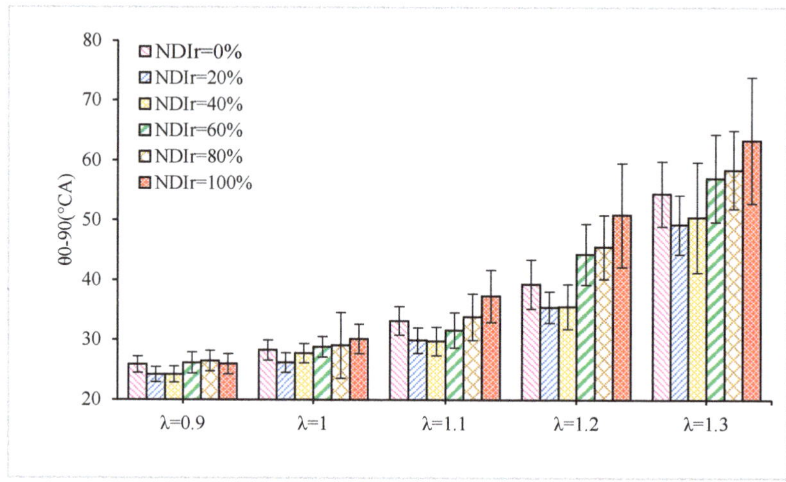

Figure 8. θ0-90 with NDIr at different values of λ.

Figure 9 shows the mean indicated pressure (IMEP) COV with NDIr under different λ. As seen from Figure 9, with the increase in NDIr, COV decreases first and then increases with NDIr, reaching a maximum at N20DI. It is interesting that COV changes significantly with DIr as λ increases. As can be seen in Figure 9, when λ is 1.2, compared with NPI and NDI, the COV of N20DI decreased by 30% and 60%, respectively. When λ is 1.3, compared with NPI and NDI, the COV of N20DI decreased by 30% and 90%, respectively. When λ is less than 1.2, the NDIr has little effect on COV under dual injection mode. However, comparing to NDI, the dual injection mode significantly improves engine stability. The reason for the above phenomenon is that the early stage of flame development is the main factor that causes the engine cyclic variation [33]. A small amount of butanol is directly injected into the cylinder, and the mixture is enriched near the spark plug to form stratified gas, which makes the ignition process more stable. In addition, it improves the flame propagation speed, reduces the influence of uncertain factors such as local misfires, and makes the engine run more smoothly. When the NDIr is relatively large, it can be seen from the previous analysis that the heterogeneity of the mixture and the fuel characteristics of butanol itself lead to the deterioration of combustion, leading to the increase in COV. The results show that the dual injection mode with NDIr of 20–40% can effectively reduce COV compared to NPI and NDI, especially under lean burn mode.

3.2. Gaseous Emissions Characteristics

Figure 10 shows the HC emissions with NDIr at λ values of 0.9, 1, 1.1, 1.2 1.3. It is shown that for a given λ. HC emissions decrease first and then increase with the increase in NDIr. This is because a small amount of n-butanol injected directly into the cylinder will cause a localized layered mixture near the spark plug, which makes it easy to form flame nuclei and makes the fuel burn faster and more completely. As the NDIr increases further, more fuel is injected into the cylinder, exacerbating the effect of preventing the formation of a homogeneous mixture. Additionally, the high latent heat of vaporization and the poor atomization of n-butanol lead to an uneven mixture. Therefore, the combustion process is incomplete, and HC emissions increase. We can also see from Figure 10 that HC emission first decreases and then increases with the increase in λ, reaching a maximum at λ = 0.9 and a minimum at λ = 1.2. This is because when λ is 0.9, more fuel is injected into the cylinder, significantly increasing localized n-butanol enrichment areas and incomplete combustion. On the other hand, the appropriate increase in λ increases the oxygen concentration in the cylinder, which is conducive to full fuel combustion. However, when λ = 1.2, the mixture is too thin, leading to misfire and incomplete combustion, and the HC emission increases accordingly. The above results show that the dual injection mode can effectively reduce HC emissions compared to NPI and NDI when the NDIr is below 60%.

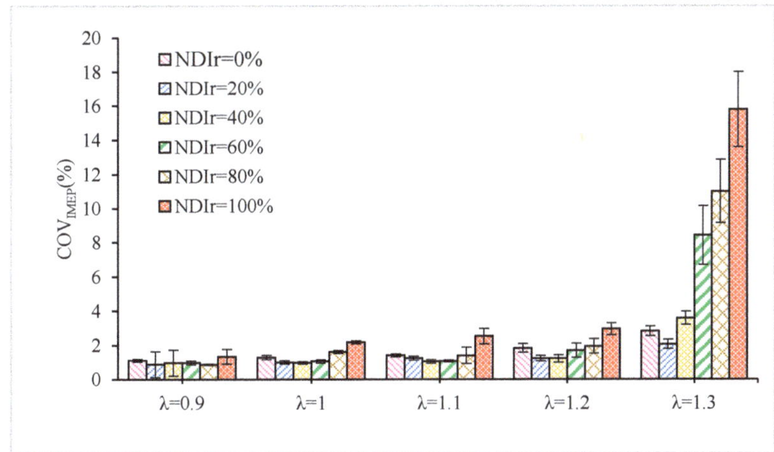

Figure 9. COV_{IMEP} with NDIr at different values of λ.

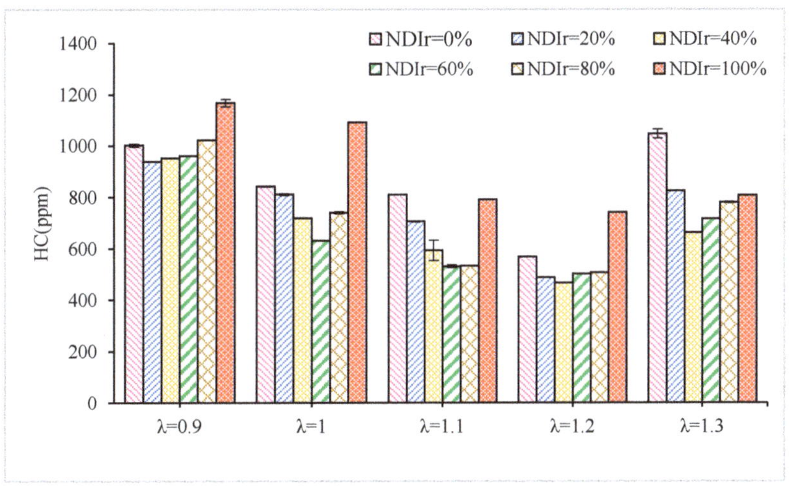

Figure 10. HC with NDIr at different values of λ.

Figure 11 shows the CO emissions with NDIr at λ values of 0.9, 1, 1.1, 1.2 1.3. Figure 11 shows that when λ is more than 0.9, CO emissions remain low and are little affected by NDIr. This may be explained by the fact that the n-butanol fuel itself contains oxygen and burns under the condition of sufficient oxygen, and the oxidation efficiency of CO is very high, so CO emissions are not affected by NDIr much. When λ is 0.9, the CO emissions are much higher than those obtained at the other four conditions. This is because when λ = 0.9, the mixture in the cylinder is rich, and the oxygen content is insufficient, leading to incomplete fuel combustion and higher CO emissions. The results indicated that compared with the in-cylinder direct injection, the dual injection mode of n-butanol can significantly reduce CO emission for the rich mixture.

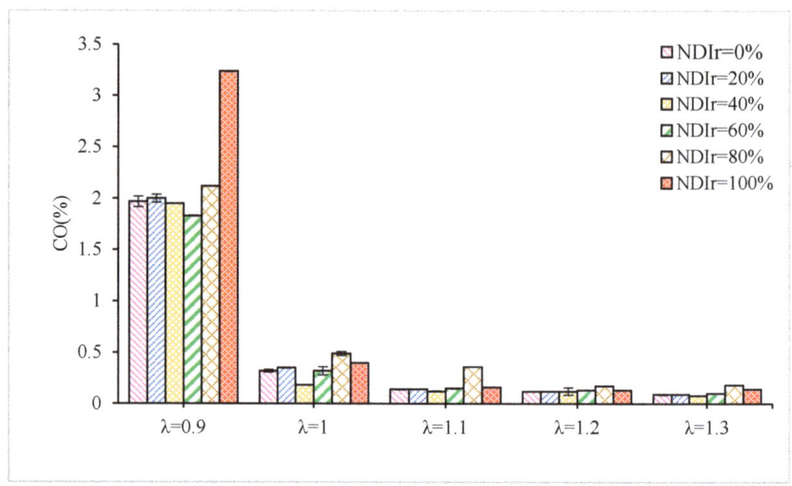

Figure 11. CO with NDIr at different values of λ.

Figure 12 shows the NOx emissions with NDIr at λ values of 0.9, 1, 1.1, 1.2 1.3. It can be seen that NOx emissions first increase slightly and then decreases s with the increasing NDIr. Especially when λ = 0.9, 1.0, and 1.1, the increase in NDIr has a significant effect on reducing NOx. NOx emission is very little at λ = 1.3. The above results may be explained by the following reasons. The formation of NOx emissions is relevant to cylinder temperature, oxygen concentration, and long residence time at high temperatures. A small amount of n-butanol direct injection promotes the combustion in the cylinder and increases the in-cylinder temperature, leading to an increase in NOx emissions at NDIr = 20%. As the NDIr increase further, NOx emissions tend to decrease. This is because a large amount of n-butanol is injected directly into the cylinder, causing the uneven mixing of the mixture because of the mixing time limitation, then leading to incomplete combustion. Secondly, a higher amount of fuel injected directly into the cylinder causes an increase in the specific heat capacity, adding to the effect of evaporation and atomization on the decrease in temperature. All of these jointly inhibit the generation of NOx emissions. So the NOx emissions first increase and then decrease. In addition, NOx emissions tend to increase initially and then decrease with λ, reaching a maximum at λ = 1. This is because when λ = 1, the oxygen content is sufficient, the combustion is sufficient, and the temperature in the cylinder is high, so the NOx emission appears the maximum. It is worth noting that the NOx emissions reduced significantly under the other four λ conditions compared with λ = 1. This indicates that the lack of oxygen and lean burn conditions can effectively reduce the NOx emission of n-butanol engines.

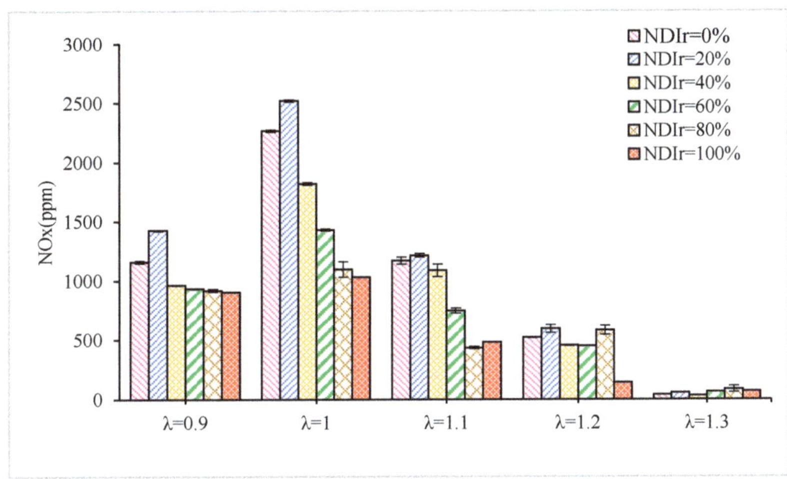

Figure 12. NOx with NDIr at different values of λ.

3.3. Particle Emissions Number

Particle emissions are important performance evaluation indexes of direct injection engines [34]. In the following section, NPN represents the number of nucleation mode particles. APN represents the number of accumulation mode particles [35]. TPN represents the total particle number [36]. NPN and APN can be used to show the number of different modes and their proportions in total particle emission, and TPN can be used to reveal the level of total particle number.

Figure 13 displays the APN with NDIr at λ values of 0.9, 1, 1.1, 1.2 1.3. When the λ is more than 0.9, the APN remains at a very low level and is little affected by NDIr. There are two possible reasons for this result. Firstly, with the increase in NDIr, the phenomenon of n-butanol spray-wall impingement increases, and the diffusion and combustion of the oil film on the wall surface intensifies, which may increase particle emission generation. Secondly, the formation of accumulation mode particles is closely related to the polycyclic aromatic hydrocarbons (PAHs). The hydroxyl radical (–OH) content produced during the combustion process of n-butanol can promote the oxidation of the PAHs [37]. Additionally, in the condition of sufficient oxygen, the formation conditions of high temperature and hypoxia of particles are inhibited. Thereby the APN remains at a very low level. At λ = 0.9, the APN increases significantly when NDIr is more than 80%. This is due to the combination of high-temperature hypoxia and direct injection mode, which leads to the increase in APN. Although the APN emission is relatively small, it is evident that when A = 0.9, compared with the in-cylinder direct injection, the dual injection mode can significantly reduce the APN.

It can be seen from Figure 14 that the NPN with NDIr at λ values of 0.9, 1, 1.1, 1.2, and 1.3. Figure 14 shows that the NPN drops first and then rises as NDIr decreases. Especially when the NDIr value is larger, NPN changes significantly. When the NDIr value ranges from 100% to 80%, NPN decreases by 73.3%, 70.9%, 65.9%, 73.5%, and 87.4% under different λ values. This can be explained in two parts. Firstly, with the decrease in NDIr, the phenomenon of n-butanol spray-wall impingement is significantly reduced, the diffusion and combustion of the oil film on the wall are improved, and the generation of particulate matter is also reduced. Secondly, according to Figure 10, HC emissions for dual injection mode with NDIr of 20% to 80% are lower than HC emissions with NDI, which should contribute to a reduction in NPN. In addition, as the λ increase, NPN decreases significantly, especially at the larger NDIr. At the condition of NDI, the increases at different λ cause a decrease in NPN by about 67.3%, 80.9%, 86.0%, and 75.9%. This is because when λ = 0.9, the fuel in the cylinder is over-rich, the mixture distribution is uneven, and the combustion is incomplete, leading to the increase in NPN. With the increase in λ, the oxygen content in the cylinder increases, and the formation conditions of hypoxia of particles are inhibited; thus, the concentration of NPN decreases. It can be concluded that, compared with the NDI, the dual injection mode can effectively reduce the NPN.

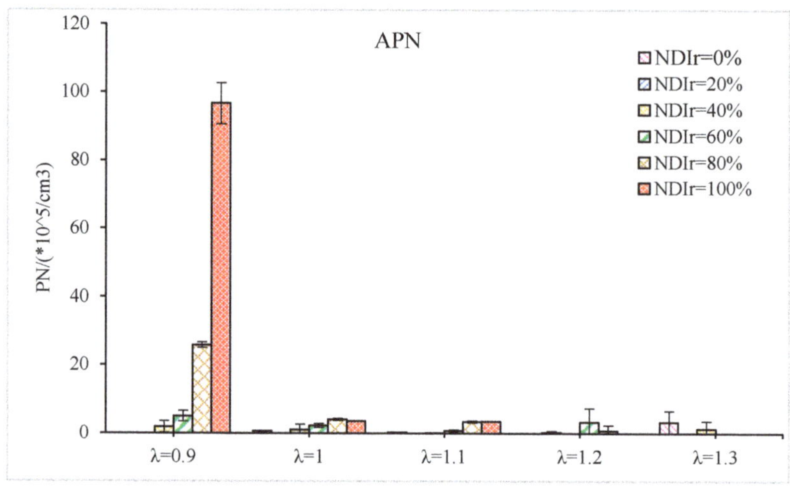

Figure 13. APN with NDIr at different values of λ.

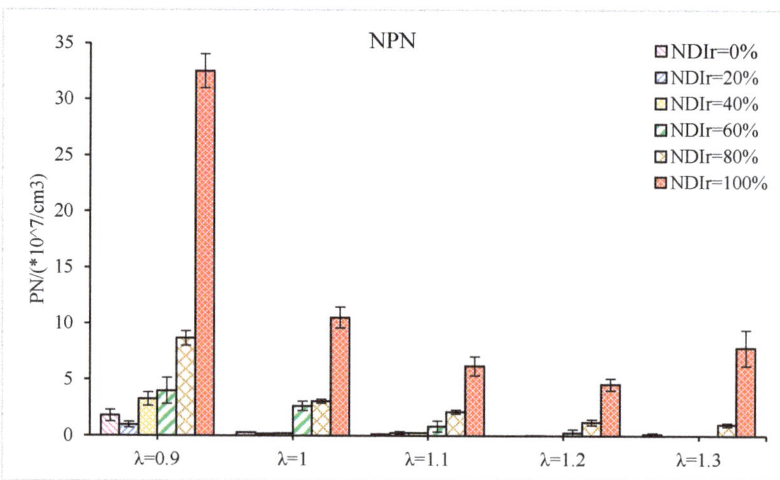

Figure 14. NPN with NDIr at different values of λ.

Figure 15 is a sum of Figures 13 and 14. The sum of NPN and APN for NDIr and different λ can be seen. The TPN shows similar trends as NPT, with an increase for increasing NDIr, and decreases initially and then increases slightly with the increasing λ. This is because APN is very low, almost negligible compared with NPN; the behavior of TNP can be explained in terms of the behavior of NPN.

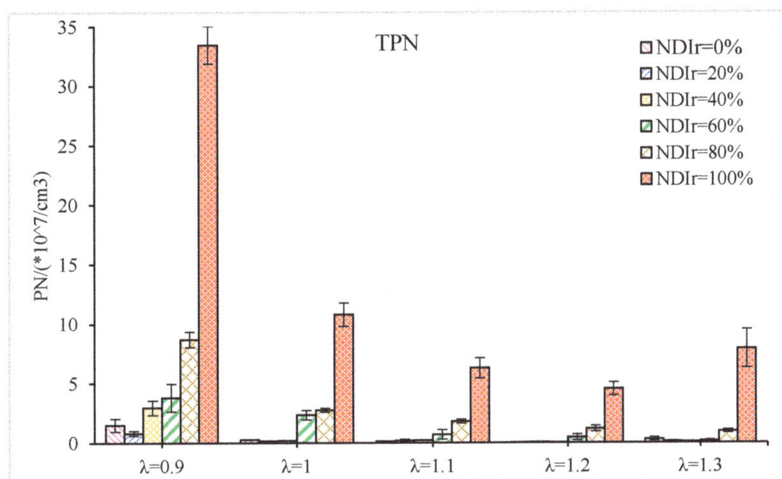

Figure 15. TPN with NDIr at different values of λ.

4. Conclusions

In this paper, the combustion and emission characteristics of the proposed combustion method were evaluated from several angles through experiments. The advanced technology of the engine and the advantages of the physicochemical properties of n-butanol were fully combined to explore a reasonable injection mode. The research in this paper will significantly contribute to butanol replacing gasoline as an effective and feasible method to reduce fossil energy consumption, obtain good combustion performance, and reduce emissions. The specific experimental conclusions are as follows:

1. Compared with the single injection mode, The dual injection of butanol engines with a smaller NDIr can form a local fuel-rich region near the spark, which is conducive to ignition and creates a more stable flame core, thus improving the comprehensive performance of butanol engines.

2. From the combustion performance index, Ttq, Pmax, Tmax, and dQmax all rise first and then drop, reaching the maximum value at NDIr = 20% and minimum value at NDIr = 100%. The AdQmax, θ0-90, and COV decrease first and then increase as NDIr increases. Thus, N20DI is considered to have an optimal combustion performance, followed by N40DI.

3. The dual injection mode has little affected on CO emission except for λ = 0.9, but can significantly reduce HC emissions than those relative to NPI and NDI. When λ is more than 0.9, the NDIr of 40–60% reduces HC emission most significantly, while the NDIr of 20% at λ = 0.9. NOx emissions increase in the dual injection mode, especially at N20DI, but not by much at NDIr of 40–80%.

4. NPN, APN, and TPN increase continuously as NDIr increases, the dual injection mode. The dual injection mode can effectively reduce particulate emissions relative to the in-cylinder direct injection mode, especially when λ = 0.9.

5. To summarize, the dual injection mode with NDIr of 20% to 40% significantly reduces HC and particulate emissions and maintains good combustion characteristics. In addition, the dual injection mode with a proper excess air ratio can effectively inhibit the increase in NOx emission caused by dual injection. Therefore, the dual injection mode with the proper NDIr can effectively optimize butanol engines' combustion and emission performance.

6. The combination of combined injection technology and butanol can reflect the excellent characteristics of butanol fuel to a greater extent, but the change of combustion state under different working conditions needs the real-time adjustment of combined

injection strategy to meet the effect of dynamic optimization under full operating conditions. Therefore, it is necessary to explore the combustion and emission characteristics of the combined injection butanol engine under more engine speed and load conditions and establish the global optimization control strategy in the next research stage.

Author Contributions: Conceptualization, W.S. and X.Y.; Data curation, W.S. and Z.G.; Formal analysis, W.S. and Z.G.; Funding acquisition, X.X.; Investigation, W.S. and Z.G.; Methodology, W.S. and X.Y.; Supervision, K.M.; Writing—original draft, W.S. and Z.G.; Writing—review and editing, X.Y., H.L. and K.M. All authors have read and agreed to the published version of the manuscript.

Funding: This research was funded by the Fundamental Research Funds for the Central Universities, grant number 2022-JCXK-23, and the Education Department of Jilin Province, grant number JJKH20220591KJ, and the Young Elite Scientists Sponsorship Program by CAST, grant number 2021QNRC001.

Institutional Review Board Statement: Not applicable.

Informed Consent Statement: Not applicable.

Data Availability Statement: Not applicable.

Conflicts of Interest: The authors declare no conflict of interest.

Nomenclature

SI	spark ignition	Pmax	peak in-cylinder pressure
NDIr	n-butanol direct injection ratio	Ttq	the engine torque
PFI	the port fuel injection	dQmax	the maximum rate of heat release
NDI	n-butanol direct injection	AdQmax	position of dQmax
NPI	port n-butanol injection	Tmax	peak in-cylinder temperature
MBT	maximum brake torque	θ0-90	total combustion duration
TDC	top dead center	COV	coefficient of variance
BTDC	before top dead center	TPN	total particle number
λ	excess air ratio	APN	accumulation mode particle number
IMEP	indicated mean effective pressure	TPN	nucleation mode particle number
HC	Hydrocarbon	CO	carbon monoxide
NOx	nitrogen monoxide or nitric oxide		

References

1. Bao, J.; Qu, P.; Wang, H.; Zhou, C.; Zhang, L.; Shi, C. Implementation of various bowl designs in an HPDI natural gas engine focused on performance and pollutant emissions. *Chemosphere* **2022**, *303*, 135275. [CrossRef] [PubMed]
2. Awad, O.I.; Mamat, R.; Ali, O.M.; Sidik, N.C.; Yusaf, T.; Kadirgama, K.; Kettner, M. Alcohol and ether as alternative fuels in spark ignition engine: A review. *Renew. Sustain. Energy Rev.* **2018**, *82*, 2586–2605. [CrossRef]
3. Algayyim, S.J.M.; Wandel, A.P. Comparative assessment of spray behavior, combustion and engine performance of ABE-biodiesel/diesel as fuel in DI diesel engine. *Energies* **2020**, *13*, 6521. [CrossRef]
4. Iliev, S.; Stanchev, H.; Mitev, E. An Experimental Investigation of a Common-rail Diesel Engine with Butanol Additives. In Proceedings of the 2020 7th International Conference on Energy Efficiency and Agricultural Engineering (EE&AE), Ruse, Bulgaria, 12–14 November 2020; pp. 1–4.
5. Kumar, B.R.; Saravanan, S. Use of higher alcohol biofuels in diesel engines: A review. *Renew. Sustain. Energy Rev.* **2016**, *60*, 84–115. [CrossRef]
6. Shi, C.; Chai, S.; Wang, H.; Ji, C.; Ge, Y.; Di, L. An insight into direct water injection applied on the hydrogen-enriched rotary engine. *Fuel* **2023**, *339*, 127352. [CrossRef]
7. Bao, J.; Wang, H.; Wang, R.; Wang, Q.; Di, L.; Shi, C. Comparative experimental study on macroscopic spray characteristics of various oxygenated diesel fuels. *Energy Sci. Eng.* **2023**, *11*, 1579–1588. [CrossRef]
8. Shang, W.; Yu, X.; Shi, W.; Xing, X.; Guo, Z.; Du, Y.; Liu, H.; Wang, S. Effect of exhaust gas recirculation and hydrogen direct injection on combustion and emission characteristics of a n-butanol SI engine. *Int. J. Hydrogen Energy* **2020**, *45*, 17961–17974. [CrossRef]

9.   Thangavel, V.; Momula, S.Y.; Gosala, D.B.; Asvathanarayanan, R. Experimental studies on simultaneous injection of ethanol–gasoline and n-butanol–gasoline in the intake port of a four stroke SI engine. *Renew. Energy* **2016**, *91*, 347–360. [CrossRef]

10.   Papagiannakis, R.G.; Rakopoulos, D.C.; Rakopoulos, C.D. Theoretical Study of the Effects of Spark Timing on the Performance and Emissions of a Light-Duty Spark Ignited Engine Running under Either Gasoline or Ethanol or Butanol Fuel Operating Modes. *Energies* **2017**, *10*, 1198. [CrossRef]

11.   Atmanli, A.; Yilmaz, N. A comparative analysis of n-butanol/diesel and 1-pentanol/diesel blends in a compression ignition engine. *Fuel* **2018**, *234*, 161–169. [CrossRef]

12.   Hergueta, C.; Bogarra, M.; Tsolakis, A.; Essa, K.; Herreros, J.M. Butanol-gasoline blend and exhaust gas recirculation, impact on GDI engine emissions. *Fuel* **2017**, *208*, 662–672. [CrossRef]

13.   Csemány, D.; DarAli, O.; Rizvi SA, H.; Józsa, V. Comparison of volatility characteristics and temperature-dependent density, surface tension, and kinematic viscosity of n-butanol-diesel and ABE-diesel fuel blends. *Fuel* **2022**, *312*, 122909. [CrossRef]

14.   Jeon, J.M.; Song, H.S.; Lee, D.G.; Hong, J.W.; Hong, Y.G.; Moon, Y.M.; Bhatia, S.K.; Yoon, J.J.; Kim, W.; Yang, Y.H. Butyrate-based n-butanol production from an engineered *Shewanella oneidensis* MR-1. *Bioprocess Biosyst. Eng.* **2018**, *41*, 1195–1204. [CrossRef]

15.   Ferreira, S.; Pereira, R.; Wahl, S.A.; Rocha, I. Metabolic engineering strategies for butanol production in *Escherichia coli*. *Biotechnol. Bioeng.* **2020**, *117*, 2571–2587. [CrossRef]

16.   Bao, T.; Feng, J.; Jiang, W.; Fu, H.; Wang, J.; Yang, S.T. Recent advances in n-butanol and butyrate production using engineered *Clostridium tyrobutyricum*. *World J. Microbiol. Biotechnol.* **2020**, *36*, 138. [CrossRef] [PubMed]

17.   Zhao, F.; Lai, M.C.; Harrington, D.L. Automotive spark-ignited direct-injection gasoline engines. *Prog. Energy Combust. Sci.* **1999**, *25*, 437–562. [CrossRef]

18.   Shang, Z.; Yu, X.; Shi, W.; Huang, S.; Li, G.; Guo, Z.; He, F. Numerical research on effect of hydrogen blending fractions on idling performance of an n-butanol ignition engine with hydrogen direct injection. *Fuel* **2019**, *258*, 116082. [CrossRef]

19.   Guo, Z.; Yu, X.; Sang, T.; Chen, Z.; Cui, S.; Xu, M.; Yu, L. Experimental study on combustion and emissions of an SI engine with gasoline port injection and acetone-butanol-ethanol (ABE) direct injection. *Fuel* **2021**, *284*, 119037. [CrossRef]

20.   Liu, H.; Wang, X.; Zhang, D.; Dong, F.; Liu, X.; Yang, Y.; Huang, H.; Wang, Y.; Wang, Q.; Zheng, Z. Investigation on blending effects of gasoline fuel with n-butanol, DMF, and ethanol on the fuel consumption and harmful emissions in a GDI vehicle. *Energies* **2019**, *12*, 1845. [CrossRef]

21.   Zhang, M.; Hong, W.; Xie, F.; Su, Y.; Han, L.; Wu, B. Experimental Investigation of Impacts of Injection Timing and Pressure on Combustion and Particulate Matter Emission in a Spray-Guided GDI Engine. *Int. J. Automot. Technol.* **2018**, *19*, 393–404. [CrossRef]

22.   Wang, Y.; Yu, X.; Ding, Y.; Du, Y.; Chen, Z.; Zuo, X. Experimental comparative study on combustion and particle emission of n-butanol and gasoline adopting different injection approaches in a spark engine equipped with dual-injection system. *Fuel* **2018**, *211*, 837–849. [CrossRef]

23.   Yang, J.; Ji, C. A comparative study on performance of the rotary engine fueled hydrogen/gasoline and hydrogen/n-butanol. *Int. J. Hydrogen Energy* **2018**, *43*, 22669–22675. [CrossRef]

24.   Ravikumar, R.; Antony, A.J. An experimental investigation to study the performance and emission characteristics of n-butanol-gasoline blends in a twin spark ignition engine. *Int. J. Mech. Prod. Eng. Res. Dev.* **2020**, *10*, 401–414.

25.   Agbro, E.; Zhang, W.; Tomlin, A.S.; Burluka, A. Experimental Study on the Influence of n-Butanol Blending on the Combustion, Autoignition, and Knock Properties of Gasoline and Its Surrogate in a Spark-Ignition Engine. *Energy Fuels* **2018**, *32*, 10052–10064. [CrossRef]

26.   Saraswat, M.; Chauhan, N.R. Comparative assessment of butanol and algae oil as alternate fuel for SI engines. *Eng. Sci. Technol. Int. J.* **2020**, *23*, 92–100. [CrossRef]

27.   Meng, F.; Yu, X.; He, L.; Liu, Y.; Wang, Y. Study on combustion and emission characteristics of a n-butanol engine with hydrogen direct injection under lean burn conditions. *Int. J. Hydrogen Energy* **2018**, *43*, 7550–7561. [CrossRef]

28.   Su, T.; Ji, C.; Wang, S.; Cong, X.; Shi, L.; Yang, J. Investigation on combustion and emissions characteristics of a hydrogen-blended n-butanol rotary engine. *Int. J. Hydrogen Energy* **2017**, *42*, 26142–26151. [CrossRef]

29.   Merola, S.S.; Irimescu, A.; Marchitto, L.; Tornatore, C.; Valentino, G. Effect of injection timing on combustion and soot formation in a direct injection spark ignition engine fueled with butanol. *Int. J. Engine Res.* **2017**, *18*, 490–504. [CrossRef]

30.   Sandhu, N.S.; Yu, X.; Leblanc, S.; Zheng, M.; Ting, D.; Li, T. *Combustion Characterization of Neat n-Butanol in an SI Engine*; SAE Technical Papers; SAE International: Warrendale, PA, USA, 2020.

31.   Kalwar, A.; Singh, A.P.; Agarwal, A.K. Utilization of primary alcohols in dual-fuel injection mode in a gasoline direct injection engine. *Fuel* **2020**, *276*, 118068. [CrossRef]

32.   Rakopoulos, D.C.; Rakopoulos, C.D.; Giakoumis, E.G.; Dimaratos, A.M.; Kyritsis, D.C. Effects of butanol–diesel fuel blends on the performance and emissions of a high-speed DI diesel engine. *Energy Convers. Manag.* **2010**, *51*, 1989–1997. [CrossRef]

33.   Heywood, J.B.; Vilchis, F.R. Comparison of flame development in a spark-ignition engine fueled with propane and hydrogen. *Combust. Sci. Technol.* **1984**, *38*, 313–324. [CrossRef]

34.   Tian, Z.; Zhen, X.; Wang, Y.; Liu, D.; Li, X. Combustion and emission characteristics of n-butanol-gasoline blends in SI direct injection gasoline engine. *Renew. Energy Int. J.* **2020**, *146*, 267–279. [CrossRef]

35.   Luo, J.; Zhang, Q.; Luo, J.; Zhang, Y. Particle Size Distributions of Butanol-Diesel and Acetone–Butanol–Ethanol (ABE)-Diesel Blend Fuels in Wick-Fed Diffusion Flames. *Energy Fuels* **2020**, *34*, 16212–16219. [CrossRef]

36. Liu, B.; Cheng, X.; Liu, J.; Pu, H. Investigation into particle emission characteristics of partially premixed combustion fueled with high n-butanol-diesel ratio blends. *Fuel* **2018**, *223*, 1–11. [CrossRef]
37. Yu, X.; Guo, Z.; He, L.; Dong, W.; Sun, P.; Shi, W.; Du, Y.; He, F. Effect of gasoline/n-butanol blends on gaseous and particle emissions from an SI direct injection engine. *Fuel* **2018**, *229*, 1–10. [CrossRef]

 *sustainability*

Article

Effect of Hydrogen-Rich Syngas Direct Injection on Combustion and Emissions in a Combined Fuel Injection—Spark-Ignition Engine

Zhen Shang [1,2], Yao Sun [2,*], Xiumin Yu [2], Ling He [2] and Luquan Ren [1]

[1] Key Laboratory of Bionic Engineering, Ministry of Education, Jilin University, Changchun 130022, China; shangzhen@jlu.edu.cn (Z.S.)
[2] State Key Laboratory of Automotive Simulation and Control, Jilin University, Changchun 130022, China
* Correspondence: syao@jlu.edu.cn; Tel.: +86-188-0430-3978

Abstract: To utilize the high efficiency of gasoline direct injection (GDI) and solve the high particulate number (PN) issue, hydrogen-rich syngas has been adopted as a favorable sustainable fuel. This paper compares and analyzes the effects of the injection configurations (GDI, gasoline port injection combined with GDI (PGDI), and gasoline port injection combined with hydrogen-rich syngas direct injection (PSDI)) and fuel properties on combustion and emissions in a spark-ignition engine. The operational points were fixed at 1800 rpm with a 15% throttle position, and the excess air ratio was 1.1. The conclusions show that PSDI gained the highest maximum brake thermal efficiency (BTE) at the MBT point, and the maximum BTE for GDI was only 94% of that for PSDI. PSDI's CoV_{IMEP} decreased by 22% compared with GDI's CoV_{IMEP}. CO and HC emissions were reduced by approximately 78% and 60% from GDI to PSDI among all the spark timings, respectively, while PSDI emitted the highest NO_X emissions. As for particulate emissions, PSDI emitted the highest nucleation-mode PN, while GDI emitted the lowest. However, the accumulation-mode PN emitted from PSDI was approximately 52% of that from PGDI and 5% of that from GDI. This study demonstrates the benefits of PSDI for sustainability in vehicle engineering.

Keywords: hydrogen-rich syngas; syngas direct injection; combined fuel injection; brake thermal efficiency; particulate number

Citation: Shang, Z.; Sun, Y.; Yu, X.; He, L.; Ren, L. Effect of Hydrogen-Rich Syngas Direct Injection on Combustion and Emissions in a Combined Fuel Injection—Spark-Ignition Engine. *Sustainability* **2023**, *15*, 8448. https://doi.org/10.3390/su15118448

Academic Editors: Cheng Shi, Jinxin Yang, Jianbing Gao and Peng Zhang

Received: 17 April 2023
Revised: 19 May 2023
Accepted: 19 May 2023
Published: 23 May 2023

1. Introduction

With the development of engine technologies, gasoline direct injection (GDI) and port fuel injection (PFI) have been widely used for modern spark-ignition engines. As direct fuel injection can be more accurately controlled both for quantity and timing, GDI generally provides better transient condition performance, which results in better fuel economy and lower CO_2 emissions [1,2]. However, previous research has concluded that GDI emits more particulate emissions than PFI because the time for mixture forming is relatively short, and locally fuel-rich regions appear more [3,4].

It is important to recognize the connection between particulate emissions and sustainability as well as take steps to reduce emissions and promote sustainable practices due to their adverse impact on air quality, human health, and the environment [5]. Therefore, to make GDI engines conform to the sustainability of vehicle engineering, GDI combined with PFI seems to be a promising method to improve efficiency and simultaneously solve the emissions issue [6–8]. Kang et al. developed a single-cylinder, four-stroke engine that adopts one direct injection system combined with one port injector. The results showed that the engine load characteristics were widened compared with a conventional spark-ignition direct injection (SIDI) engine, and knock reduction and engine flexibility can also be found in a dual-fuel dual-injection engine [9]. Sun et al. investigated the particulate number (PN) reduction and size distribution in a combined dimethyl ether/gasoline injection SI engine.

They concluded that by increasing the proportion of dimethyl ether direct injection, both the nucleation and accumulation modes of PN emissions drop remarkably [10].

On the other side, finding alternative fuels for internal combustion engines is also a feasible pathway for sustainable development to meet stringent emission regulations and solve the shortage of conventional fuel [11]. Synthesis gas (syngas) is considered an attractive substitute energy due to its abundant sources and clean combustion characteristics [12]. The feedstock for syngas can be biomass, coal, refinery coke, or even landfill waste, whilst the manufacturing methods include gasification, fuel reforming, and fermentation [13,14]. The combustible species of syngas are H_2, CO, and CH_4, and the inert diluents of syngas are mainly N_2 and CO_2. The high hydrogen content in syngas determines that syngas belongs to clean energy [15]. The production method affects the specific composition of syngas, for example, applying catalytic gasification technology to gasify biomass could yield a hydrogen-enriched synthetic gas with hydrogen and CO contents of up to 50% and 17% by volume [16]. Harun et al. reported that the CO contained in syngas could increase the knock limit, and the combustion duration was also prolonged [17]. Concerning hydrogen-rich syngas, the physicochemical characteristics of hydrogen can be inherited to some extent, while the sources of syngas are much more convenient than hydrogen [18,19]. Hydrogen has some excellent physicochemical characteristics, such as wide flammability, low minimum ignition energy, a high laminar burning velocity, and a small quenching distance [20]. Therefore, injecting syngas directly into a cylinder may have a better effect on reducing emissions than gasoline direct injection because of the fuel properties.

To figure out the environmental issues and sustainable development, the influences of syngas on gaseous emissions have been widely investigated [15]. Huang et al. studied how the ignition timing affects emissions from a syngas internal combustion engine containing hydrogen by using a spark plug reformer system [21]. They successfully developed a spark plug reformer system that can reduce power consumption and operate under a low operating temperature [22]. The experiment showed that when the spark timing was adjusted to the MBT, HC and NO_X emissions decreased, while CO_2 and CO emissions slightly increased with the use of syngas. Grzegorz et al. also concluded that hydrogen-rich syngas and high equivalence ratios cause a higher reaction temperature that favors NO_X emissions [23]. Harun et al. compared the content ratio of hydrogen and CO in syngas and obtained that the emissions were greatly related to the syngas composition. In particular, the NO level with a H_2/CO ratio of 2.36 was lower than that with a H_2/CO ratio of 0.62 even though it had a high exit temperature and hydrogen content [24]. Similarly, Ouimette et al. reported a different NO_X emission tendency of syngas in partially premixed combustion conditions. They indicated that NO_X emissions remained stable for syngas mixtures with a H_2/CO ratio of 0–1.3, whereas NO_X emissions exhibited a clear downward trend with higher ratios (>1.3) [25].

There are few published papers that have studied the particulate matter (PM) in syngas combustion, but the effect of hydrogen on particulate emissions has been investigated. Singh A. P. et al. compared the PM emissions from hydrogen-, CNG-, HCNG-, gasoline-, and diesel-fueled engines. They reported that hydrogen emitted the lowest PN and the lowest amount of PM among the researched fuels. Moreover, hydrogen enrichment of CNG reduced the total PM emissions [26]. Zhao et al. analyzed PM emissions from a GDI engine using a hydrogen and gasoline mixture. The findings revealed that under a low load, blending 5% hydrogen into the stoichiometric mixture can lower the total PM mass and PM number by up to 90%, and a further reduction in the total mass to 95% as well as in the total number to 97% can be achieved with 10% hydrogen. Nevertheless, under a high load, although hydrogen addition decreased the number of smaller particles, it encouraged the generation of accumulation-mode particles [27].

Thus, in this study, to realize both high thermal efficiency and low engine emissions, the adoption of hydrogen-rich syngas direct injection in a combined fuel injection–SI engine was evaluated. Two sets of experiments were formulated. The first one was about the comparison between GDI and gasoline port injection combined with GDI (PGDI) and aimed to analyze the configuration characteristic of the combined injection system. The second set of experiments was about the comparison between PGDI and gasoline port injection combined with syngas direct injection (PSDI), aiming to further analyze the fuel properties of syngas. In addition, from our previous research, a comparison between the injection modes containing GDI, PGDI, and gasoline port injection combined with hydrogen direct injection was carried out, and the experiments showed that the combination of an injection mode with fuel had a great influence on engine behavior [28–30], which means this specific investigation on the combustion, gaseous emission, and PN emission characteristics of syngas, which have not been studied yet, is novel and will be useful to meet sustainable development requirements.

2. Materials and Methods

2.1. Experimental Setup

The prototype engine was an in-line four-cylinder water-cooled spark-ignition direct injection (SIDI) engine, which originally had two injection systems, one direct injection system, and one port injection system. The cylinder head was furnished with centrally mounted spark plugs and direct injectors situated between intake valves. Port injectors were arranged on intake manifolds. Table 1 lists the detailed specifications of the prototype engine. The injection timing and duration for both gasoline and syngas were controlled with a self-developed ECU that can also adjust the throttle position and ignition timing. The syngas used in this experiment was hydrogen-rich syngas with fixed 85% H_2 and 15% CO by volume to eliminate interferences [31,32]. The specifications of the gasoline and syngas are shown in [28] and [33]. To maintain the operating state with a fixed excess air ratio (λ), the requested fuel distribution ratio was obtained through the precise monitoring of the airflow, syngas flow, and gasoline flow. Moreover, due to the wide flammability of hydrogen, a flame arrestor was installed within the hydrogen supply line to avoid backfire. Figure 1 illustrates the overall schematic layout of the experimental system.

Table 1. Original engine specifications.

Parameter	Unit	Value
Compression ratio	-	9.6
Total displacement	L	1.984
Stroke	mm	92.8
Bore	mm	82.5
Rated torque	Nm, rpm	350, 1500–4500
Rated power	kW, rpm	160, 4500–6200

During the experiments, the engine's speed and torque were measured using an ECD CW160 that was controlled with an FST-OPEN system. A GU13Z-24 piezoelectric pressure sensor was used to measure the cylinder pressure, and the crank angle was measured with a Kistler-2614B4 encoder. Both sets of data were transmitted to a combustion analyzer to calculate the real-time cylinder pressure and analyze the combustion process through the combustion analysis software (DS 0928). The excess air ratio was recorded with a Meter LA4 lambda sensor, and a wideband oxygen transducer was mounted onto the exhaust pipe. The syngas mass flow was monitored with a DMF-1-1 AB gas flow meter. In addition, the regular gas emissions could be measured simultaneously with a DiCom 4000 analyzer. Particulate emissions were recorded using a DMS500, which can measure the particulate number and provide the PN size distribution. The specific information on the testing instruments including the measuring error is shown in Table 2.

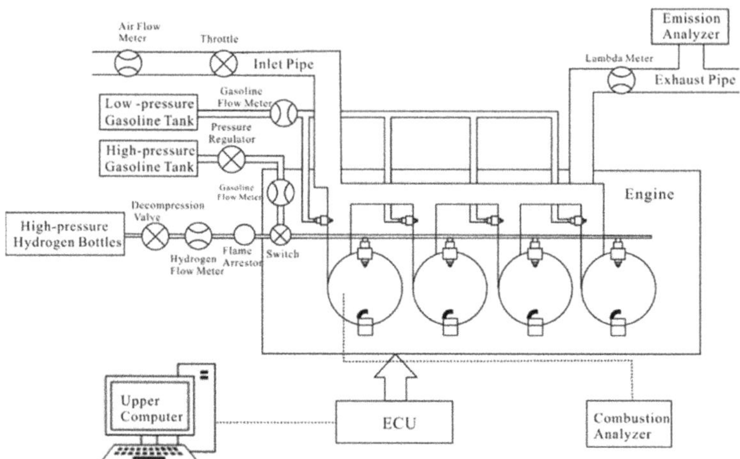

Figure 1. Schematic layout of the experimental system.

Table 2. Information on testing instruments.

Apparatus	Parameter	Manufacturer	Type	Uncertainty
Dynamometer	Engine speed Torque	LY Nanfeng	CW160	$\leq\pm1$ rpm $\leq\pm0.28$ Nm
Pressure sensor	Cylinder pressure	AVL	GU 13Z-24	$\leq\pm0.3$ bar
Lambda analyzer	Excess air ratio	ETAS	LAMBDA LA4	$\leq\pm0.1$
Gas flowmeter	Syngas quantity	Beijing SINCERITY	DMF-1-1 AB	$\leq\pm0.01$ g/s
Fuel flowmeter	Gasoline quantity	ONO SOKKI	DF-2420	$\leq\pm0.01$ g/s
Emission analyzer	CO HC NO_X	AVL	DiCom 4000	$\leq\pm0.01\%$ $\leq\pm30$ ppm $\leq\pm20$ ppm
Fast particulate analyzer	Particulate emissions	CAMBUSTION	DMS500	$\leq\pm1\%$

2.2. Experimental Procedures

In all the experiments, the engine speed was constant at 1800 rpm, representing a normal urban situation. Some research studies have indicated that different injection modes could affect volumetric efficiency [34]. To highlight the influence of volumetric efficiency, a low engine load associated with a small throttle position was chosen. Therefore, the throttle position was fixed at 15% for all the experiments, and other loads will be investigated in future work. The port fuel injection was aimed to form a homogenous mixture, and the port injection time was set at 300 crank angle degrees before the top dead center (CAD BTDC). Both the syngas and gasoline direct injection pressures were set at 3 MPa to unify the standards. In previous studies, when hydrogen was injected at 100 CAD BTDC, the engine exhibited the best efficiency and the highest power output, so the direct injection timings in this study were set at 100 CAD BTDC (both for syngas and gasoline) [35]. As low-temperature combustion achieved via lean-burn has been proven to be beneficial for increasing engine efficiency and decreasing emissions [36,37], and syngas can extend the lean-burn limit, the experiments were conducted under lean-burn conditions. The λ was fixed at 1.1, and the definition of λ is shown as follows:

$$\lambda = \frac{V_{air}\rho_{air}}{m_s AF_s + m_g AF_g} \tag{1}$$

In Equation (1), V_{air} represents the air volumetric flow rate, and ρ_{air} represents the density of air. m_s and m_g denote the measured mass flow rates of syngas and gasoline. Especially for the PDI mode, m_g equals the total gasoline mass flow rate including both direct injection and port injection. AF_s and AF_g are equal to 29.52 and 14.6 as the stoichiometric air–fuel ratios of syngas and gasoline, respectively.

The definition of the fuel distribution ratio is indicated in Equation (2) representing the energy ratio:

$$\Phi_D = \frac{q_D}{q_D + q_P} \tag{2}$$

In Equation (2), Φ_D denotes the fuel distribution ratio, q_D and q_P represent the heat released by the direct injection fuel (the gasoline direct injection portion in PGDI and that of syngas in PSDI) and port injection fuel, respectively. The heat is calculated from the mass flow rate and the low heating value (LHV). As the onboard syngas production amount was limited, the direct injection portion was regarded as an improver to fit the practical application [38]. Therefore, the Φ_D for the PGDI mode was fixed at 20%, while the PSDI mode had a lower Φ_D of 10%.

To find the effect on the combustion and emissions characteristics at various sparking timings, a range of spark timings from 10 CAD BTDC to 30 CAD BTDC were employed. The combustion parameters were measured and averaged in 200 continuous cycles at all test points. Concretely, the cylinder pressure, brake thermal efficiency (BTE), combustion durations (CA 0–10 and CA 10–90), indicated mean effective pressure (IMEP), and coefficient of variation in IMEP (CoV$_{IMEP}$) were measured to analyze the combustion process. Moreover, the CO, HC, NO$_X$, and PN emissions were recorded in the experiments to study the emission characteristics. Notably, the comparisons of the three injection modes are novel and will be useful for future vehicle engineering sustainability.

3. Results and Discussion

3.1. Cylinder Pressure and Brake Thermal Efficiency

Figures 2 and 3 plot the in-cylinder pressure for the three kinds of injection modes with spark timings at 10 CAD BTDC and 25 CAD BTDC. The cylinder pressures for PSDI show the highest cylinder pressure, while GDI presents the lowest value in both figures. Firstly, investigating the injection configuration and then analyzing the fuel properties indicated some detailed effects among the three injection modes. Port fuel injection has more time to form a homogenous mixture beneficially, and the direct injection portion can achieve stable and reliable ignition, so the combustion efficiency and cylinder pressure can be increased with combined fuel injection. This conclusion is similar to that in [39], which mainly focuses on fuel consumption and half-load performance. Furthermore, as syngas has a high laminar flame speed that can enhance the constant volume combustion degree, and its small quenching distance also facilitates more complete combustion, the cylinder pressure of PSDI performs higher than that of PGDI.

With 2 spark timings, the relevant crank angles for the maximum cylinder pressure of PSDI were only 16 CAD ATDC and 1 CAD ATDC, respectively. The relevant phasing was much more advanced than that of the GDI and PGDI modes, mainly because of the dramatically shortened combustion period. As syngas has a high laminar burning velocity, the combustion is more concentrated in the top dead center (TDC) with a specific spark timing. The cylinder pressure is an integrated result of the piston motion and combustion process, and the piston motion produces the highest pressure around the TDC (due to the smallest displacement). The concentrated combustion moves the highest cylinder pressure closer to the TDC, which enlarges the integrated pressure. Furthermore, advancing ignition lets more fuel burn before the TDC, and the accumulated heat before the TDC gradually increases, which results in higher in-cylinder pressure. However, the combustion deteriorates, as the relevant crank angle for the highest cylinder pressure is too early with an over-advanced spark timing.

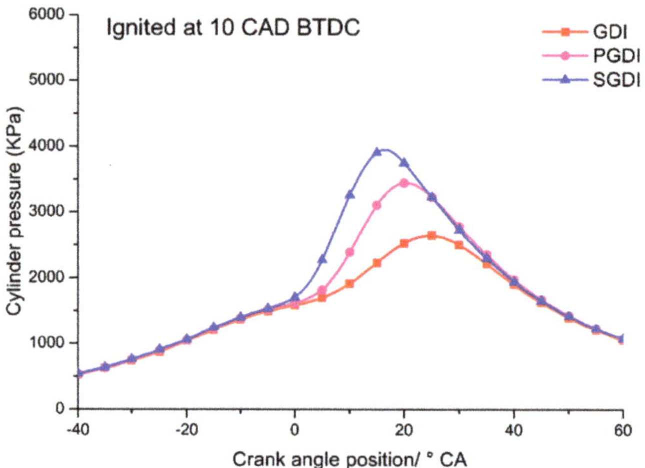

Figure 2. In-cylinder pressure. Spark timing set at 10 CAD BTDC for 3 kinds of injection modes.

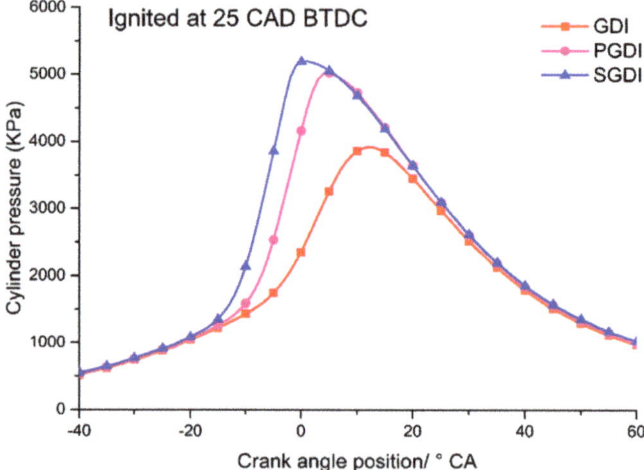

Figure 3. In-cylinder pressure. Spark timing set at 25 CAD BTDC for 3 kinds of injection modes.

Figure 4 plots the brake thermal efficiency (BTE) at various spark timings for three kinds of injection modes. As the minimum spark advance for best torque (MBT) for GDI and PGDI is 20 CAD BTDC, while the MBT for PSDI is only 10 CAD BTDC, the investigated ranges of ignition timings had a slight difference. When the spark timing was set at the MBT, PSDI demonstrated the highest maximum BTE (30.3%), while GDI showed the least maximum value (28.7%), and the maximum BTE of PGDI was 29.8%. The maximum BTE increased by 1.6% from GDI to PSDI. PGDI can form a more homogeneous mixture, producing complete combustion and higher BTE. As syngas has high diffusive efficiency and low ignition energy, the homogenous condition was further enhanced, and the requirement for combustion declined. These factors are all beneficial for complete combustion, which is critically beneficial for BTE. Moreover, a high laminar burning velocity shortens the combustion period and reduces heat transfer loss. As such, PSDI has higher BTE than GDI because of more complete combustion, higher combustion efficiency, and less heat transfer loss.

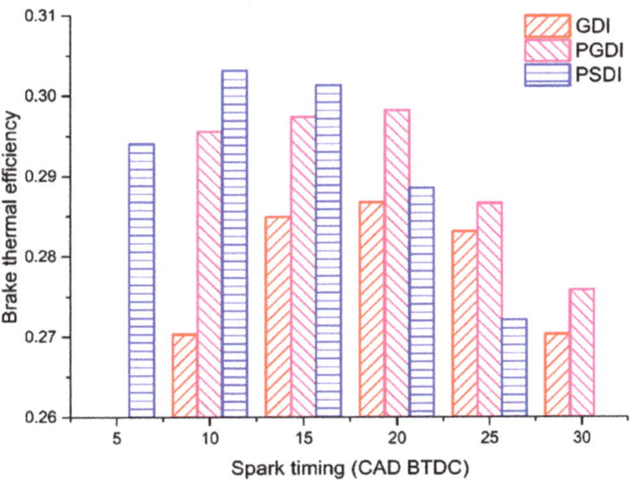

Figure 4. BTE versus spark timings for three kinds of injection modes.

Figure 4 also shows that the BTE first rises and then descends with an advancing spark timing. This is because an over-advanced spark timing makes more fuel burn during the compression stroke which produces more negative compression work and decreases the BTE. Comparatively, over-retarding the spark timing would enlarge the extent of post-combustion, and more fuel would burn during the expansion stroke, resulting in lower BTE. It is worth noting that when the spark advance angle was larger than 15 CAD, the BTE of PSDI decreased quickly and was even worse than that of GDI. This is because the combustion duration for PSDI is relatively short, and over-advancing ignition causes most of the combustion to be completed before the TDC, and the negative compression work increases. Subsequently, the BTE for PSDI declined dramatically. As the combustion speeds are lower in GDI and PGDI, the influences by over-advancing ignition are relatively small for these two modes. Thus, PSDI is more sensitive to spark timings because of the high flame speed.

Figure 5 shows the cylinder temperature at the exhaust valve opening (T_{EVO}) versus the spark timings for the three kinds of injection modes. The T_{EVO} reflects the engine post-combustion and exhaust loss to some extent [40]. The cam phasing was fixed just as the original engine and the exhaust valve opened at 26 CAD before the bottom dead center (BBDC). The T_{EVO}s for the GDI and PGDI modes are relatively similar, and the T_{EVO} for the PSDI mode is substantially higher than those for the other modes. The exhaust temperature for PSDI shows the highest value because of the fast burning velocity and higher maximum cylinder pressure. The T_{EVO} mainly affected the emissions characteristics, and the details will be discussed in Section 3.3.

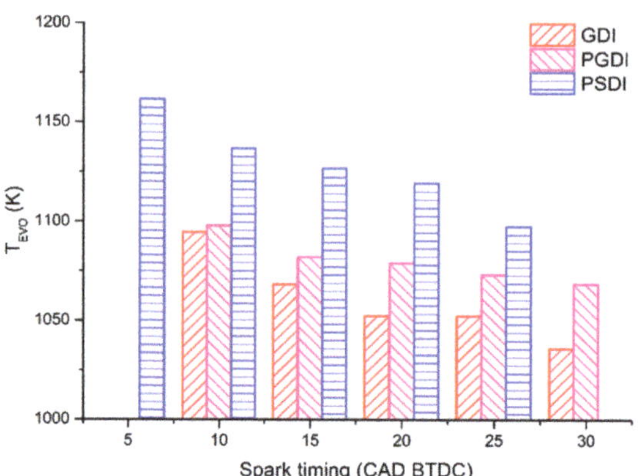

Figure 5. T_{EVO} versus spark timings for three kinds of injection modes.

3.2. Combustion Analysis

Figure 6 plots the flame development duration (0–10% mass fraction burned; CA 0–10) at various spark timings for the 3 kinds of injection modes, and Figure 7 plots the flame propagation duration (10–90% mass fraction burned; CA 10–90). PGDI shows a shorter duration for both CA 0–10 and CA 10–90 than GDI, and PSDI demonstrates the shortest durations for both parts. The combined injection mode can form a more stratified mixture than the pure GDI mode, which eases the formation of the flame kernel and accelerate the whole combustion period, so PGDI indicates shorter combustion durations. According to previous studies, adding hydrogen to methane and ethanol stimulates the formation of O, OH, and H radicals, which are kinds of improvers for chain reactions [41,42]. The syngas used in this investigation was hydrogen-rich, so it inherited some hydrogen characteristics, and gasoline is a kind of hydrocarbon fuel that is chemically similar to ethanol and methane. Thus, the formation of radicals stimulated via syngas accelerates combustion and shortens the combustion period. A shorter CA 0–10 means more stable combustion, while a shorter CA 10–90 means the combustion process is closer to the constant volume process, allowing higher thermal efficiency [43]. Hence, it is further certified that PSDI has the most stable combustion and the highest thermal efficiency among the three injection modes. On the other hand, by advancing the ignition timing, CA 0–10 was prolonged while CA 10–90 was shortened for all three injection modes. Advancing ignition deteriorates the initial condition for forming the flame kernel, and CA 0–10 is prolonged. Retarding ignition expands the degree of post-combustion, and the combustion at this stage is mainly diffusion combustion with a low burning rate. Thus, the CA 10–90 duration is extended by retarding the ignition timing.

Figure 8 plots the coefficients of variation (CoVs) in the IMEP at various spark timings for the three kinds of injection modes. The CoV_{IMEP} of GDI is dramatically higher than those of the other two modes, and it is very sensitive to spark timings. Specifically, the minimum CoV_{IMEP} of PSDI decreases by 22% from the minimum CoV_{IMEP} of GDI. Many investigations have indicated that a short combustion duration helps to ease an engine's cyclic variations [44]. It is seen in Figures 6 and 7 that the combustion duration of GDI is obviously long, leading to a high CoV_{IMEP}. Figure 8 also indicates that the CoV_{IMEP} reaches its lowest value at around the MBT point.

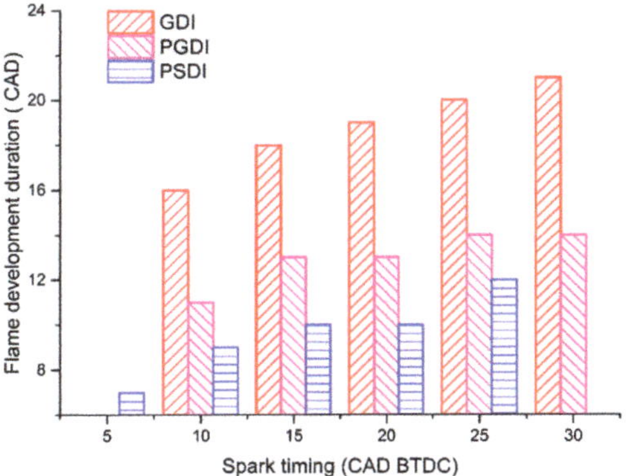

Figure 6. CA 0–10 versus spark timings for three kinds of injection modes.

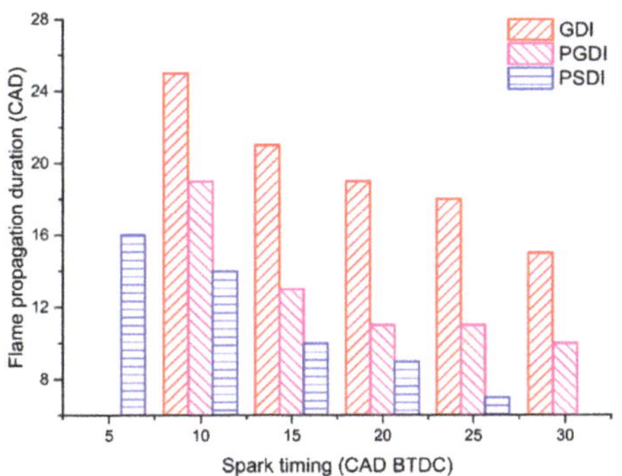

Figure 7. CA 10–90 versus spark timings for three kinds of injection modes.

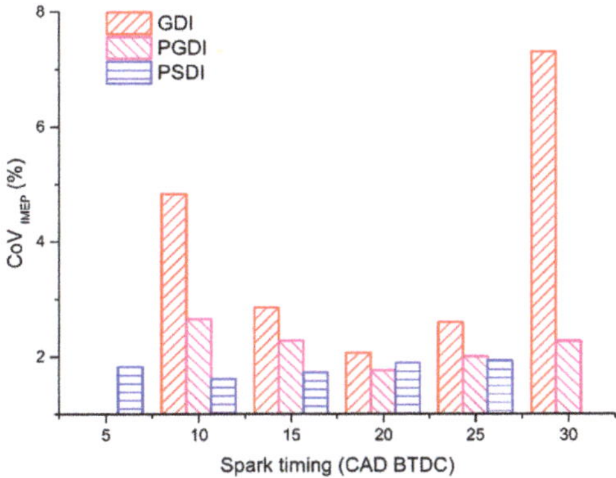

Figure 8. CoV_{IMEP} versus spark timings for three kinds of injection modes.

3.3. Gaseous Emissions and PN Emissions

Figure 9 plots CO emission at various spark timings for the three kinds of injection modes. GDI emits the highest CO emission, while PSDI emits the least. CO emissions were reduced by approximately 78% from GDI to PSDI among the whole range of spark timings. However, a 15% volumetric fraction of the syngas used in this experiment was CO, and the CO emission still shows the least value for PSDI, which further certifies that PSDI is beneficial for complete combustion and shows better stability. Ji C. et al. concluded that adjusting the spark timing has no clear effect on the CO emission of a hybrid hydrogen–gasoline engine [45], which agrees with the PSDI profile. CO emission decreased by retarding ignition for the GDI and PGDI modes, as the prolonged post-combustion induced by retarding ignition enhances the oxidation of CO.

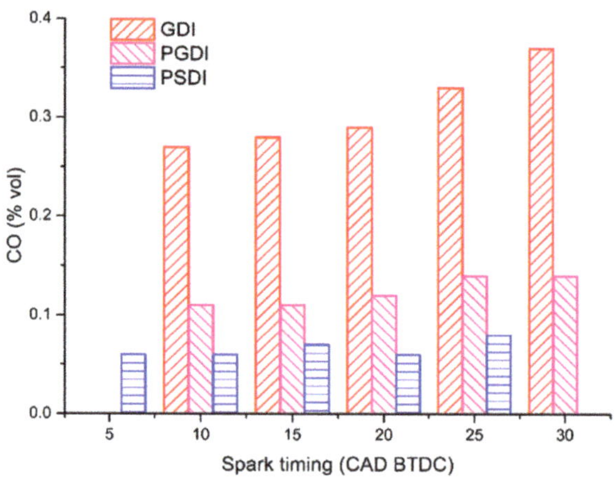

Figure 9. CO emissions versus spark timings for three kinds of injection modes.

As is shown in Figure 10, GDI emits the highest HC emission, and PGDI and PSDI only emit small amounts. HC emissions were reduced by approximately 60% from GDI to PSDI with a 20 CAD BTDC spark timing. This is because the combined injection mode improves combustion completeness, which reduces the sources of HCs. For the GDI mode, HC emission decreased by retarding ignition, the reason for which could be ascribed to the enhanced oxidizing process caused by prolonged post-combustion and increased T_{EVO}.

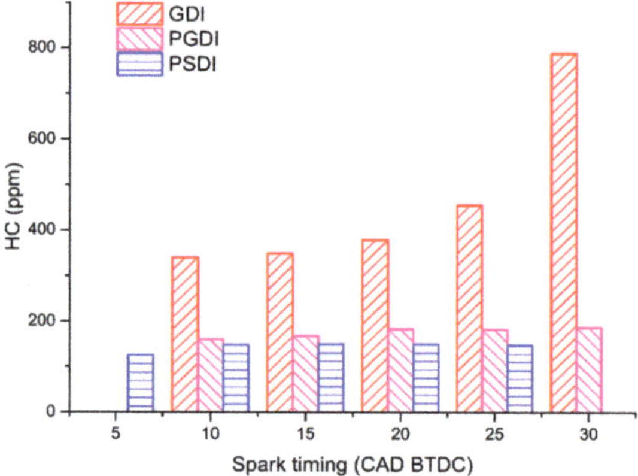

Figure 10. HC emissions versus spark timings for three kinds of injection modes.

Figure 11 plots the NO_X emissions at various spark timings for the three kinds of injection modes. PSDI produced the highest NO_X emission, whereas GDI produced the lowest NO_X emission. Specifically, the NO_X emission from GDI was 44% and 32% of that of PGDI and PSDI, respectively. The main factors affecting NO_X emissions include the oxygen fraction, cylinder temperature, and reserved time for high-temperature reactions [46]. As the data in Figure 11 show, PSDI still has a high NO_X emission problem. However, as syngas direct injection avails stable combustion, lean-burn combustion and exhaust gas recirculation (EGR) technologies could be used to reduce NO_X emissions without causing much damage to the engine performance according to [47]. Figure 11 also indicates that retarding ignition linearly reduces NO_X emissions. This is because retarding ignition decreases the maximum cylinder temperature, which is a critical factor for NO_X emission.

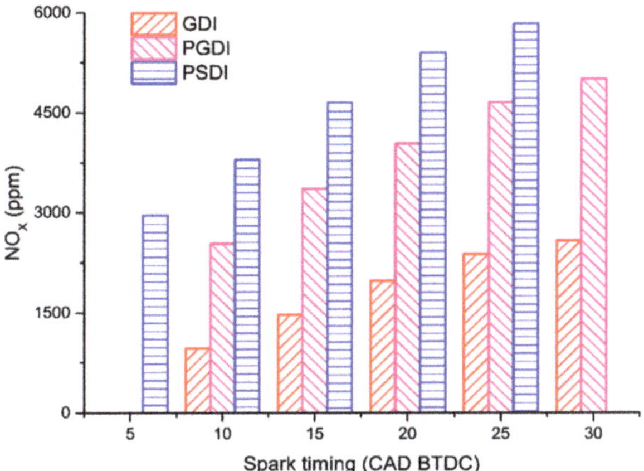

Figure 11. NO_X emissions versus spark timings for three kinds of injection modes.

Particulate emissions are a common issue for SIDI engines, and SIDI engines always have high particulate numbers relative to diesel engines, so the particulate emission numbers will be focused on herein. Figure 12 plots the total nucleation-mode PNs at various spark timings for the three kinds of injection modes, and Figure 13 shows the total accumulation-mode PNs for the three kinds of injection modes. Particles in nucleation mode are typically composed of volatile organic and sulfur compounds that form during exhaust dilution and cooling. Meanwhile, particles in accumulation mode are mainly made up of carbonaceous agglomerates and associated adsorbed materials [48–50].

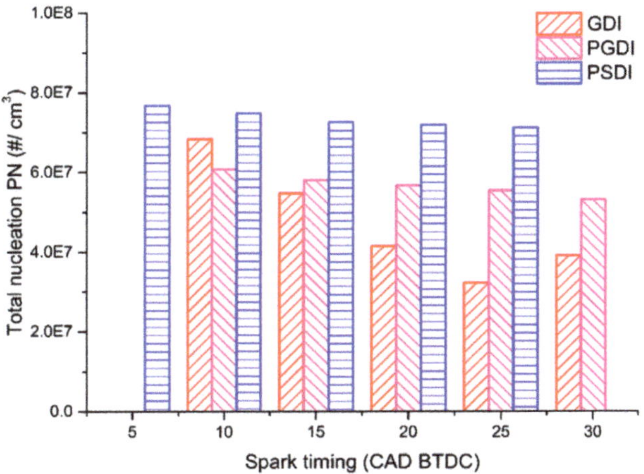

Figure 12. Total nucleation-mode PNs versus spark timings for three kinds of injection modes.

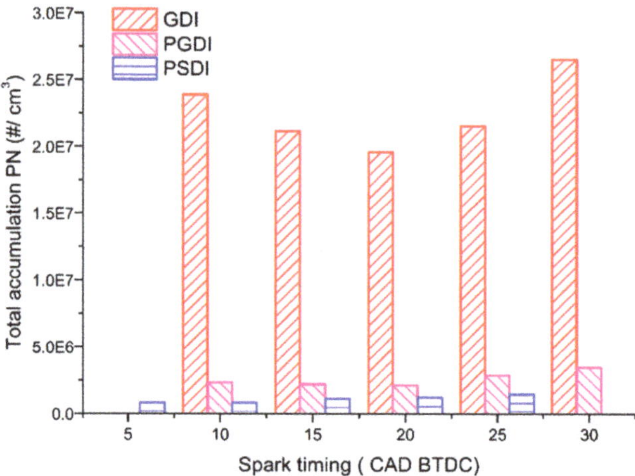

Figure 13. Total accumulation-mode PNs versus spark timings for three kinds of injection modes.

Regarding the nucleation-mode PN, PSDI generally emits the highest value while GDI emits the lowest value. The nucleation-mode PN for GDI is only 45% of that for PSDI at a 20 CAD BTDC spark timing. As nucleation-mode particles are mainly from exhaust dilution and cooling, a higher exhaust temperature will enhance the nucleation process and result in more nucleation-mode particles. From the results shown in Figure 5, the T_{EVO} of PSDI is the highest among the three injection modes, which is conducive to the formation of nucleation-mode particles. The three injection modes present different ignition characteristics in Figure 12, and GDI emits the lowest nucleation-mode PN at the MBT point, while PGDI and PSDI show slight variations with the spark timings. This is ascribed to the relatively high exhaust temperature and low unburned hydrocarbons in PSDI and PGDI, which decrease the propensity to oxidize the particles. Consequently, the PGDI and PSDI modes respond less sensitively to spark timings. Although PSDI increases the nucleation-mode PN, improving the dilution condition in the exhaust pipe and eliminating temperature differences will dramatically reduce the nucleation-mode PN.

Regarding the accumulation-mode PN, GDI shows the highest value, and PSDI indicates the lowest value. The accumulation-mode PN of PSDI is approximately 52% of PGDI's and 5% of GDI's, respectively. Accumulation particles are mainly formed during the combustion process of an inhomogeneous mixture [51]. PGDI dramatically enhances the homogenous situation in a cylinder via the port fuel injection, and the accumulation particle number obviously decreases from that of GDI. Particle growth and augmentation are mainly enacted through H atom loss and continue through the addition of the acetylene (HACA) mechanism [52]. Adding syngas increases the H atom amount in the cylinder, and the HACA mechanism is suppressed. Therefore, PSDI further reduces the accumulation-mode PN compared with PGDI.

4. Conclusions

This paper experimentally investigated the effect of hydrogen-rich syngas direct injection on combustion and emissions in a combined fuel injection—spark-ignition engine. Direct comparisons between the GDI, PGDI, and PSDI modes with cylinder pressure, BTE, CA0–10, CA10–90, CoV_{IMEP}, CO, HC, NO_X, and PN emissions versus spark timing were performed to analyze the effects of the injection configurations and the fuel properties to meet the requirements of sustainable development. The main conclusions are listed as follows:

1. When the spark timing was fixed at the MBT for the three injection modes, PSDI gained the highest maximum BTE, while the maximum BTE of GDI was only 94% of

PSDI's. In addition, the BTE of PSDI was much more sensitive than that of the other two modes due to the high burning rate of syngas.

2. PSDI performed the shortest durations, and GDI showed the longest duration for both CA0–10 and CA 10–90. The CoV_{IMEP} of GDI was dramatically higher than that of the other two modes, and the variations were very sensitive to spark timings in GDI. The minimum CoV_{IMEP} of PSDI decreased by 22% from the minimum value of GDI.

3. CO emissions were reduced by approximately 78% from GDI to PSDI among the whole range of spark timings, and HC emissions were reduced by approximately 60% from GDI to PSDI. However, PSDI showed the highest NO_X emissions, and GDI showed the lowest value. Specifically, the NO_X emissions from GDI were 44% and 32% of that from PGDI and PSDI, respectively. Retarding ignition linearly reduced NO_X emissions for the three injection modes.

4. PSDI generally emitted the highest nucleation PN while GDI emitted the lowest. The nucleation-mode PN for GDI was only 45% of that for PSDI at a 20 CAD BTDC spark timing. Improving the exhaust conditions and eliminating temperature differences will dramatically reduce the nucleation-mode PN.

5. GDI showed the highest accumulation-mode PN and PSDI indicated the lowest. The accumulation-mode PN for PSDI was approximately 52% of that for PGDI and only 5% of that for GDI. The small amount of accumulation-mode particles certifies the effect of hydrogen-rich syngas on reducing particles. Thus, PSDI is a feasible method to solve the high particulate emission issue in DISI engines and also improve engine performance.

In conclusion, the method of PSDI can exhibit the dual advantages of combined injection and syngas fuel properties to achieve high BTE and low CO, HC, and particulate emissions. In future research, we may develop a special exhaust after-treatment system for syngas or investigate the performance of syngas direct injection engines under various load conditions so as to better adapt to the sustainable development goal of vehicles.

Author Contributions: Conceptualization, Y.S. and X.Y.; validation, L.H.; formal analysis, Y.S.; investigation, Z.S. and Y.S.; resources, X.Y.; data curation, Z.S. and Y.S.; writing—original draft preparation, Y.S.; writing—review and editing, Z.S.; supervision, X.Y.; project administration, L.R.; funding acquisition, L.R. All authors have read and agreed to the published version of the manuscript.

Funding: This research was funded by the National Natural Science Foundation of China (NSFC, 62103160), the Jilin Province Science and Technology Development Plan Project (no. 20210508058RQ), and the Dingxin Scholar Support Program of Jilin University (no. BD0097).

Institutional Review Board Statement: Not applicable.

Informed Consent Statement: Not applicable.

Data Availability Statement: The data presented in this study are available on request from the corresponding author. The data are not publicly available due to privacy restrictions.

Conflicts of Interest: The authors declare no conflict of interest.

Abbreviations

SI, spark ignition; GDI, gasoline direct injection; PGDI, gasoline port injection combined with GDI; PSDI, gasoline port injection combined with hydrogen-rich syngas direct injection; λ, excess air ratio; Φ_D, fuel distribution ratio; BTE, brake thermal efficiency; CA 0–10, flame development duration; CA 10–90, flame propagation duration; IMEP, indicated mean effective pressure; CoV_{IMEP}, coefficient of variation in IMEP; TEVO, cylinder temperature at exhaust valve opening; MBT, minimum spark advance for best torque; TDC, top dead center; BDC, bottom dead center; PN, particulate number; PM, particulate matter.

References

1. Mon Oo, H.; Karin, P.; Charoenphonphanich, C.; Chollacoop, N.; Hanamura, K. Physicochemical characterization of direct injection Engines's soot using TEM, EDS, X-ray diffraction and TGA. *J. Energy Inst.* **2021**, *96*, 181–191.
2. Bao, J.; Qu, P.; Wang, H.; Zhou, C.; Zhang, L.; Shi, C. Implementation of various bowl designs in an HPDI natural gas engine focused on performance and pollutant emissions. *Chemosphere* **2022**, *303*, 135275. [CrossRef] [PubMed]
3. Wang, X.; Chen, W.; Huang, Y.; Wang, L.; Zhao, Y.; Gao, J. Advances in soot particles from gasoline direct injection engines: A focus on physical and chemical characterization. *Chemosphere* **2023**, *311*, 137181. [CrossRef] [PubMed]
4. Catapano, F.; Di Iorio, S.; Magno, A.; Vaglieco, B.M. Effect of fuel quality on combustion evolution and particle emissions from PFI and GDI engines fueled with gasoline, ethanol and blend, with focus on 10–23 nm particles. *Energy* **2022**, *239*, 122198. [CrossRef]
5. Akasha, H.; Ghaffarpasand, O.; Pope, F. Climate change, air pollution and the associated burden of disease in the Arabian peninsula and neighbouring regions: A critical review of the literature. *Sustainability* **2023**, *15*, 3766. [CrossRef]
6. Huang, Y.; Hong, G.; Huang, R. Effect of injection timing on mixture formation and combustion in an ethanol direct injection plus gasoline port injection (EDI + GPI) engine. *Energy* **2016**, *111*, 92–103. [CrossRef]
7. Melaika, M.; Herbillon, G.; Dahlander, P. Spark ignition engine performance, standard emissions and particulates using GDI, PFI-CNG and DI-CNG systems. *Fuel* **2021**, *293*, 120454. [CrossRef]
8. Yin, X.; Xu, L.; Duan, H.; Wang, Y.; Wang, X.; Zeng, K.; Wang, Y. In-depth comparison of methanol port and direct injection strategies in a methanol/diesel dual fuel engine. *Fuel Process. Technol.* **2023**, *241*, 107607. [CrossRef]
9. Purayil, S.T.P.; Hamdan, M.O.; Al-Omari, S.A.B.; Selim, M.Y.E.; Elnajjar, E. Review of hydrogen–gasoline SI dual fuel engines: Engine performance and emission. *Energy Rep.* **2023**, *9*, 4547–4573. [CrossRef]
10. Sun, P.; Feng, J.; Yang, S.; Wang, C.; Cui, K.; Dong, W.; Du, Y.; Yu, X.; Zhou, J. Particulate number and size distribution of dimethyl ether/gasoline combined injection spark ignition engines at medium engine speed and load. *Fuel* **2022**, *313*, 122645. [CrossRef]
11. Zhang, B.; Wang, H.; Wang, S. Computational investigation of combustion, performance, and emissions of a diesel-hydrogen dual-fuel engine. *Sustainability* **2023**, *15*, 3610. [CrossRef]
12. Ghulamullah, M.; Imran, A.; Kashif, H.M.; Safdar, A.; Hubdar, A.M.; Imran, N.U.; Abdul, M.P. Thermochemical conversion of biomass for syngas production: Current status and future trends. *Sustainability* **2022**, *14*, 2596.
13. Kravos, A.; Seljak, T.; Opresnik, S.; Katrasnik, T. Operational stability of a spark ignition engine fuelled by low H2 content synthesis gas: Thermodynamic analysis of combustion and pollutants formation. *Fuel* **2020**, *261*, 116457. [CrossRef]
14. Tartakovsky, L.; Sheintuch, M. Fuel reforming in internal combustion engines. *Prog. Energy Combust. Sci.* **2018**, *67*, 88–114. [CrossRef]
15. Paykani, A.; Chehrmonavari, H.; Tsolakis, A.; Alger, T.; Northrop, W.F.; Reitz, R.D. Synthesis gas as a fuel for internal combustion engines in transportation. *Prog. Energy Combust. Sci.* **2022**, *90*, 100995. [CrossRef]
16. Ji, M.; Wang, J. Review and comparison of various hydrogen production methods based on costs and life cycle impact assessment indicators. *Int. J. Hydrog. Energy* **2021**, *46*, 38612–38635. [CrossRef]
17. Harun, Y.; Ilker, Y. Combustion and emission characteristics of premixed $CNG/H_2/CO/CO_2$ blending synthetic gas flames in a combustor with variable geometric swirl number. *Energy* **2019**, *172*, 117–133.
18. Praveen, K.G.; Amit, K.S.; Arun, J.; Gopinath, K.P. Hydrogen-rich syngas production from the lignocellulosic biomass by catalytic gasification: A state of art review on advance technologies, economic challenges, and future prospectus. *Fuel* **2023**, *342*, 127800.
19. Matamba, T.; Lglauer, S.; Keshavarz, A. A progress insight of the formation of hydrogen rich syngas from coal gasification. *J. Energy Inst.* **2022**, *105*, 81–102. [CrossRef]
20. Alexandru, C.; Constantin, P.; Niculae, N.; Gheorghe, L.; Cristian, N.; Dinu, F. Hydrogen—An alternative fuel for automotive diesel engines used in transportation. *Sustainability* **2020**, *12*, 9321.
21. Huang, D.; Jang, J.; Lin, P.; Chen, B. Effect of ignition timing on the emission of internal combustion engine with syngas containing hydrogen using a spark plug reformer system. *Energy Procedia* **2017**, *105*, 1570–1575. [CrossRef]
22. Huang, D.; Lin, B.; Jang, J. Emission of internal combustion with low temperature plasma reformer. *Energy Procedia* **2015**, *75*, 3036–3041. [CrossRef]
23. Grzegorz, P.; Willian, C.N. Emissions and efficiency of pure CO and H_2-rich simulated gases combustion with peach and rice biomasses syngas characteristics: A renewable energy pathway for the agroindustrial sector in Rio Grande do Sul, Brazil. *Energy Convers. Manag.* **2023**, *282*, 116375.
24. Harun, Y.; Omer, C.; Ilker, Y. A comparison study on combustion and emission characteristics of actual synthetic gas mixtures. *Fuel* **2020**, *263*, 116712.
25. Ouimette, P.; Seers, P. NOx emission characteristics of partially premixed laminar flames of H2/CO/CO2 mixtures. *Int. J. Hydrog. Energy* **2009**, *34*, 9603–9610. [CrossRef]
26. Singh, A.P.; Pal, A.; Agarwal, A.K. Comparative particulate characteristics of hydrogen, CNG, HCNG, gasoline and diesel fueled engines. *Fuel* **2016**, *185*, 491–499. [CrossRef]
27. Zhao, H.; Stone, R.; Zhou, L. Analysis of the particulate emissions and combustion performance of a direct injection spark ignition engine using hydrogen and gasoline mixtures. *Int. J. Hydrog. Energy* **2010**, *35*, 4676–4686. [CrossRef]
28. Shang, Z.; Yu, X.; Ren, L. Comparative study on effects of injection mode on combustion and emission characteristics of a combined injection n-butanol/gasoline SI engine with hydrogen direct injection. *Energy* **2020**, *213*, 118903. [CrossRef]

29. Sun, Y.; Yu, X.; Jiang, L. Effects of direct hydrogen injection on particle number emissions from a lean burn gasoline engine. *Int. J. Hydrog. Energy* **2016**, *41*, 18631–18640. [CrossRef]
30. He, F.; Li, S.; Yu, X. Comparison study and synthetic evaluation of combined injection in a spark ignition engine with hydrogen-blended at lean burn condition. *Energy* **2018**, *157*, 1053–1062. [CrossRef]
31. Zhang, G.; Li, G.; Li, H.; Lv, J. Effect of diluent gas on propagation and explosion characteristics of hydrogen-rich syngas laminar premixed flame. *Energy* **2022**, *246*, 123370. [CrossRef]
32. Kim, S.; Kim, J. Feasibility assessment of hydrogen-rich syngas spark-ignition engine for heavy-duty long-distance vehicle application. *Energy Convers. Manag.* **2022**, *252*, 115048. [CrossRef]
33. Chacartegui, R.; Sanchez, D.; Escalona, J.M.; Sanchez, A.M.T. Gas and steam combined cycles for low calorific syngas fuels utilisation. *Appl. Energy* **2013**, *101*, 81–92. [CrossRef]
34. Zareei, J.; Ghadamkheir, K.; Farkhondeh, S.; Abed, A.M.; Opulencia, M.; Alvarez, J. Numerical investigation of hydrogen enriched natural gas effects on different characteristics of a SI engine with modified injection mechanism from port to direct injection. *Energy* **2022**, *255*, 124445. [CrossRef]
35. Yu, X.; Du, Y.; Sun, P.; Liu, L.; Wu, H.; Zuo, X. Effects of hydrogen direct injection strategy on characteristics of lean-burn hydrogen–gasoline engines. *Fuel* **2017**, *208*, 602–611. [CrossRef]
36. Hong, B.; Lius, A.; Mahendar, S.; Mihaescu, M.; Cronhjort, A. Energy and exergy characteristics of an ethanol-fueled heavy-duty SI engine at high-load operation using lean-burn combustion. *Appl. Therm. Eng.* **2023**, *224*, 120063. [CrossRef]
37. Chaurasiya, R.; Krishnasamy, A. A single fuel port and direct injected low temperature combustion strategy to reduce regulated pollutants from a light-duty diesel engine. *Fuel* **2023**, *335*, 127114. [CrossRef]
38. Shi, C.; Chai, S.; Di, L.M.; Ji, C.; Ge, Y.; Wang, H. Combined experimental-numerical analysis of hydrogen as a combustion enhancer applied to Wankel engine. *Energy* **2023**, *263*, 125896. [CrossRef]
39. Gong, C.; Yu, J.; Liu, F. Combined impact of excess air ratio and injection strategy on performances of a spark-ignition port- plus direct-injection dual-injection gasoline engine at half load. *Fuel* **2023**, *340*, 127605. [CrossRef]
40. Zhang, B.; Ji, C.; Wang, S. Combustion analysis and emissions characteristics of a hydrogen-blended methanol engine at various spark timings. *Int. J. Hydrog. Energy* **2015**, *40*, 4707–4716. [CrossRef]
41. Atelge, M.R. Experimental study of a blend of Diesel/Ethanol/n-Butanol with hydrogen additive on combustion and emission and exegetic. *Fuel* **2022**, *325*, 124903. [CrossRef]
42. Liu, X.; Zhao, M.; Feng, M.; Zhu, Y. Study on mechanisms of methane/hydrogen blended combustion using reactive molecular dynamics simulation. *Int. J. Hydrog. Energy* **2023**, *48*, 1625–1635. [CrossRef]
43. Li, J.; Zhang, R.; Pan, J.; Wei, H.; Shu, G.; Chen, L. Ammonia and hydrogen blending effects on combustion stabilities in optical SI engines. *Energy Convers. Manag.* **2023**, *280*, 116827. [CrossRef]
44. Duan, X.; Deng, B.; Liu, Y.; Zou, S.; Liu, J.; Feng, R. An experimental study the impact of the hydrogen enrichment on cycle-to-cycle variations of the large bore and lean burn natural gas spark-ignition engine. *Fuel* **2020**, *282*, 118868. [CrossRef]
45. Ji, C.; Wang, S.; Zhang, B. Effect of spark timing on the performance of a hybrid hydrogen-gasoline engine at lean conditions. *Int. J. Hydrog. Energy* **2010**, *35*, 2203–2212. [CrossRef]
46. Sun, S.; Xu, D.; Liang, Y.; Jiang, G.; Diao, Y.; Gong, X.; Wang, B. Effect of temperature, oxygen concentration, and CaO addition on SO_2 and NOx emissions during oxygen-fuel combustion of municipal sludge. *J. Energy Inst.* **2022**, *105*, 424–432. [CrossRef]
47. Zhao, L.; Su, X.; Wang, X. Comparative study of exhaust gas recirculation (EGR) and hydrogen-enriched EGR employed in a SI engine fueled by biobutanol-gasoline. *Fuel* **2020**, *268*, 117194. [CrossRef]
48. Wang, C.; Pei, Y.; Peng, Z.; Li, X. Insights into the effects of direct injection timing on the characteristics of deposited fuel film and Particulate Matter (PM) emissions from a Dual-Fuel Spark Ignition (DFSI) engine. *Fuel Process. Technol.* **2022**, *238*, 107515. [CrossRef]
49. Thawko, A.; Tartakovsky, L. The mechanism of particle formation in non-premixed hydrogen combustion in a direct-injection internal combustion engine. *Fuel* **2022**, *327*, 125187. [CrossRef]
50. Kurien, C.; Srivastava, A.K.; Lesbats, S. Experimental and computational study on the microwave energy based regeneration in diesel particulate filter for exhaust emission control. *J. Energy Inst.* **2020**, *93*, 2133–2147. [CrossRef]
51. Su, Y.; Zhang, Y.; Xie, F.; Duan, J.; Li, X.; Liu, Y. Influence of ethanol blending ratios on in-cylinder soot processes and particulate matter emissions in an optical direct injection spark ignition engine. *Fuel* **2022**, *308*, 121944. [CrossRef]
52. Anna, A.D.; Sirignano, M.; Kent, J. A model of particle nucleation in premixed ethylene flames. *Combust. Flame* **2010**, *157*, 2106–2115.

 *sustainability*

A Migration Learning Method Based on Adaptive Batch Normalization Improved Rotating Machinery Fault Diagnosis

Xueyi Li [1,2,*], Tianyu Yu [1], Daiyou Li [1], Xiangkai Wang [1], Cheng Shi [3,*], Zhijie Xie [1] and Xiangwei Kong [2,4]

[1] College of Mechanical and Electrical Engineering, Northeast Forestry University, Harbin 150040, China
[2] Key Laboratory of Vibration and Control of Aero-Propulsion System, Ministry of Education, Northeastern University, Shenyang 110819, China
[3] School of Vehicle and Energy, Yanshan University, Qinhuangdao 066004, China
[4] School of Mechanical Engineering and Automation, Northeastern University, Shenyang 110819, China
* Correspondence: lixueyiphm@163.com (X.L.); shicheng@ysu.edu.cn (C.S.)

Abstract: Sustainable development has become increasingly important as one of the key research directions for the future. In the field of rotating machinery, stable operation and sustainable performance are critical, focusing on the fault diagnosis of component bearings. However, traditional normalization methods are ineffective in target domain data due to the difference in data distribution between the source and target domains. To overcome this issue, this paper proposes a bearing fault diagnosis method based on the adaptive batch normalization algorithm, which aims to enhance the generalization ability of the model in different data distributions and environments. The adaptive batch normalization algorithm improves the adaptability and generalization ability to better respond to changes in data distribution and the real-time requirements of practical applications. This algorithm replaces the statistical values in a BN with domain adaptive mean and variance statistics to minimize feature differences between two different domains. Experimental results show that the proposed method outperforms other methods in terms of performance and generalization ability, effectively solving the problems of data distribution changes and real-time requirements in bearing fault diagnosis. The research results indicate that the adaptive batch normalization algorithm is a feasible method to improve the accuracy and reliability of bearing fault diagnosis.

Keywords: fault diagnosis; AdaBN; transfer learning; rotating machinery

Citation: Li, X.; Yu, T.; Li, D.; Wang, X.; Shi, C.; Xie, Z.; Kong, X. A Migration Learning Method Based on Adaptive Batch Normalization Improved Rotating Machinery Fault Diagnosis. *Sustainability* **2023**, *15*, 8034. https://doi.org/10.3390/su15108034

Academic Editor: Talal Yusaf

Received: 17 March 2023
Revised: 30 April 2023
Accepted: 10 May 2023
Published: 15 May 2023

1. Introduction

Against the backdrop of global development, sustainability has become an important topic in various fields. Whether it is in terms of economics, society, or the environment, sustainability is a goal we should strive for. Over the past few decades, human over-exploitation of natural resources and environmental damage have brought irreversible impacts to the earth. Therefore, promoting sustainable development has become one of the most important tasks of the current era. For the sustainable development of energy, many scholars have made many contributions [1–4]. At the same time, industrial production also needs to promote the sustainable development of energy. Rotating machinery is an essential part of industrial production, and rolling bearings are important parts of rotating machinery. Once the bearing faults in the mechanical equipment, it is likely to cause serious safety accidents such as mechanical jamming, resulting in economic losses [5–7]. In order to avoid economic losses, more and more scholars pay attention to the fault diagnosis method of bearings.

Bearing fault diagnosis methods mainly include methods based on signal processing, traditional machine learning, and deep learning [8]. The fault diagnosis method based on traditional machine learning is mainly divided into two steps: 1. Feature processing on the collected signal to extract useful fault features. The main methods include wavelet transform (WT) [9], Empirical Mode Decomposition (EMD) [10], Singular Value Decomposition (SVD) [11], and Short-Time Fourier Transform (STFT) [12]. 2. Distinguish the fault types, including the fault size of the same fault type. The main methods include Support Vector Machines (SVMs) [13], Artificial Neural Networks (ANNs) [14], K-Nearest [15], etc. The traditional machine-learning-based fault diagnosis methods described above usually require manual extraction of fault features and expert experience. For example, the wavelet transform needs to find a suitable wavelet basis function, the Short-Time Fourier Transform needs to adjust the length and width of the required window function, and the decision tree needs to analyze the independent features of the sample extraction [16].

At present, deep learning theory has been gradually applied to the field of bearing fault diagnosis. Although traditional machine learning methods can diagnose bearing faults, manual feature extraction relies mainly on manual labor. In addition, traditional machine learning model generalization is less capable. Deep learning is a new topic in the field of machine learning, and research on neural networks began in the 1980s [17], such as Restricted Boltzmann Machines (RBM) [18] and Convolutional Neural Networks (CNN) [19]. These theories have advanced the development of bearing fault diagnosis, including methods based on the Deep Belief Network (DBN) [20], Stacked Autoencoders (SAE), CNN, and ResNet [21] methods. The deep learning method automatically learns fault features from the collected data, providing an end-to-end bearing fault diagnosis model.

Although deep learning methods have made some progress in the field of rotating machinery fault diagnosis, there are still the following problems: 1. Due to the involvement of multiple devices and scenarios, the distribution of the training and test datasets for bearing fault detection may change in practical applications. If the model is trained and tested only on specific datasets, its performance may decrease on other datasets. 2. In practical applications, bearing fault detection needs to be performed in real time, so the model needs to have high generalization and adaptability and be able to perform accurate fault diagnosis in different data distributions and environments. Therefore, domain generalization can make the model more adaptive and have better generalization, which can better deal with changes in data distribution and real-time requirements in practical applications and improve the accuracy and reliability of bearing fault diagnosis. In this paper, the AdaBN algorithm is used to solve the problem of domain generalization. Specifically, the AdaBN algorithm replaces the mean and variance statistics in a BN with domain-adaptive mean and variance statistics, which can be obtained by minimizing the feature differences between two different domains. In this way, AdaBN can effectively solve the problem of insufficient generalization of the model when the distribution of training and test data is different.

Pan et al. [18] introduced a component analysis method for domain adaptation by transfer that reduces the distance between the source and target domains. However, this method assumes that the conditional distributions of the source and target domain data are approximately equal. Long et al. [19] proposed a transfer feature learning approach with joint distribution adaptation, which aims to simultaneously reduce the marginal and conditional distributions between domains. Zhong et al. [22] trained the model on enough normal samples, and then passed the SVM to replace fully connected layers. Zhao et al. [23] proposed a multi-scale convolutional transfer learning network pretrained on the source domain; then, they transferred the model to other different but similar domains for fine-tuning. Balanced Distribution Adaptation (BDA) is used in [24,25] to adaptively balance marginal and conditional distribution differences between feature domains learned by deep neural networks. Qian et al. [24] considered higher-order moments and proposed using Kullback–Leibler (KL) divergence to adjust the fault diagnosis domain distribution of rotating machinery. Wang et al. [25] aligned marginal and conditional distributions in multiple layers by using a conditional Maximum Mean Discrepancy (MMD) based on estimated pseudolabels. Yang et al. [26] proposed using a polynomial kernel instead of a Gaussian kernel in MMD for better alignment of the domain distribution. Han et al. [27] and Qian et al. [28] used Joint Distribution Adaptation (JDA) [29] to align conditional and marginal distributions, They used MMD and domain adversarial training to train two feature extractors and classifiers, respectively. Sheng et al. [30] proposed a linear combination of multiple Gaussian kernels to reduce the variance between domain distributions.

This paper proposes an end-to-end unsupervised method for domain-adaptive bearing fault diagnosis based on an improved BN. The preprocessed data are directly input into the model using the Convolutional Neural Network. The fault features are automatically extracted, and the parameter quantity of the model is significantly reduced compared with that of the Artificial Neural Network, which is beneficial to prevent the model from overfitting. In this paper, the AdaBN algorithm is used to realize the fault diagnosis of bearings of the same fault type and different working conditions. The method proposed can be extended from the source domain to the target domain [31], and the diagnosis accuracy reaches 100% on the CWRU dataset. Compared with the traditional BN method, it shows better fault diagnosis results. The results show that the method proposed in this paper has better unsupervised domain adaptation diagnosis accuracy for bearing faults. The innovative summary of the method proposed in this paper is as follows:

1. Introduction of AdaBN layer: The AdaBN [32] method introduces the AdaBN layer in the deep neural network, which dynamically adjusts the parameters of the BN layer according to the input data to adapt to different data distributions. This dynamic adjustment mechanism enables the AdaBN method to better adapt to complex and changing data distributions, thus improving the performance of deep neural networks.
2. Consideration of different data distributions: The AdaBN method considers the different data distributions and dynamically adjusts the BN layer parameters for each batch of data, enabling the model to better adapt to changes in data distribution. This adjustment mechanism tailored to different data distributions can improve the performance of the model on many datasets.
3. Effectively addressing the limitations of the BN layer on small batch data: The BN layer performs poorly on small batch data, while the AdaBN method can adapt to small batch data by dynamically adjusting the parameters of the BN layer, thus improving the performance of the model. This method can effectively address the limitations of the BN layer on small batch data and also make the model training more efficient.

The structure of this paper is as follows: Section 2 introduces the basic definition of unsupervised transfer learning, Section 3 elaborates on the bearing fault diagnosis method proposed in this paper, and Section 4 describes the proposed method in the Western Reserve University dataset and laboratory simulation data. It is validated on the set and compared with the results of the BN model without optimization. Section 5 concludes this paper and looks forward to future research directions.

2. Unsupervised Deep Transfer Learning

Existing transfer learning mainly focuses on the study of closed sets. Specifically, the fault categories in the source and target domains are the same, which is obviously only an ideal transfer learning scenario. In a real transfer learning environment, the source domain and the target domain often only share some categories of information, even if there is no common category between the source domain and the target domain. A scenario where the categories of the source and target domains completely overlap is called a closed set. A scenario where the source and target domains share a part of the categories is called an open set. A scenario where the source and target domains do not share any categories at all is called a fully open set. The main content of this paper is based on a closed set.

Unsupervised deep transfer learning with overlapping categories is defined as follows: it is assumed that the source domain data are labeled, and the target domain data are unlabeled. Unsupervised deep transfer learning refers to the source domain data without labels. The fault types of the domain and the target domain are the same, which is also the situation studied in this paper, but the actual situation may be different. First, the mechanical equipment is usually in a normal working state, and it is difficult to collect data on bearing faults with labels. The data are relatively small, and the fault under the real operating condition of the bearing can only be approximated using electric discharge machining (EDM) of the bearing in the laboratory. However, this method has two disadvantages: 1. It is difficult to grasp the size of the fault type processed, and it is difficult to simulate the real type of fault. 2. There is an inconsistency between the processed fault type and the real fault type. Therefore, the research is based on the same fault type in the source domain and the target domain. Then, the research on the state of different fault types in the source domain and the target domain is carried out. Assuming that the label of the source domain is available, the definition of the source domain is as follows:

$$\mathcal{D}_s = \{(x_i^s, y_i^s)\}_{i=1}^{n_s} \ x_i^s \in \mathbf{X_s}, \ y_i^s \in \mathbf{Y_s} \tag{1}$$

where $\mathcal{D}_s$ represents the source domain, $x_i^s \in \mathbb{R}^d$ is the i-th sample, $\mathbf{X_s}$ is the union of all samples, y_i^s is the i-th label of the i-th sample, $\mathbf{Y_s}$ is the union of all different labels, and n_s is the total number of samples in the source domain. In addition, assuming that the label of the target domain is not available, the definition of the source domain is as follows:

$$\mathcal{D}_t = \{(x_i^t)\}_{i=1}^{n_t} \ x_i^t \in \mathbf{X_t} \tag{2}$$

where $\mathcal{D}_t$ represents the target domain, $x_i^t \in \mathbb{R}^d$ is the i-th sample, $\mathbf{X_t}$ is the union of all samples, and n_t is the total number of target samples.

3. The Proposed Method

3.1. Batch Normalization

The BN [33] is for $\mathrm{x} = \left(x^{(1)} \ldots x^{(d)}\right)$ with d-dimensional input, and the features of each dimension were normalized.

$$\hat{x}^{(k)} = \frac{x^{(k)} - \mathrm{E}\left[x^{(k)}\right]}{\sqrt{Var\left[x^{(k)}\right]}} \tag{3}$$

where $x^{(k)}$ and $y^{(k)}$ are input/output scalars that respond to a neuron in a data sample. The data normalization method above may change the data distribution of the layers. For example, normalizing the inputs of the sigmoid will restrict them to a nonlinear state. To solve this problem, this paper sets the value $x^{(k)}$ for each activation.

$$y^{(k)} = \gamma^{(k)} \hat{x}^{(k)} + \beta^{(k)} \tag{4}$$

to introduce a pair of parameters $\gamma^{(k)}$ and $\beta^{(k)}$, which shift and scale the standard value. These parameters are learned at the same time as the original model and have the ability to restore the network. In fact, the original value x^k can be restored by setting $\gamma^{(k)} = \sqrt{Var\left[x^{(k)}\right]}$ and $\gamma^{(k)} = \sqrt{Var\left[x^{(k)}\right]}$ for the stochastic gradient descent method optimization algorithm. Stable input distribution can greatly promote the convergence of the model, reduce the training time, and allow the use of a relatively large learning rate. It is helpful to slow down gradient disappearance and gradient explosion. Many experiments have demonstrated that the BN can significantly reduce the number of iterations while improving the final model performance. The BN is already a necessary part of many top-level architectures such as ResNet [34] and Inception V3 [35].

3.2. Domain-Adaptive AdaBN Algorithm

Figure 1 shows the flowchart of the AdaBN algorithm proposed for fault diagnosis. The model obtains parameters through training samples and can extract fault features. This is generally only applicable to the source domain, and the accuracy on the source domain is relatively high, but the accuracy rate will be relatively low for fault migration.

The main reason is the data distribution is not the same. This paper proposes a simple and effective method called the improved AdaBN [36] algorithm for bearing fault diagnosis, and Table 1 shows the algorithm flow chart. The algorithm uses the μ_t and σ_t^2 of each BN layer of the target domain samples instead of the μ_t and σ_t^2 calculated by the samples of the active domain in the original BN layer. Domain adaptation via the BN. The weights with fault feature extraction ability learned by the model in the training set are frozen. The domain-related knowledge is represented by the statistical data of the BN layer. Therefore, the trained model can be easily applied to related fields by modeling the statistical data in the BN layer, thereby reducing the training time and computing cost of the model.

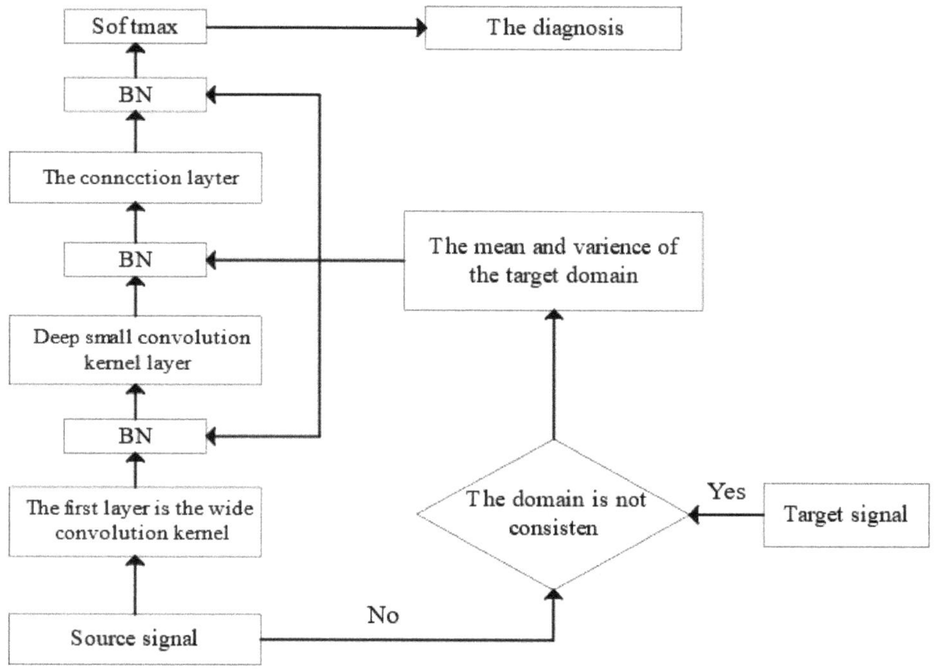

Figure 1. The flowchart of the AdaBN algorithm proposed for fault diagnosis.

Table 1. DCNN algorithm based on AdaBN domain adaptation.

Algorithm	DCNN Algorithm Based on AdaBN Domain Adaptive
Enter	Signal p of the target domain, expressed in the i neuron of the BN layer of the DCNN $x_t^{(i)}(p) \in \mathbf{x}_t^{(i)}$ of which $\mathbf{x}_t^{(i)} = \left\{ x_t^{(i)}(1), \ldots, x_t^{(i)}(n) \right\}$, for the i neuron, has been trained to scale with parallel parameters $\gamma_s^{(i)}$ and $\beta_s^{(i)}$.
Output	The adjusted DCNN network
For	For each neuron i and each signal p in the target domain, compute the mean and variance of all samples in the target domain: $$\mu_t^{(i)} \leftarrow \mathbf{E}\left[\mathbf{x}_t^{(i)}\right]$$ $$\sigma_t^{(i)} \leftarrow Var\left[\mathbf{x}_t^{(i)}\right]$$ Calculate the output of the BN layer: $$\hat{\mathbf{x}}_t^{(i)}(p) = \frac{x_t^{(i)}(p) - \mu_t^{(i)}}{\sigma_t^{(i)}}$$ $$\mathbf{y}_t^{(i)}(p) = \gamma_s^{(i)}\hat{\mathbf{x}}_t^{(i)}(p) + \beta_s^{(i)}$$
End for	

3.3. DCNN Model Based on AdaBN Algorithm

Figure 2 is the network architecture diagram of the Deep Convolutional Neural Networks with the Wide First-Layer Kernel (DCNN) model, and Table 2 shows the one-dimensional neural network structure parameters. The DCNN model obtains the parameters of the model through the training samples and can learn fault features. When the DCNN model faces the target domain data, the accuracy of the diagnosis model will decrease compared with the source domain data. In order to reduce the performance degradation of the model, the AdaBN was used to improve the domain adaptation ability of the DCNN model.

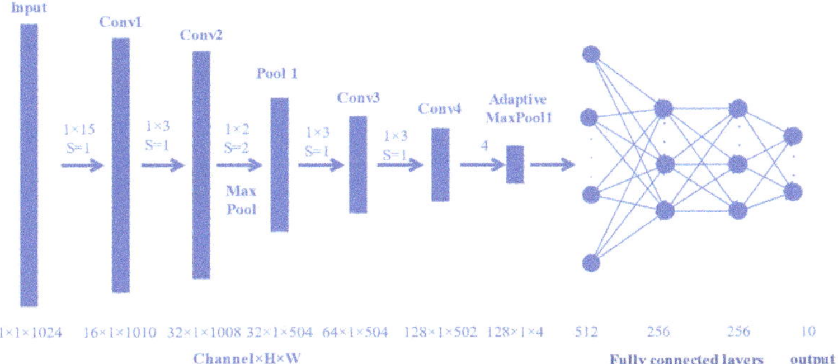

Figure 2. DCNN network architecture diagram.

Table 2. One-dimensional neural network structure parameters.

Number	Network Layer	Kernel Size/Step	Number of Kernel	Output Size (Width × Depth)
1	Conv1	$1 \times 15/1$	16	1×1010
2	Conv2	$1 \times 3/1$	32	1×1008
3	Pool1	$1 \times 2/2$	32	1×504
4	Conv3	$1 \times 3/1$	64	1×504
5	Conv4	$1 \times 3/1$	128	1×502
6	AdaptiveMaxpool	4	128	1×4
7	Fc 1	——	——	512
8	Fc 2	——	——	256
9	Fc 3	——	——	256
10	Fc 4	——	——	10

3.4. Discussion on AdaBN

The ultimate goal of standardization in the AdaBN algorithm is to make the data received by each layer come from a similar data distribution to alleviate the impact of domain offset. The AdaBN was used to distribute alignment. For example, MMD [32] in Equation (5) is commonly used to measure the degree of offset between the source and target domains.

$$MMD[\mathcal{F}, p, q] := \sup_{f \in \mathcal{F}} \left(\mathbf{E}_{x \sim p}[f(x)] - \mathbf{E}_{y \sim q}[f(y)] \right) \tag{5}$$

where sup is the upper bound, E : is the expectation, and $x \sim p$: x is the sample space of p.

Actually, an MMD with a Gaussian kernel can be viewed as minimizing the distance between the weighted sum of all moments. This advantage also makes it possible for AdaBN to be applied in the whole network, since AdaBN performs an explicit matching of the secondary moments and does not require very time-consuming kernel computation. The simplicity of AdaBN is in stark contrast to the complexity of the domain migration problem.

Consider a simple neural network with input $\mathbf{X} \in \mathbb{R}^{p_1 \times 1}$, which has a BN layer with a mean and variance of μ_i and $\sigma_i^2 (i \in \{1 \ldots p_2\})$ for each feature, a fully connected layer with weight matrix $\mathbf{W} \in \mathbb{R}^{p_1 \times p_2}$ and bias $\mathbf{b} \in \mathbb{R}^{p_2 \times 1}$, and a nonlinear transformation layer $f(\cdot)$, where p_1 and p_2 correspond to the feature sizes of the input and output. If there is no BN, the output of the network is $f(\mathbf{W}_a \mathbf{x} + \mathbf{b}_a)$.

$$
\begin{aligned}
\mathbf{W}_a &= \mathbf{W}^T \mathbf{\Sigma}^{-1} \\
\mathbf{b}_a &= -\mathbf{W}^T \mathbf{\Sigma}^{-1} \mu + \mathbf{b} \\
\mathbf{\Sigma} &= diag(\sigma_1, \ldots, \sigma_{p_1}) \\
\mu &= (\mu_1, \ldots, \mu_{p_1})
\end{aligned}
\tag{6}
$$

It can be seen that the transformation is not very simple, even for a simple computational layer. As the CNN architecture goes deeper, it can gain more capabilities to represent complex nonlinear transformations [37].

4. Model Validation

In order to verify the method proposed in this paper, CWRU and laboratory simulation bench data are used for verification.

1. Validation on the CWRU dataset

The CWRU bearing center data acquisition system is shown in Figure 3. The experimental object of this experiment is the drive end bearing shown in the figure. The diagnosed bearing model is the SKF6205 deep groove ball bearing, and the fault bearing is made using electric discharge machining. The sampling frequency of the system is 12 kHz. There are three types of defects in the diagnosed bearing: rolling element damage, outer race damage, and inner race damage, with defect diameters of 0.007 inch, 0.014 inch, and 0.021 inch, respectively, resulting in a total of nine damage states. In the experiment, 1024 data points were used for diagnosis each time.

Figure 3. CWRU rolling bearing data acquisition system.

The improved BN algorithm proposed in this paper is experimentally verified using the CWRU dataset. Dividing the data into four groups, the rotation speed corresponds to 1797 rpm, 1772 rpm, 1750 rpm, and 1730 rpm, and the corresponding labels are 0, 1, 2, and 3, respectively. Each group contains 10 pieces of data, and these 10 pieces of data all include the original vibration signal of a normal bearing, the original vibration signal of an outer ring faulty bearing, the original vibration signal of an inner ring faulty bearing, and the original vibration signal of a rolling element faulty bearing. Figure 3 shows the accuracy rate of the model that migrated from the 0th group to the 1st group on the CWRU dataset; that is, the model migrated from the speed of 1797 rpm to 1772 rpm on the source domain training set and the source domain test set, respectively. Accuracy on the test set was found in the target domain. From Figure 3, we can see that the domain-adaptive diagnosis results based on AdaBN in the target domain are faster and more stable in the 0–150 epoch than the domain-adaptive diagnosis results without AdaBN, and the variance and mean are relatively larger. Domain-adaptive diagnosis results using AdaBN on epochs 150–300 have more stable convergence, relatively small variance, and higher accuracy. In order to verify the effectiveness and robustness of the method proposed in this paper, experiments were carried out for each transfer learning model under different rotational speeds, and a total of six groups of experiments were performed, namely: 0→1, 0→2, 0→3, 1→2, 1→3, and 2→3. Figure 4 is a line chart of six transition states, the abscissas correspond to six groups of experiments, and each experiment corresponds to the source training set (SDT) of the source domain, the source test set (SDV) of the source domain, and the target test set (TDV) of the target domain and target test set (TDV). The validation indicators include the mean and variance of the first 150 epochs and the mean and variance of the last 150 epochs. The model throughout the training process is trained on the training set of the source domain, the test set of the source domain, and the validation set of the target domain.

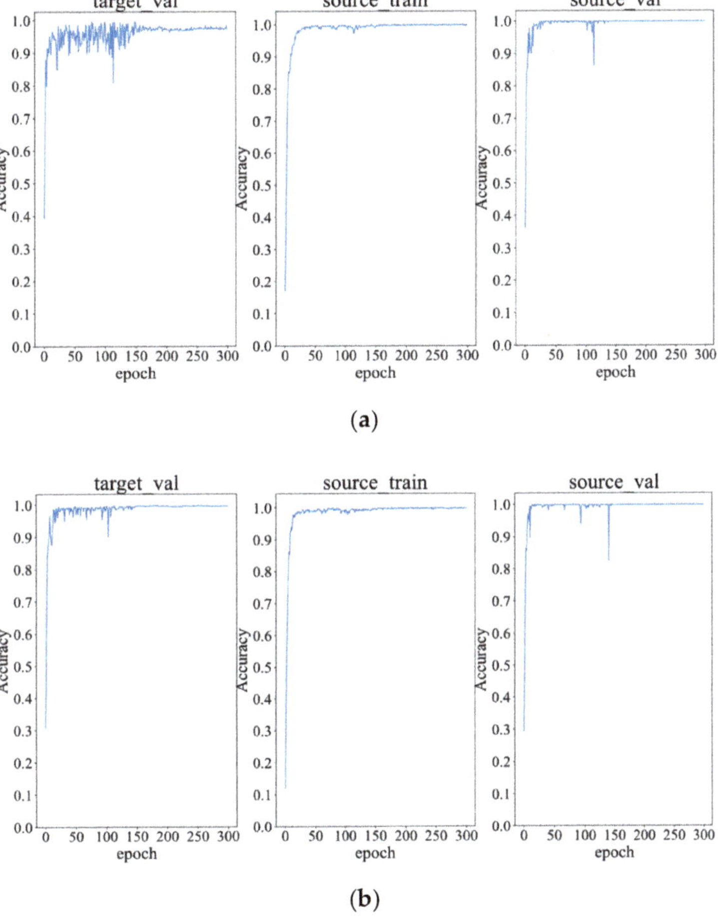

Figure 4. (a) Accuracy rate in each domain using the AdaBN method. (b) Accuracy rate in each domain without using the AdaBN method.

In this experiment, the authors computed the mean value of the first 150 epochs to evaluate whether using AdaBN could improve the initial accuracy of the model in the early stage of training. Additionally, authors measured the variance of the first 150 epochs and the last 150 epochs to investigate the stability of model at the beginning and end of training. The authors compared the variance of the two models when the average accuracy rate was similar to determine if the AdaBN model could provide better stability during the early stage of training. The results in Figure 4 demonstrate that using the AdaBN method generally resulted in higher accuracy than not using it in many training epochs.

It can also be observed from Figure 5 that the variance with the AdaBN method is smaller than that without the AdaBN method. Variance is an important parameter that reflects the stability of the data, so, whether in the source domain or the target domain, the accuracy of transfer learning after using the AdaBN method is more stable. Figure 6 shows the mean and maximum values of the training set, the validation set of the source domain, and the validation set of the target domain under different transfer states on CWRU. Figure 7a shows the confusion matrix without AdaBN, and Figure 7b shows the confusion matrix with AdaBN. From the comparison of the two figures, it can be seen that the model without the AdaBN confusion matrix has a large number of misjudgments in Category 8. As shown in Table 3, our proposed method was first compared with traditional machine learning methods with six transfer conditions in detail. The results show that our proposed method outperforms the traditional SVM method by nearly 30%, the traditional MLP method by about 15%, and also has a stable improvement compared with our proposed method without AdaBN optimization. The AdaBN method proposed in this paper improves the accuracy of direct migration under different working conditions, which proves the effectiveness of the AdaBN method.

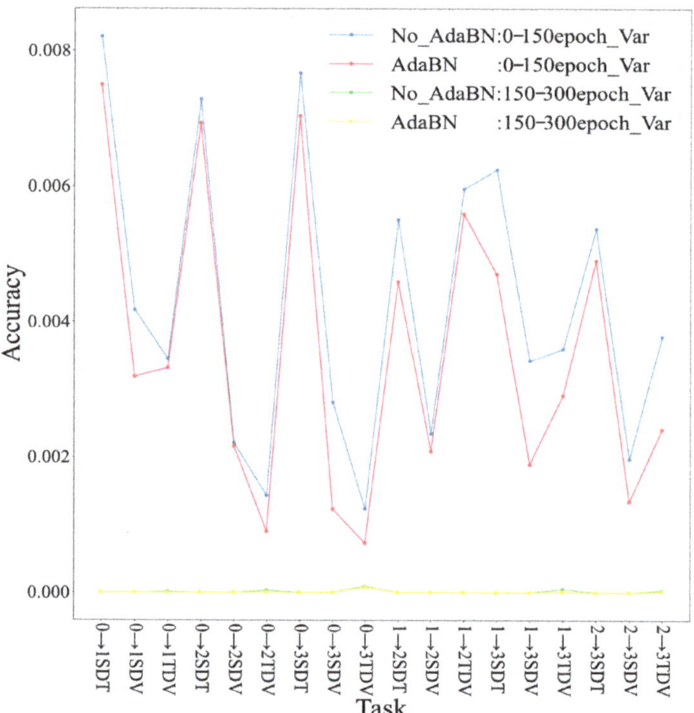

Figure 5. The variance on the CWRU dataset.

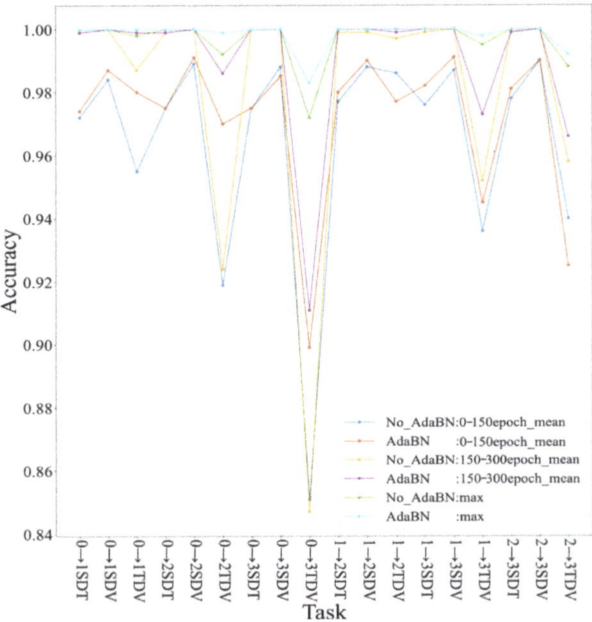

Figure 6. The accuracy of the mean and maximum value in different domains on the CWRU dataset.

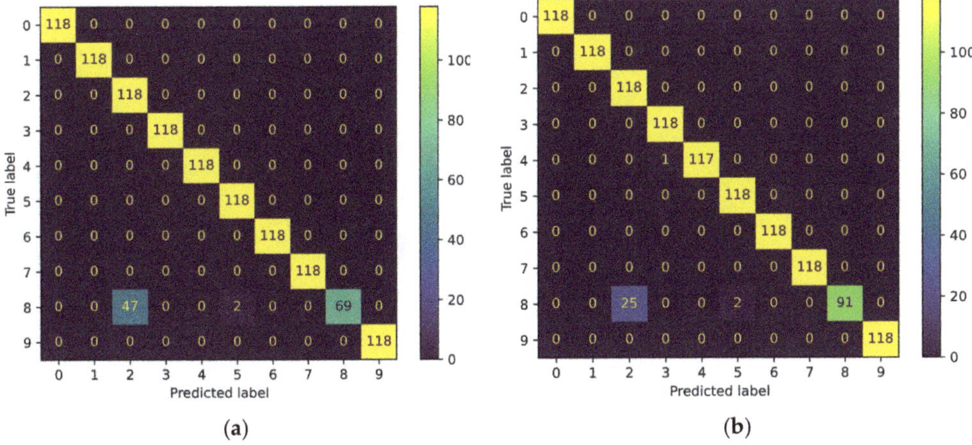

(a) (b)

Figure 7. (**a**) Confusion matrix without AdaBN on CWRU dataset. (**b**) Confusion matrix with AdaBN on CWRU dataset.

Table 3. Comparison of accuracy of each algorithm in six migration states.

Task	0→1	0→2	0→3	1→2	1→3	2→3
SVM	70.34%	74.23%	71.23%	68.45%	73.12%	68.49%
MLP	85.24%	82.93%	80.98%	78.21%	84.82%	88.49%
DCNN	99.12%	98.89%	97.53%	99.59%	99.53%	98.53%
DCNN (AdaBN)	99.89%	99.85%	98.83%	99.59%	99.82%	99.12%

The main sources of error in this experiment are data collection, data preprocessing, model selection, and parameter tuning. In this experiment, parameter tuning is the main source of error, and different hyperparameters have a significant impact on the error of the model. Therefore, it is necessary to choose appropriate hyperparameters based on experience and actual conditions [38,39]. The key hyperparameters used in training the neural network in this paper are batch_size: 64, optim: Adam, learning_rate: 1×10^{-3}, moment: 0.9, weight-decay: 1×10^{-5}, lr_scheduler: Step, and epoch: 600.

2.   Validation on a laboratory testbed dataset

(1)   Introduction to the dataset

The tapered roller bearing used in this experiment was NUP205. The inner diameter was 25 mm, the outer diameter was 52 mm, and the width was 15 mm. The data were collected at seven different rotational speeds, and each rotational speed included normal bearing data, inner ring faulty bearing data, and outer ring faulty bearing data. Among them, 11 types of outer ring faults were designed, and 6 types of inner ring faults were designed. Two channels of data were collected in the horizontal and vertical directions for each specific type. In this paper, the vibration signal data in the horizontal direction at four different speeds of 900 rpm, 1200 rpm, 1500 rpm, and 1650 rpm were selected for the experiment. Figure 8 shows the experimental bench for simulating bearing data in the laboratory.

Figure 8. The data acquisition test bench for the laboratory dataset.

(2)   Analysis of results

It can be seen from Table 4 that whether it is at a 0–150 epoch or 150–300 epoch, the accuracy of using the AdaBN method is obviously higher than that of using AdaBN on the laboratory test bench. It can be seen from Figure 9 that the variance of the AdaBN method is dominant in most transfer models, and it is better than the method without AdaBN in most cases. Therefore, using the AdaBN method in the bearing fault diagnosis migration model can improve the stability and accuracy of the model. The confusion matrix for never using AdaBN is shown in Figure 10a, and the confusion matrix for using AdaBN is shown in Figure 10b. Further, there are five more misclassifications other than those in Categories 5 and 9, namely, Categories 2, 3, 4, and 6. The exact number of use for the AdaBN method in Category 7 is much higher than that without using the AdaBN method. The effectiveness of the AdaBN method proposed in this paper is confirmed.

Table 4. The mean and maximum values under different migration states.

Task	No_AdaBN 0–150 Epoch Mean	AdaBN 0–150 Epoch Mean	No_AdaBN 150–300 Epoch Mean	AdaBN 150–300 Epoch Mean	No_AdaBN 0–300 Epoch Max	AdaBN 0–300 Epoch Max
0→1	0.833	**0.869**	0.862	**0.883**	0.951	**0.952**
0→2	0.942	**0.948**	0.996	**0.998**	**0.999**	**0.999**
0→3	0.904	**0.905**	0.942	**0.948**	0.946	**0.956**
1→2	0.744	**0.811**	0.765	**0.824**	0.853	**0.900**
1→3	0.936	**0.944**	0.997	**0.999**	0.999	**1.000**
2→3	0.902	**0.904**	**0.951**	0.950	**0.956**	0.955
1→0	0.723	**0.781**	0.730	**0.780**	0.838	**0.890**
2→1	**0.966**	0.960	**0.999**	**0.999**	**1.000**	**1.000**

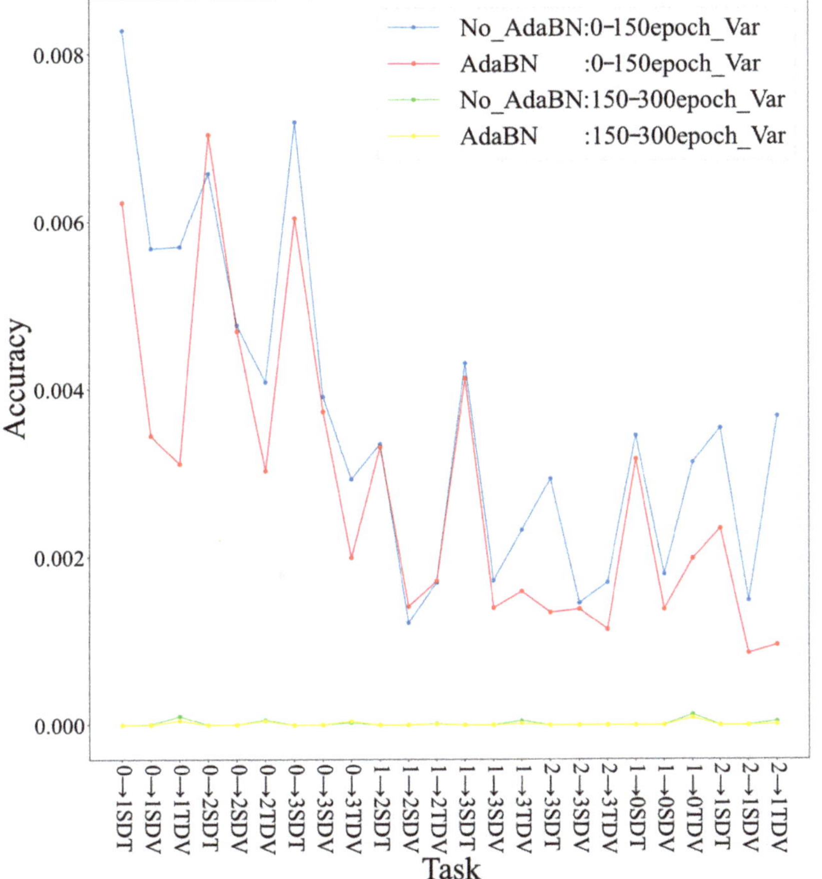

Figure 9. The variance on the laboratory dataset.

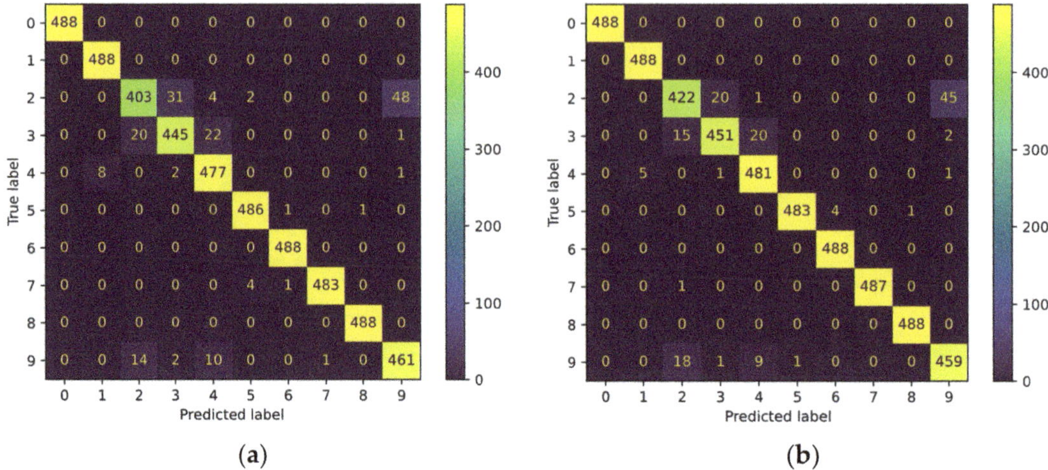

Figure 10. (**a**) Confusion matrix without AdaBN on laboratory dataset. (**b**) Confusion matrix with AdaBN on laboratory dataset.

5. Conclusions

Sustainable energy is one of the most important research directions across various disciplines. In order to achieve timely detection of faults in rotary machinery during operation, this paper proposes a rotating machinery fault diagnosis method based on AdaBN adaptive domain generalization, which effectively improves the convergence speed and stability of the model. Compared with traditional machine learning methods, this method has higher accuracy and is beneficial for the timely diagnosis of faults in rotating machinery, thus promoting the sustainable development of energy. In the field of sustainability, adaptive batch normalization can also help detect faults in other mechanical equipment in the energy sector, thus promoting sustainable energy development. In addition to bearing fault diagnosis, adaptive batch normalization can also be used for tasks such as image classification, speech recognition, natural language processing, and sustainability. Adaptive batch normalization can enhance the generalization ability of different datasets, thus improving classification accuracy and enhancing the generalization ability of the model. In the future, the authors will continue to apply the AdaBN algorithm to the energy sector to promote sustainable energy development.

Author Contributions: Investigation, X.W.; resources, D.L.; writing—original draft, X.L. and T.Y.; visualization, C.S.; project administration, Z.X.; funding acquisition, X.K. All authors have read and agreed to the published version of the manuscript.

Funding: This work is supported in part by the Fundamental Research Funds for the Central Universities (No. 2572022BF07) and in part by the Key Laboratory of Vibration and Control of Aero-Propulsion System, Ministry of Education, Northeastern University (VCAME202209).

Institutional Review Board Statement: Not applicable.

Informed Consent Statement: Not applicable.

Data Availability Statement: Not applicable.

Conflicts of Interest: The authors declare no conflict of interest.

Abbreviation

Adaptive Batch Normalization	AdaBN
Wavelet Transform	WT
Empirical Mode Decomposition	EMD
Singular Value Decomposition	SVD
Short-Time Fourier Transform	STFT
Support Vector Machines	SVM
Artificial Neural Network	ANN
Restricted Boltzmann Machines	RBM
Convolutional Neural Networks	CNN
Deep Belief Network	DBN
Stacked Autoencoders	SAE
Balanced Distribution Adaptation	BDA
Kullback–Leibler	KL
Maximum Mean Discrepancy	MMD
Joint Distribution Adaptation	JDA
Electric Discharge Machining	EDM
Batch Normalization	BN
Deep Convolutional Neural Networks	DCNN
Source training set	SDT
Source test set	SDV
Target test set	TDV
Case Western Reserve University	CWRU

References

1. Singh, N.; Hamid, Y.; Juneja, S.; Srivastava, G.; Dhiman, G.; Gadekallu, T.R.; Shah, M.A. Load balancing and service discovery using Docker Swarm for microservice based big data applications. *J. Cloud Comput.* **2023**, *12*, 1–9. [CrossRef]
2. Slathia, S.; Kumar, R.; Lone, M.; Viriyasitavat, W.; Kaur, A.; Dhiman, G. A Performance Evaluation of Situational-Based Fuzzy Linear Programming Problem for Job Assessment. In *Proceedings of Third International Conference on Advances in Computer Engineering and Communication Systems*; ICACECS 2022; Springer: Berlin/Heidelberg, Germany, 2023; pp. 655–667.
3. Tan, D.; Meng, Y.; Tian, J.; Zhang, C.; Zhang, Z.; Yang, G.; Cui, S.; Hu, J.; Zhao, Z. Utilization of renewable and sustainable diesel/methanol/n-butanol (DMB) blends for reducing the engine emissions in a diesel engine with different pre-injection strategies. *Energy* **2023**, *269*, 126785. [CrossRef]
4. Tan, D.; Wu, Y.; Lv, J.; Li, J.; Ou, X.; Meng, Y.; Lan, G.; Chen, Y.; Zhang, Z. Performance optimization of a diesel engine fueled with hydrogen/biodiesel with water addition based on the response surface methodology. *Energy* **2023**, *263*, 125869. [CrossRef]
5. Hu, J.; Zhang, L. Risk based opportunistic maintenance model for complex mechanical systems. *Expert Syst. Appl.* **2014**, *41*, 3105–3115. [CrossRef]
6. Bao, J.; Qu, P.; Wang, H.; Zhou, C.; Zhang, L.; Shi, C. Implementation of various bowl designs in an HPDI natural gas engine focused on performance and pollutant emissions. *Chemosphere* **2022**, *303*, 135275. [CrossRef]
7. Shi, C.; Chai, S.; Di, L.; Ji, C.; Ge, Y.; Wang, H. Combined experimental-numerical analysis of hydrogen as a combustion enhancer applied to Wankel engine. *Energy* **2023**, *263*, 125896. [CrossRef]
8. Lei, Y.; Yang, B.; Jiang, X.; Jia, F.; Li, N.; Nandi, A.K. Applications of machine learning to machine fault diagnosis: A review and roadmap. *Mech. Syst. Signal Process.* **2020**, *138*, 106587. [CrossRef]
9. Peng, T.; Gui, W.; Wu, M.; Xie, Y. A fusion diagnosis approach to bearing faults. In Proceedings of the International Conference on Modeling and Simulation in Distributed Applications, Changsha, China, 25–27 September 2001; pp. 759–766.
10. Lei, Y.; Zuo, M.J.; He, Z.; Zi, Y. A multidimensional hybrid intelligent method for gear fault diagnosis. *Expert Syst. Appl.* **2010**, *37*, 1419–1430. [CrossRef]
11. Stewart, G.W. On the early history of the singular value decomposition. *SIAM Rev.* **1993**, *35*, 551–566. [CrossRef]
12. Griffin, D.; Lim, J. Signal estimation from modified short-time Fourier transform. *IEEE Trans. Acoust. Speech Signal Process.* **1984**, *32*, 236–243. [CrossRef]
13. Cortes, C.; Vapnik, V. Support-vector networks. *Mach. Learn.* **1995**, *20*, 273–297. [CrossRef]
14. Schmidhuber, J. Deep learning in neural networks: An overview. *Neural Netw.* **2015**, *61*, 85–117. [CrossRef] [PubMed]
15. Cover, T.; Hart, P. Nearest neighbor pattern classification. *IEEE Trans. Inf. Theory* **1967**, *13*, 21–27. [CrossRef]
16. Song, Y.-Y.; Ying, L. Decision tree methods: Applications for classification and prediction. *Shanghai Arch. Psychiatry* **2015**, *27*, 130.
17. Glorot, X.; Bengio, Y. Understanding the difficulty of training deep feedforward neural networks. In Proceedings of the Thirteenth International Conference on Artificial Intelligence and Statistics, Sardinia, Italy, 13–15 May 2010; JMLR Workshop and Conference Proceedings. pp. 249–256.

18. Rumelhart, D.E.; Hinton, G.E.; Williams, R.J. Learning representations by back-propagating errors. *Nature* **1986**, *323*, 533–536. [CrossRef]
19. Albawi, S.; Mohammed, T.A.; Al-Zawi, S. Understanding of a convolutional neural network. In Proceedings of the 2017 International Conference on Engineering and Technology (ICET), Antalya, Turkey, 21–23 August 2017; IEEE: Piscataway, NJ, USA, 2017; pp. 1–6.
20. Hua, Y.; Guo, J.; Zhao, H. Deep belief networks and deep learning. In Proceedings of the 2015 International Conference on Intelligent Computing and Internet of Things, Harbin, China, 17–18 January 2015; IEEE: Piscataway, NJ, USA, 2015; pp. 1–4.
21. Wu, Z.; Shen, C.; Van Den Hengel, A. Wider or deeper: Revisiting the resnet model for visual recognition. *Pattern Recognit.* **2019**, *90*, 119–133. [CrossRef]
22. Zhong, S.; Fu, S.; Lin, L. A novel gas turbine fault diagnosis method based on transfer learning with CNN. *Measurement* **2019**, *137*, 435–453. [CrossRef]
23. Zhao, B.; Zhang, X.; Zhan, Z.; Pang, S. Deep multi-scale convolutional transfer learning network: A novel method for intelligent fault diagnosis of rolling bearings under variable working conditions and domains. *Neurocomputing* **2020**, *407*, 24–38. [CrossRef]
24. Wang, K.; Wu, B. Power equipment fault diagnosis model based on deep transfer learning with balanced distribution adaptation. In Proceedings of the International Conference on Advanced Data Mining and Applications, Nanjing, China, 16–18 November 2018; Springer: Berlin/Heidelberg, Germany, 2018; pp. 178–188.
25. Wang, Y.; Wang, C.; Kang, S.; Xie, J.; Wang, Q.; Mikulovich, V. Network-combined broad learning and transfer learning: A new intelligent fault diagnosis method for rolling bearings. *Meas. Sci. Technol.* **2020**, *31*, 115013. [CrossRef]
26. Yang, B.; Lei, Y.; Jia, F.; Li, N.; Du, Z. A polynomial kernel induced distance metric to improve deep transfer learning for fault diagnosis of machines. *IEEE Trans. Ind. Electron.* **2019**, *67*, 9747–9757. [CrossRef]
27. Han, T.; Liu, C.; Yang, W.; Jiang, D. Deep transfer network with joint distribution adaptation: A new intelligent fault diagnosis framework for industry application. *ISA Trans.* **2020**, *97*, 269–281. [CrossRef] [PubMed]
28. Qian, W.; Li, S.; Yi, P.; Zhang, K. A novel transfer learning method for robust fault diagnosis of rotating machines under variable working conditions. *Measurement* **2019**, *138*, 514–525. [CrossRef]
29. Li, Y.; Song, Y.; Jia, L.; Gao, S.; Li, Q.; Qiu, M. Intelligent fault diagnosis by fusing domain adversarial training and maximum mean discrepancy via ensemble learning. *IEEE Trans. Ind. Inform.* **2020**, *17*, 2833–2841. [CrossRef]
30. Zhang, Y.; Ren, Z.; Zhou, S. A new deep convolutional domain adaptation network for bearing fault diagnosis under different working conditions. *Shock. Vib.* **2020**, *2020*, 8850976. [CrossRef]
31. Shao, H.; Zhang, X.; Cheng, J.; Yang, Y. Intelligent fault diagnosis of bearing using enhanced deep transfer auto-encoder. *J. Mech. Eng.* **2020**, *56*, 84–91.
32. Li, C.-L.; Chang, W.-C.; Cheng, Y.; Yang, Y.; Póczos, B. Mmd gan: Towards deeper understanding of moment matching network. *Adv. Neural Inf. Process. Syst.* **2017**, *30*. [CrossRef]
33. Ioffe, S.; Szegedy, C. Batch normalization: Accelerating deep network training by reducing internal covariate shift. In Proceedings of the International Conference on Machine Learning, Lille, France, 6–11 July 2015; PMLR: New York, NY, USA, 2015; pp. 448–456.
34. He, K.; Zhang, X.; Ren, S.; Sun, J. Deep residual learning for image recognition. In Proceedings of the IEEE Conference on Computer Vision and Pattern Recognition, Las Vegas, NV, USA, 27–30 June 2016; pp. 770–778.
35. Li, X.; Su, K.; He, Q.; Wang, X. Research on fault diagnosis of highway bi-lstm based on attention mechanism. *Eksploat. I Niezawodn.-Maint. Reliab.* **2023**, *25*. [CrossRef]
36. Li, J.; He, D. A Bayesian optimization AdaBN-DCNN method with self-optimized structure and hyperparameters for domain adaptation remaining useful life prediction. *IEEE Access* **2020**, *8*, 41482–41501. [CrossRef]
37. Li, Y.; Wang, N.; Shi, J.; Liu, J.; Hou, X. Revisiting batch normalization for practical domain adaptation. *arXiv* **2016**, arXiv:1603.04779.
38. Huang, L. Normalization ináTask-Specific Applications. In *Normalization Techniques in Deep Learning*; Springer: Berlin/Heidelberg, Germany, 2022; pp. 87–100.
39. Jin, T.; Yan, C.; Chen, C.; Yang, Z.; Tian, H.; Guo, J. New domain adaptation method in shallow and deep layers of the CNN for bearing fault diagnosis under different working conditions. *Int. J. Adv. Manuf. Technol.* **2023**, *124*, 3701–3712. [CrossRef]

 *sustainability*

Article

Computational Investigation of Combustion, Performance, and Emissions of a Diesel-Hydrogen Dual-Fuel Engine

Bo Zhang [1], Huaiyu Wang [2],* and Shuofeng Wang [3]

[1] CRRC Academy Corporation Limited, Beijing 100160, China
[2] School of Mechanical Engineering, Beijing Institute of Technology, Beijing 100081, China
[3] Beijing Lab of New Energy Vehicles and Key Lab of Regional Air Pollution Control, College of Energy and Power Engineering, Beijing University of Technology, Beijing 100124, China
* Correspondence: huaiyu.wang@bit.edu.cn

Abstract: This paper aims to expose the effect of hydrogen on the combustion, performance, and emissions of a high-speed diesel engine. For this purpose, a three-dimensional dynamic simulation model was developed using a reasonable turbulence model, and a simplified reaction kinetic mechanism was chosen based on experimental data. The results show that in the hydrogen enrichment conditions, hydrogen causes complete combustion of diesel fuel and results in a 17.7% increase in work capacity. However, the increase in combustion temperature resulted in higher NOx emissions. In the hydrogen substitution condition, the combustion phases are significantly earlier with the increased hydrogen substitution ratio (*HSR*), which is not conducive to power output. However, when the *HSR* is 30%, the CO, soot, and THC reach near-zero emissions. The effect of the injection timing is also studied at an HSR of 90%. When delayed by 10°, IMEP improves by 3.4% compared with diesel mode and 2.4% compared with dual-fuel mode. The NOx is reduced by 53% compared with the original dual-fuel mode. This study provides theoretical guidance for the application of hydrogen in rail transportation.

Keywords: diesel-hydrogen dual-fuel engines; hydrogen substitution ratio; injection timing; rail transportation

Citation: Zhang, B.; Wang, H.; Wang, S. Computational Investigation of Combustion, Performance, and Emissions of a Diesel-Hydrogen Dual-Fuel Engine. *Sustainability* **2023**, *15*, 3610. https://doi.org/10.3390/su15043610

Academic Editors: Cheng Shi, Jinxin Yang, Jianbing Gao and Peng Zhang

Received: 23 January 2023
Revised: 9 February 2023
Accepted: 13 February 2023
Published: 15 February 2023

1. Introduction

Adjusting the industrial and energy structures is inevitable to realize an emission peak and carbon neutrality [1,2]. The gaseous and particle emissions from internal combustion engines (ICEs) contribute an important proportion of total atmospheric pollutants [3,4]. Therefore, searching for low-carbon or zero-carbon fuels has become an important research direction for developing ICEs [5–7]. Hydrogen is regarded as the clean energy with the most development potential in the 21st century because of many advantages, such as diverse sources, being clean and low-carbon, being flexibile and efficient, and having various application scenarios [8–10]. Hydrogen energy has become the preferred direction for the new round of carbon emission reduction and carbon neutrality worldwide [11,12]. It has been incorporated into the energy strategy deployment by many countries [13,14]. From the strategic point of view of energy security and sustainable development, China has considered hydrogen energy a new strategic industry for development [15,16].

There are two main ways to utilize hydrogen energy: fuel cells [17] and ICEs [18]. Fuel cells have the advantages of high efficiency and zero-emissions, but they are technically complex, costly, and dependent on the construction of supporting systems [19]. The hydrogen-fueled ICEs can use industrial by-product hydrogen to convert energy by the combustion method to achieve similar thermal efficiency as fuel cells, which has the significant advantage of low cost [20]. Hydrogen-fueled ICEs retain the main structure and system of traditional ICEs [21–23]. Based on traditional ICEs, hydrogen-fueled ICEs can be realized by simply replacing the hydrogen supply and injection system, hydrogen-specific cold spark plugs, matching a new turbocharger, and adapting the lubrication and crankcase ventilation accordingly [24–26]. Therefore, hydrogen-fueled ICEs are an essential technology direction to promote the upgrading and transformation of various application fields of traditional ICEs, which help achieve peak emissions and carbon neutrality [27–30].

For the spark-ignited hydrogen-fueled ICEs, the higher laminar flame speed and larger dilute combustion limits of hydrogen allow the engine to operate over a wide range of equivalence ratios, in which the thermal efficiency typically equals or exceeds that of gasoline-fueled engines [31]. It has been shown that spark-ignited hydrogen-fueled ICEs can achieve near-zero NOx emissions under lean-burn operating conditions [32]. However, hydrogen-fueled ICEs face problems such as backfire, premature ignition, and detonation at high loads, which severely limit their rapid development [33]. The heat transfer loss through the wall increases rapidly as the equivalent ratio increases [34]. The problems of detonation and premature ignition are challenging for heavy engines, which has prompted research on hydrogen compression ignition (CI) engines [35]. Therefore, dual-fuel technology was developed for engines [36]. The diesel fuel is designed to assist the ignition of the hydrogen fuel, known as the diesel pilot ignition mode [37,38]. In fact, this concept has long been widely used in natural gas-fueled ICEs [39] and extensively used in the marine field [40,41]. Tripathi et al. [42] investigated the performance and emissions of a diesel-hydrogen dual-fuel engine using numerical simulations. The results showed that the combination of two injection strategies can simultaneously reduce NOx emissions and improve IMEP. Sharma et al. [43] investigated the performance and emissions of dual-fuel engines with different compression ratios and hydrogen fractions through a similar methodology. The results show that compression-ignition mode is not suitable for compression ratios less than 14.5. Köse et al. [44] experimentally investigated that hydrogen enrichment reduced pollutant emissions except for NOx and increased brake thermal efficiency (BTE) and exhaust temperature. When operated at 1750 rpm, 40.4% BTE was achieved at 2.5% hydrogen enrichment, while in diesel mode, the BTE was 33%. Ramsay et al. [45] studied the effect of the constant volume combustion phase on the performance and emissions of a dual-fuel engine under various load and hydrogen energy share conditions. The results demonstrated that this method could improve thermal efficiency with far lower carbon-based emissions under all conditions. Taghavifar et al. [46], through a 1-D model, parametrically investigated the effects of levels of diesel and hydrogen, compressor pressure ratio, and combustion duration on energy, exergy, and performance in a diesel-hydrogen dual-fuel engine. The results indicated that supercharging can significantly improve thermal efficiency and reduce fuel consumption. Wu et al. [47] optimized the operation parameters of a dual-fuel engine based on the Taguchi method. The results revealed that for NOx, using EGR technology reduces more than 60.5% at various loads, and BSFC can reduce it by 14.52%.

From the above literature, it is clear that hydrogen can significantly improve thermal efficiency and performance. It is essential to improve thermal efficiency and reduce carbon emissions for rail transportation. Therefore, in this paper, hydrogen enrichment and hydrogen substitution are investigated separately by using numerical simulation. In addition, injection timing studies are carried out to optimize performance and emissions. This paper explored the potential of hydrogen in ICEs for rail transportation and provided a theoretical basis for practical applications.

2. Materials and Methods

2.1. Numerical Methodology

In this work, the prototype engine is a high-speed diesel engine with a total displacement of 87.54 L. The cylinder bore, stroke, and compression ratio were 180 mm, 215 mm, and 17, respectively. The engine is designed to meet future emissions regulations by taking into account compactness, power-to-weight ratio, economy, and reliability. The CONVERGE code was applied to calculate the flow motion and combustion phenomena in the combustion chamber [48–50]. The SolidWorks software was used to establish the 3-D geometric model. The model (*.stl) was then imported into CONVERGE to calculate the combustion. To simulate the turbulence, spray, and combustion, the mathematical models adopted in CFD calculations are summarized in Table 1 [51]. The dual-fuel reaction mechanism with 76 species and 464 reactions was chosen to simulate the combustion of the hydrogen and diesel mixture [52]. This dual-fuel reaction mechanism was coupled with the GRI3.0 mechanism, which can accurately simulate hydrogen combustion. This mechanism was used in the simulation of pure diesel and dual-fuel, which avoids the influence of mechanism differences on the results [53]. In addition, at 1800 rpm, the temperatures of the piston, cylinder wall, and cylinder head were set to 553, 433, and 523 K, respectively.

Table 1. Mathematical models adopted in this research.

Region	Type
Turbulence	RNG $k - \varepsilon$ model
Wall heat transfer	O'Rourke and Amsden model
Spray breakup	KH-RT model
Evaporation	Frossling model
Droplet collision	O'Rourke's model
Combustion	SAGE
NOx formation	Extended Zel'dovich mechanism
Soot formation	Hiroyasu model

2.2. Model Validation

To reduce calculation time, only 1/8 of the domain was recorded and analyzed since the injection is located in the center of the chamber with eight nozzles. As illustrated in Figure 1a, the cylindrical domain was classified into angular sectors such that one injector falls at the center of each sector. Figure 1b shows the computational mesh at TDC, in which the adaptive mesh refinement and fixed embedding were activated to guarantee the calculation accuracy. Figure 2 shows the predicted pressure profile under different meshes. It can be seen from Figure 2 that the 4 mm basic grid can meet the calculation accuracy. As the dual-fuel mode is still in the development stage, only the diesel mode was tested. To validate the combustion and turbulence model, the model was verified at speeds of 600, 1400, and 1800 rpm. Figure 3 shows the comparison between experimental and simulated cylinder pressure. The calculation accuracy was high and met the engineering requirements for the next step of research. Since the reaction kinetics mechanism used includes hydrogen and diesel, the verified model was used for the next study. In addition, Tripathi et al. [42] was validated using the same turbulence model and mechanism in diesel-hydrogen conditions, which is another indication that the accuracy of the model in this paper can be studied in the next part [54].

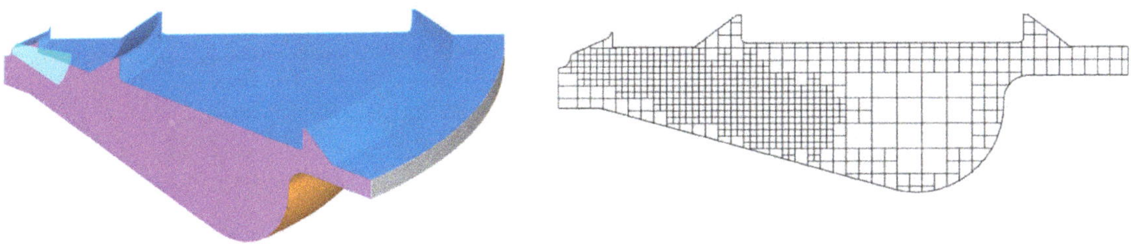

(a) Engine sector domain   (b) Computational mesh

Figure 1. Schematic sector domain and computational mesh.

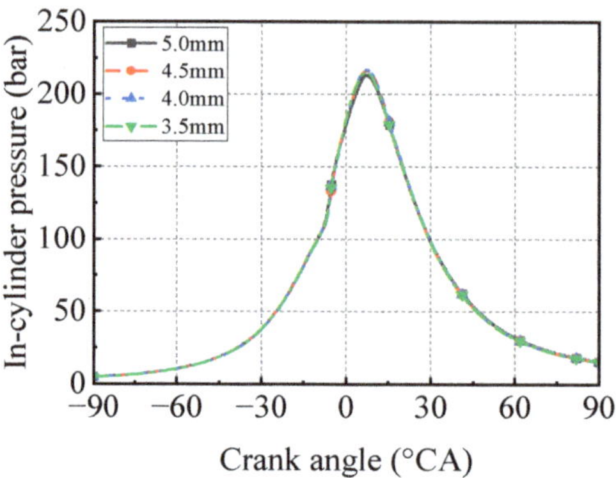

Figure 2. Comparison of predicted pressure profile under different meshes.

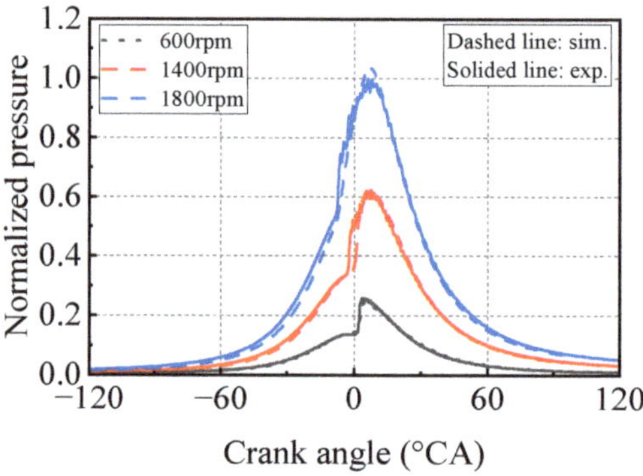

Figure 3. Model validation of in-cylinder pressure at various speeds.

2.3. Research Schemes

The present work aims to evaluate the potential of hydrogen in diesel engines and to explore the maximum hydrogen substitution ratio for the purpose of reducing carbon emissions. In all the simulations, the speed was kept at 1800 rpms, and the total ejected energy was same as with the 18-bar BMEP. In the hydrogen enrichment test, the source diesel was kept the same, and the hydrogen was the added energy. The hydrogen enrichment ratio (*HER*) was between 10 and 20%. In the hydrogen substitution test, the total energy was kept the same as with the origin diesel, and the substitution ratios (*HSR*) were 30, 60, and 90%, respectively. In the relative injection timing test, the high *HSR* was tested with varing injections. The summary of the test conditions is listed in Table 2.

Table 2. Summary of different calculated schemes.

Case Group	Hydrogen Energy Ratio/%	Relative Injection Timing/°CA
Hydrogen enrichment	0% [1]	0
	10% [2]	0
	20% [2]	0
Hydrogen substitution	30%	0
	60%	0
	90%	0

Table 2. *Cont.*

Case Group	Hydrogen Energy Ratio/%	Relative Injection Timing/°CA
Relative injection timing	0%	0
	90%	0
	90%	5
	90%	10
	90%	15

[1] pure diesel, [2] *HER*.

The definition of *HER* and *HSR* is as follows:

$$HER = \frac{m_{H_2} \cdot LHV_{H_2}}{m_{Diesel} \cdot LHV_{Diesel}} \tag{1}$$

$$HSR = \frac{m_{H_2} \cdot LHV_{H_2}}{m_{Diesel} \cdot LHV_{Diesel} + m_{H_2} \cdot LHV_{H_2}} \tag{2}$$

3. Results and Discussion

3.1. Effect of Hydrogen Enrichment

The in-cylinder pressure is an important parameter to deeply understand the combustion process and directly affects performance and emissions. The comparison of in-cylinder pressures is shown in Figure 4. In this section, D and H represent diesel and hydrogen, respectively. D100 means the diesel energy fraction is 100%, and H10 represents the hydrogen energy fraction at 10%. As shown in Figure 4, a larger *HER* leads to a higher in-cylinder pressure [42]. The corresponding CA position (PFP_CA) is delayed with the increasing *HER*, as shown in Table 3. The IMEP of D100 + H10 and D100 + H20 increased by 8.7 and 17.7%, respectively. The temperature contour is shown in Table 4. It is obvious that the temperature increases with the *HER*. This is because diesel has a dominant effect on combustion, while hydrogen only promotes combustion. The comparison of HRR and combustion phases is shown in Figure 5. The profiles of HRR and combustion phases are similar to each other. The phases are delayed with increasing *HER*, but the level is minor. This is because the overall energy of the additional hydrogen enrichment condition is higher than the original engine. Although the hydrogen enrichment could enhance the flame speed, the ratio is relatively small.

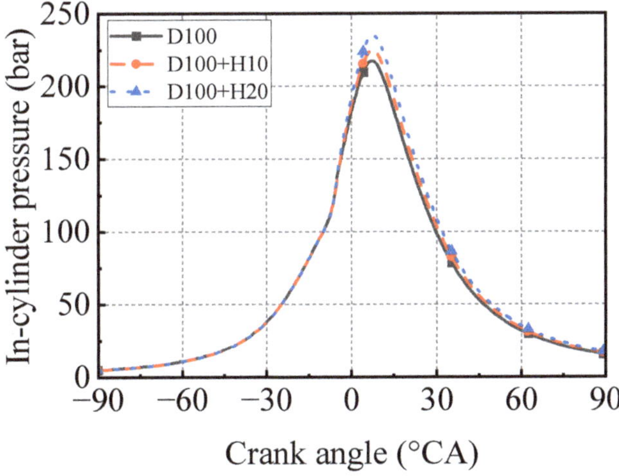

Figure 4. Comparison of in-cylinder pressure at hydrogen additional enrichment.

Table 3. Comparison of PFP, corresponding CA position, and IMEP at hydrogen additional enrichment.

Parameter	PFP (bar)	PFP_CA (°CA, aTDC)	IMEP (bar)
D100	21.75	7.26	21.53
D100 + H10	22.49	7.43	23.40
D100 + H20	23.51	7.73	25.33

Table 4. Contour of temperature distribution at hydrogen additional enrichment.

CAD	D100 (K)	D100 + H10 (K)	D100 + H20 (K)
−4 °CA aTDC			
TDC			
4 °CA aTDC			
8 °CA aTDC			
16 °CA aTDC			
24 °CA aTDC			

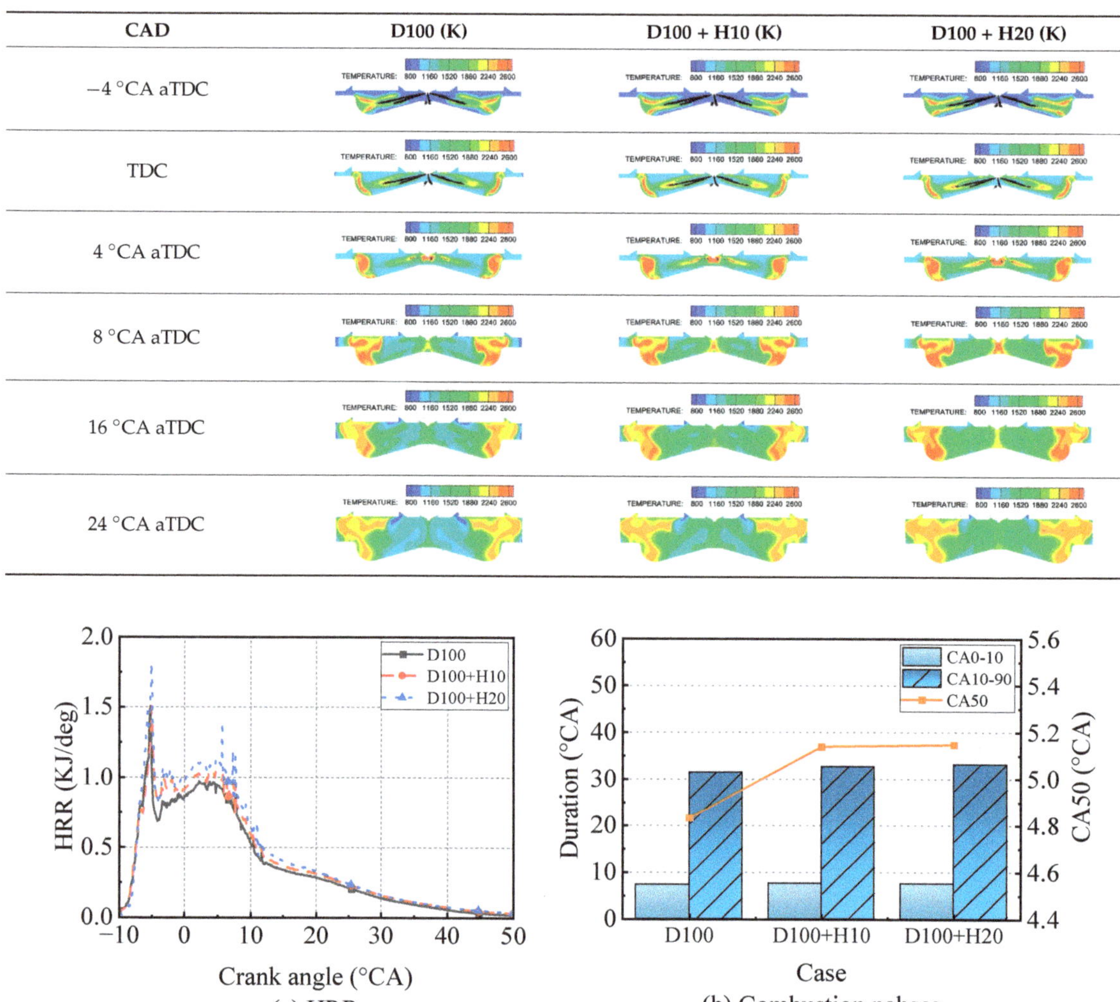

Figure 5. Comparison of combustion phases at hydrogen additional enrichment.
(a) HRR
(b) Combustion pahses

Figure 5. Comparison of combustion phases at hydrogen additional enrichment.

The comparison of emissions is shown (Figure 6) at hydrogen additional enrichment, where the value is recorded at the exhaust opening timing. The data are scaled so that it can be displayed on a single figure. In general, all pollutant emissions except NOx and CO_2 decrease with the increase of *HER*. This is because soot, THC, and CO can all be further oxidized and burn more completely as the *HER* increases. The higher *HAR* results in a higher in-cylinder temperature, which is mainly responsible for the complete oxidation of $CO + OH \Leftrightarrow CO_2 + H$ and $CO + O + M \Leftrightarrow CO_2 + M$, and improving the degree of complete combustion consequently [55]. The NOx formation is chiefly determined by mean temperature. For the case D100 + H2, the NOx emissions increased by 45%. It is recommended to combine after-treatment equipment for further optimization.

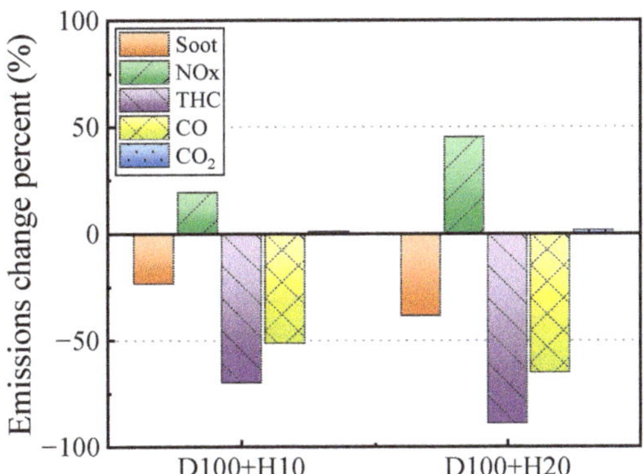

Figure 6. Comparison of emissions at hydrogen additional enrichment.

3.2. Effect of Hydrogen Substitution

In order to investigate the effect of hydrogen substitution enrichment on combustion, performance, and emissions, the conditions with HSR of 0, 30, 60, and 90% are selected for further simulation. In all the simulations, the total energy is kept the same, where D and H represent diesel and hydrogen, respectively. The D10-H90 means the diesel energy fraction is 10% and the hydrogen energy fraction is 90%. The in-cylinder pressure is shown in Figure 7. The comparison of PFP, corresponding CA, and IMEP is listed in Table 5. The PFP increases with the increase of HSR. As the HSR increases, the corresponding CA advances. This is because the higher burning velocity of hydrogen accelerates the flame speed and results in a higher PFP and an advanced CA. The IMEP of D10-H90 is lower than D40-H60, although the PFP is higher. However, too high a PFP is also not conducive to the modification of the original engine and even requires a redesign of the overall strength.

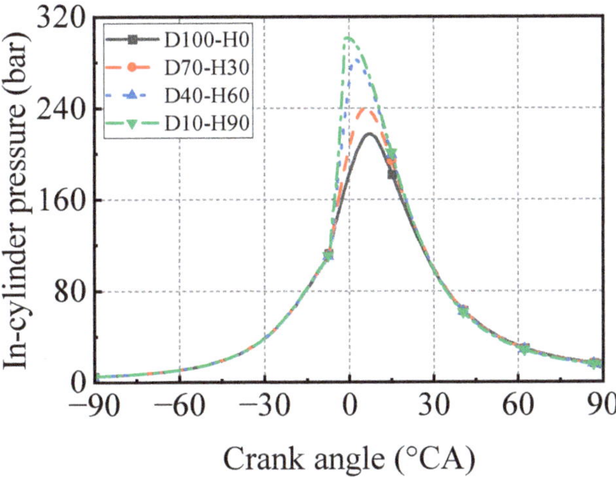

Figure 7. Comparison of in-cylinder pressure at hydrogen substitution enrichment.

Table 5. Comparison of PFP, corresponding CA position, and IMEP at hydrogen additional enrichment.

Parameter	PFP (bar)	PFP_CA (°CA, aTDC)	IMEP (bar)
D100-H0	21.75	7.26	21.53
D70-H30	23.93	5.92	22.00
D40-H60	28.33	1.85	22.01
D10-H90	30.13	−0.48	21.76

As shown in Table 6, the mean temperature is raised with the increase of HSR. The high-temperature zone appears earlier as the HSR increases. When the mixing ratio is 90%, the high-temperature zone appears inside the combustion chamber, while the high-temperature zone at other ratios is beyond the combustion chamber near the cylinder head. Therefore, increasing HSR is beneficial to reduce the heat load on the cylinder head. The high temperature zone inside the cylinder is more uniform with the increase of HSR. This is due to the fact that as the HSR increases, the combustion mode gradually shifts from diffusion combustion to premixed combustion mode. This transition can be visualized from the HRR, as shown in Figure 8a. As the HSR increases, the HRR gradually shows a dual trend. The first peak is caused by diesel igniting hydrogen, and the second peak is caused by the simultaneous ignition of hydrogen at multiple points of the combustion flame. When the HSR is 60%, the two values are similar, while the peak of hydrogen combustion exotherm is much higher than that of diesel when the HSR is 90%. It can also be seen from Table 6 that diffusion combustion generally starts from the tail, and the high temperature area is small. Because the laminar flame of hydrogen is faster, when premixed, the flame quickly spreads throughout the combustion chamber. Therefore, when the HSR is higher, the second peak of heat release is also higher. The comparison of combustion phases is also shown in Figure 8b. Due to the faster burning rate of hydrogen and the shift in the combustion mode, the combustion phases are advanced. However, when the HSR is 90%, the CA50 is already before the TDC, which is not conducive to power output.

Table 6. Contour of temperature distribution at hydrogen substitution enrichment.

CAD	D100-H0 (K)	D70-H30 (K)	D40-H60 (K)	D10-H90 (K)
−4 °CA aTDC	TEMPERATURE: 800 1160 1520 1880 2240 2600	TEMPERATURE: 800 1160 1520 1880 2240 2600	TEMPERATURE: 800 1160 1520 1880 2240 2600	TEMPERATURE: 800 1160 1520 1880 2240 2600
TDC	TEMPERATURE: 800 1160 1520 1880 2240 2600	TEMPERATURE: 800 1160 1520 1880 2240 2600	TEMPERATURE: 800 1160 1520 1880 2240 2600	TEMPERATURE: 800 1160 1520 1880 2240 2600
4 °CA aTDC	TEMPERATURE: 800 1160 1520 1880 2240 2600	TEMPERATURE: 800 1160 1520 1880 2240 2600	TEMPERATURE: 800 1160 1520 1880 2240 2600	TEMPERATURE: 800 1160 1520 1880 2240 2600
12 °CA aTDC	TEMPERATURE: 800 1160 1520 1880 2240 2600	TEMPERATURE: 800 1160 1520 1880 2240 2600	TEMPERATURE: 800 1160 1520 1880 2240 2600	TEMPERATURE: 800 1160 1520 1880 2240 2600
24 °CA aTDC	TEMPERATURE: 800 1160 1520 1880 2240 2600	TEMPERATURE: 800 1160 1520 1880 2240 2600	TEMPERATURE: 800 1160 1520 1880 2240 2600	TEMPERATURE: 800 1160 1520 1880 2240 2600

For the emissions, as shown in Figure 9, all pollutant emissions except NOx are reduced, especially THC and CO, which have become small in order of magnitude at HSR >30%. It can be analyzed in two ways: first, the increased HSR reduces the diesel fuel, resulting in lower carbon for THC and CO, i.e., the total fuel carbon content. Secondly, due to the higher combustion temperature of hydrogen enrichment, hydrogen increases the free radical content of the reaction, which promotes the combustion process and makes the combustion more adequate. For soot, when the HSR is higher, it leads to more homogeneous fuel mixing due to the lower mixture density. In addition, the hydrogen raises the temperature of the compression process, resulting in a lower in-cylinder temperature gradient, which deteriorates the soot generation environment. The increase in combustion temperature provides a good environment for soot oxidation. The synergistic effect of the two leads to extremely low soot. For CO_2, because hydrogen is a carbon-free fuel, CO_2 emissions are dramatically reduced at higher *HSR* [56].

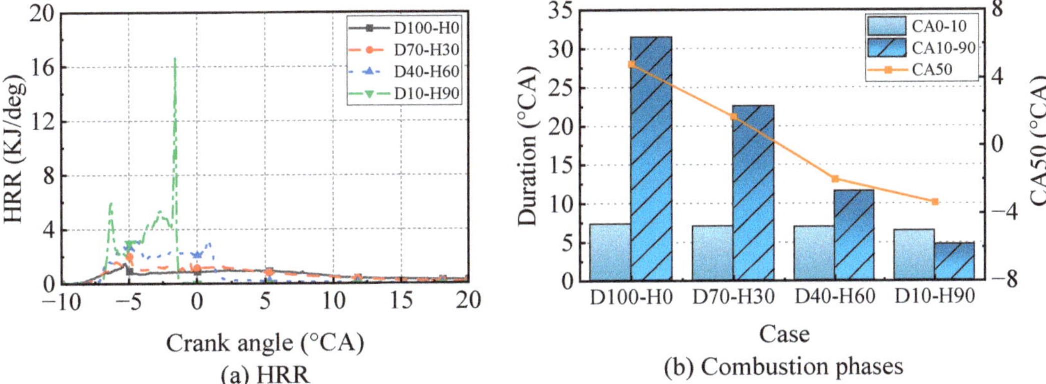

(a) HRR

(b) Combustion phases

Figure 8. Comparison of HRR and combustion phases at hydrogen substitution enrichment.

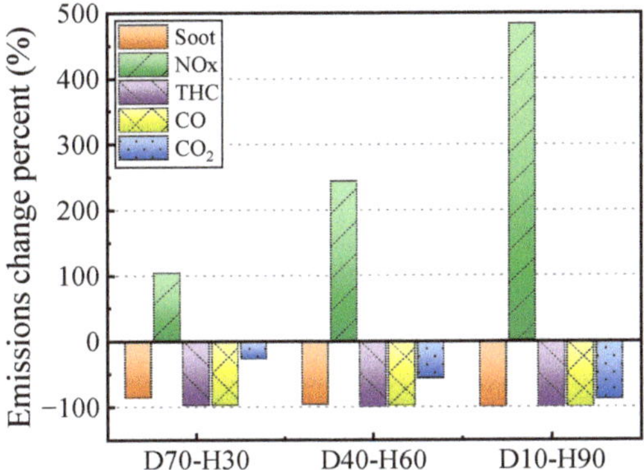

Figure 9. Comparison of emissions at hydrogen substitution enrichment.

3.3. Effect of Pilot Injection Timing

From the above study, it can be concluded that hydrogen enrichment improves the combustion process and reduces pollutant emissions except NOx. For performance, the optimization of the injection strategy is necessary because the combustion phase at high *HSR* conditions is too advanced [57]. In addition, at high *HSR*, the PFP is too high, resulting in excessive mechanical load. In addition, considering the goal of carbon reduction and the common operating conditions of rail internal combustion engines, this section investigates the effect of the pilot fuel injection timing on the performance, combustion, and emissions of dual-fuel engines.

The "I" represents the relative injection timing, i.e., H0-I0 is the injection timing of the original engine in diesel mode (the relative injection timing is 0 °CA). The H90-I5 represents 90% hydrogen energy and 10% diesel energy, and the diesel injection timing is delayed by 5 °CA. Figure 10 shows the comparison of in-cylinder pressure at different injection timings. In all cases, the PFP is higher than the diesel mode. The pressure rise rate is also increased in dual-fuel mode. As listed in Table 7, the CA position of PFP decreases with the delayed injection timing. According to the above study, in dual-fuel mode, when at the maximum *HSR*, the delayed injection timing facilitates an increase in engine power output. The IMEP exceeds the original diesel condition. When delayed by 10 °CA, IMEP improves by 3.4% compared with diesel mode and 2.4% compared with dual-fuel mode.

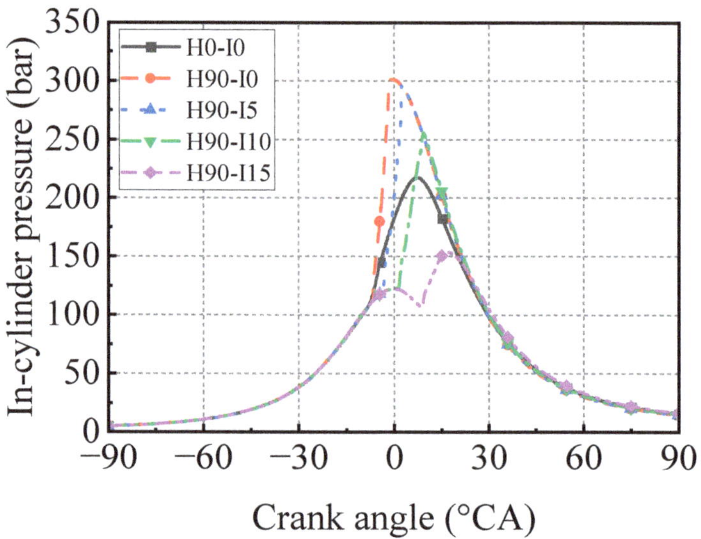

Figure 10. Comparison of in-cylinder pressure with various pilot injection timings.

Table 7. Comparison of PFP, corresponding CA position, and IMEP with various pilot injection timings.

Parameter	PFP (bar)	PFP_CA (°CA, aTDC)	IMEP (bar)
H0-I0	21.75	7.26	21.53
H90-I0	30.13	−0.48	21.76
H90-I5	29.57	2.52	22.00
H90-I10	25.55	9.32	22.27
H90-I15	15.37	17.03	21.20

The contour of temperature distribution is illustrated in Table 8. The appearance of the high temperature zone is delayed with the delay in injection timing. The mean temperature in the dual-fuel mode is higher than the diesel model, which undoubtedly causes higher NOx emissions. The trend of peak temperature is the same as the injection timing. When the injection timing is early, two separate high-temperature zones appear at the beginning of combustion. This is due to lower temperatures and incomplete fuel atomization. Therefore, the fuel in the combustion chamber and near the injector ignited hydrogen gas separately. When in the dual-fuel mode, the diesel is already atomized and broken at the end of the injection compared with the pure-diesel mode. This is due to the reduced in-cylinder ambient density, which facilitates fuel atomization. The HRR and combustion phases are shown in Figure 11. The combustion center is advanced with the early injection timing. This is because as the injection timing is delayed, the in-cylinder pressure and temperature decrease, which delay the combustion phases [58]. As the injection timing is delayed, the second peak of heat release rises initially before decreasing. When the injection timing is delayed by 15 °CA, the second peak of heat release dissipates, resulting in a shift to premixed combustion.

Table 8. Contour of temperature distribution with various pilot injection timings.

CAD	H90-I0 (K)	H90-I5 (K)	H90-I10 (K)	H90-I15 (K)
−4 °CA aTDC	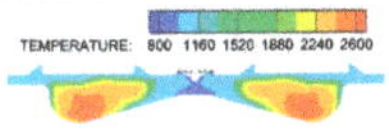	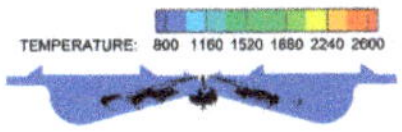	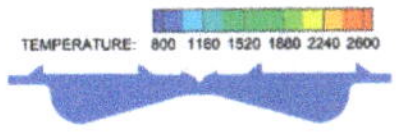	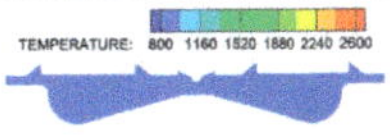
TDC	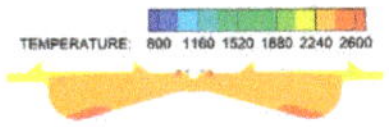	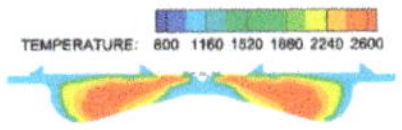	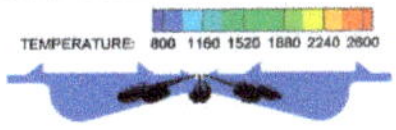	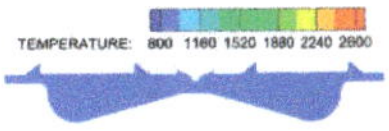
8 °CA aTDC	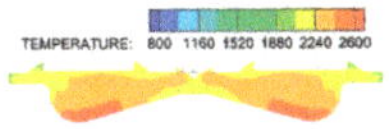	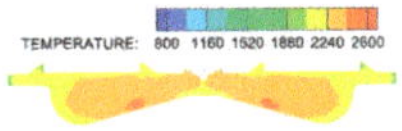	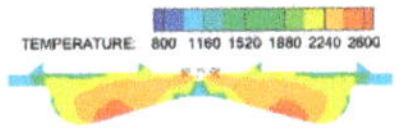	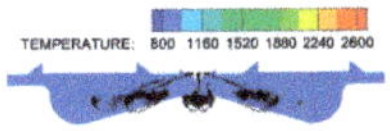
16 °CA aTDC	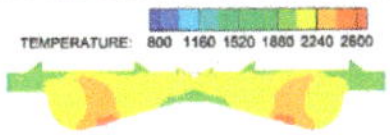	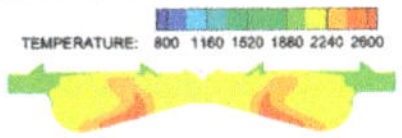	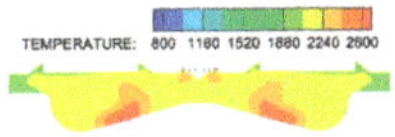	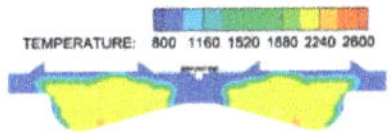
24 °CA aTDC	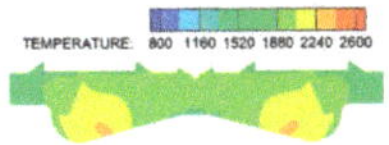	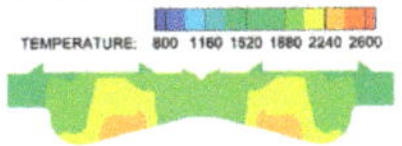	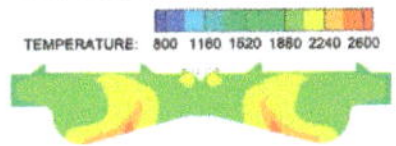	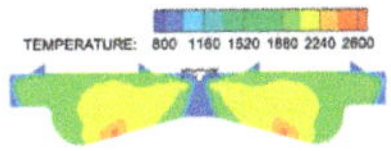

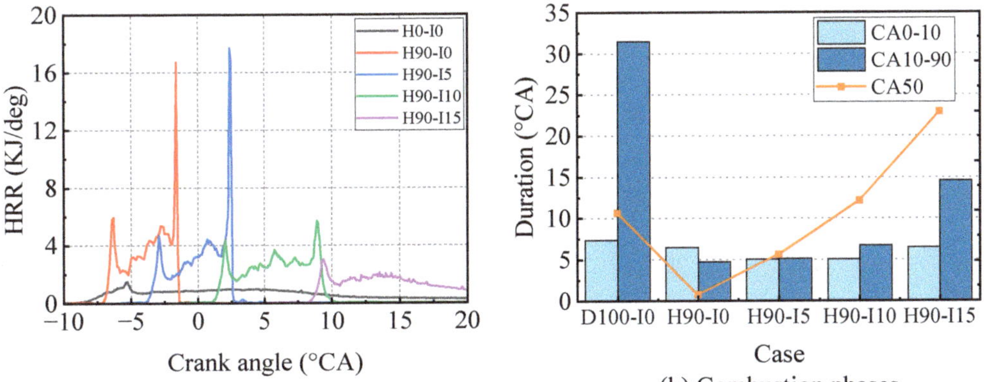

Figure 11. Comparison of HRR and combustion phases with various pilot injection timings.

The effect of the injection timing on the emissions is shown in Figure 12. The soot, THC, CO, and CO_2 are much smaller than the original engine after hydrogen enrichment conditions. Alternatively, the carbon content of the fuel is reduced to 10% of the original engine because the pilot diesel is 1/10 of the original engine. For CO_2, it is reduced by nearly 90%. Conversely, due to the higher combustion temperature in dual-fuel mode, it leads to a dramatic reduction in the production of unburned emissions. At high *HSR*, the lower in-cylinder temperature gradient due to less diesel injection mass and more homogeneous mixing is not conducive to soot production. In addition, due to the higher temperature, a good environment for soot oxidation is provided. When the injection timing was delayed by 15 °CA, a small amount of THC appeared. For NOx generation, the NOx generation decreases with the delayed injection timing. This is because the delayed injection timing reduces the in-cylinder combustion temperature, which is not conducive to NOx generation [42]. Compared with the original injection timing, NOx is reduced by 53% when delaying 10 °CA.

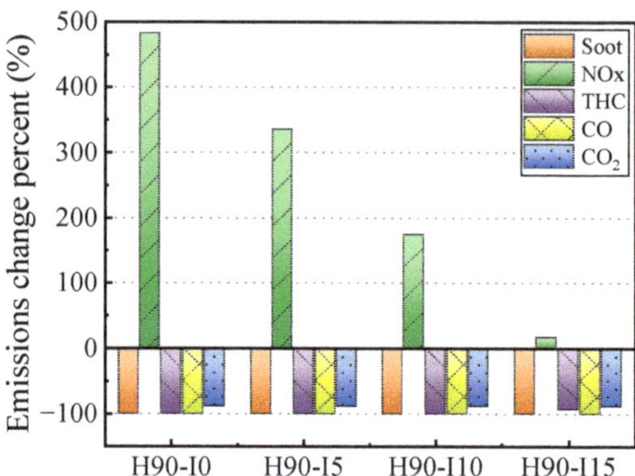

Figure 12. Comparison of emissions with various pilot injection timings.

4. Conclusions

In this paper, the effect of hydrogen on diesel combustion is investigated using a numerical method. The effects of additional hydrogen enrichment, hydrogen substitution, and pilot injection timing on combustion, performance, and emissions are investigated. The main findings are as follows:

(1) The hydrogen enrichment was used for the study at very high loads. With the increase of HER, the diesel fuel atomized better and burned more fully. When in a small HER, the combustion phase of the engine had a small range of variation and a consistent shape of the HRR. Due to the increase in temperature after hydrogen enrichment, it leads to higher NOx. However, CO, soot, and THC emissions were reduced due to more complete combustion.

(2) The hydrogen substitution was also studied in order to reduce carbon emissions and increase the HSR. As HSR increases, the peak cylinder pressure increases, and the combustion phase advances. The higher combustion temperature of hydrogen leads to more NOx. When the HSR was 90%, the center of combustion was located unfavorably in front of the upper stop, resulting in a lower IMEP.

(3) Since the center of gravity of combustion is too advanced at a higher HSR, the study of injection timing was carried out. With the delay in injection timing, the in-cylinder pressure decreases and the combustion temperature decreases. In addition, the work capacity increases and the NOx decreases.

(4) The implementation of hydrogen injectors necessitates increased control standards, as the low energy density of hydrogen may cause limitations in larger-scale applications. In addition, hydrogen safety must be taken into account.

In conclusion, this paper investigates the effect of hydrogen on diesel engine combustion and emissions and provides theoretical guidance for practical design optimization. Subsequent work will couple EGR with an injection strategy to further increase power and reduce NOx emissions.

Author Contributions: Formal analysis, B.Z. and H.W.; Funding acquisition, S.W.; Investigation, H.W.; Methodology, B.Z.; Project administration, S.W.; Resources, S.W.; Software, B.Z. and H.W.; Supervision, S.W.; Writing—original draft, B.Z.; Writing—review & editing, H.W. All authors have read and agreed to the published version of the manuscript.

Funding: This work was supported by CRRC Academy Corporation Limited and the BIT Research and Innovation Promoting Project (Grant No.2022YCXZ004). The authors also acknowledge all students involved in tests and calculations for this work.

Institutional Review Board Statement: Not applicable.

Informed Consent Statement: Not applicable.

Data Availability Statement: The data presented in this study are available on request from the corresponding author. The data are not publicly available due to privacy restrictions.

Conflicts of Interest: The authors declare no conflict of interest.

References

1. Gong, C.; Li, Z.; Sun, J.; Liu, F. Evaluation on combustion and lean-burn limit of a medium compression ratio hydrogen/methanol dual-injection spark-ignition engine under methanol late-injection. *Appl. Energy* **2020**, *277*, 115622. [CrossRef]
2. Bakar, R.A.; Widudo; Kadirgama, K.; Ramasamy, D.; Yusaf, T.; Kamarulzaman, M.K.; Sivaraos; Aslfattahi, N.; Samylingam, L.; Alwayzy, S.H. Experimental analysis on the performance, combustion/emission characteristics of a DI diesel engine using hydrogen in dual fuel mode. *Int. J. Hydrog. Energy* **2022**. [CrossRef]
3. Gong, C.; Si, X.; Liu, F. Effects of injection timing and CO_2 dilution on combustion and emissions behaviors of a stoichiometric GDI engine under medium load conditions. *Fuel* **2021**, *303*, 121262. [CrossRef]
4. Wang, H.; Ji, C.; Shi, C.; Ge, Y.; Meng, H.; Yang, J.; Chang, K.; Wang, S. Comparison and evaluation of advanced machine learning methods for performance and emissions prediction of a gasoline Wankel rotary engine. *Energy* **2022**, *248*, 123611. [CrossRef]
5. Gong, C.; Yi, L.; Zhang, Z.; Sun, J.; Liu, F. Assessment of ultra-lean burn characteristics for a stratified-charge direct-injection spark-ignition methanol engine under different high compression ratios. *Appl. Energy* **2020**, *261*, 114478. [CrossRef]
6. Duan, X.; Xu, L.; Jiang, P.; Lai, M.-C.; Sun, Z. Parallel perturbation analysis of combustion cycle-to-cycle variations and emissions characteristics in a natural gas spark ignition engine with comparison to consecutive cycle method. *Chemosphere* **2022**, *308*, 136334. [CrossRef]
7. Li, Y.; Gao, W.; Zhang, P.; Ye, Y.; Wei, Z. Effects study of injection strategies on hydrogen-air formation and performance of hydrogen direct injection internal combustion engine. *Int. J. Hydrog. Energy* **2019**, *44*, 26000–26011. [CrossRef]
8. Ma, Y.; Wang, X.; Li, T.; Zhang, J.; Gao, J.; Sun, Z. Hydrogen and ethanol: Production, storage, and transportation. *Int. J. Hydrog. Energy* **2021**, *46*, 27330–27348. [CrossRef]
9. Sun, X.; Liu, H.; Duan, X.; Guo, H.; Li, Y.; Qiao, J.; Liu, Q.; Liu, J. Effect of hydrogen enrichment on the flame propagation, emissions formation and energy balance of the natural gas spark ignition engine. *Fuel* **2022**, *307*, 121843. [CrossRef]
10. Seelam, N.; Gugulothu, S.K.; Reddy, R.V.; Bhasker, B.; Kumar Panda, J. Exploration of engine characteristics in a CRDI diesel engine enriched with hydrogen in dual fuel mode using toroidal combustion chamber. *Int. J. Hydrog. Energy* **2022**, *47*, 13157–13167. [CrossRef]
11. Liu, W.; Wan, Y.; Xiong, Y.; Gao, P. Green hydrogen standard in China: Standard and evaluation of low-carbon hydrogen, clean hydrogen, and renewable hydrogen. *Int. J. Hydrog. Energy* **2022**, *47*, 24584–24591. [CrossRef]
12. Yu, X.; Li, G.; Du, Y.; Guo, Z.; Shang, Z.; He, F.; Shen, Q.; Li, D.; Li, Y. A comparative study on effects of homogeneous or stratified hydrogen on combustion and emissions of a gasoline/hydrogen SI engine. *Int. J. Hydrog. Energy* **2019**, *44*, 25974–25984. [CrossRef]
13. Duan, X.; Liu, Y.; Liu, J.; Lai, M.-C.; Jansons, M.; Guo, G.; Zhang, S.; Tang, Q. Experimental and numerical investigation of the effects of low-pressure, high-pressure and internal EGR configurations on the performance, combustion and emission characteristics in a hydrogen-enriched heavy-duty lean-burn natural gas SI engine. *Energy Convers. Manag.* **2019**, *195*, 1319–1333. [CrossRef]
14. Arat, H.T. Alternative fuelled hybrid electric vehicle (AF-HEV) with hydrogen enriched internal combustion engine. *Int. J. Hydrog. Energy* **2019**, *44*, 19005–19016. [CrossRef]
15. Shi, X.; Zheng, Y.; Lei, Y.; Xue, W.; Yan, G.; Liu, X.; Cai, B.; Tong, D.; Wang, J. Air quality benefits of achieving carbon neutrality in China. *Sci. Total Environ.* **2021**, *795*, 148784. [CrossRef] [PubMed]
16. Zhang, R.; Hanaoka, T. Deployment of electric vehicles in China to meet the carbon neutral target by 2060: Provincial disparities in energy systems, CO2 emissions, and cost effectiveness. *Resour. Conserv. Recycl.* **2021**, *170*, 105622. [CrossRef]
17. Li, J.; Liang, M.; Cheng, W.; Wang, S. Life cycle cost of conventional, battery electric, and fuel cell electric vehicles considering traffic and environmental policies in China. *Int. J. Hydrog. Energy* **2021**, *46*, 9553–9566. [CrossRef]
18. Gao, J.; Tian, G.; Ma, C.; Balasubramanian, D.; Xing, S.; Jenner, P. Numerical investigations of combustion and emissions characteristics of a novel small scale opposed rotary piston engine fuelled with hydrogen at wide open throttle and stoichiometric conditions. *Energy Convers. Manag.* **2020**, *221*, 113178. [CrossRef]
19. Hames, Y.; Kaya, K.; Baltacioglu, E.; Turksoy, A. Analysis of the control strategies for fuel saving in the hydrogen fuel cell vehicles. *Int. J. Hydrog. Energy* **2018**, *43*, 10810–10821. [CrossRef]
20. Duan, X.; Lai, M.; Jansons, M.; Guo, G.; Liu, J. A review of controlling strategies of the ignition timing and combustion phase in homogeneous charge compression ignition (HCCI) engine. *Fuel* **2021**, *285*, 119142. [CrossRef]
21. Luo, Q.; Sun, B. Experiments on the effect of engine speed, load, equivalence ratio, spark timing and coolant temperature on the energy balance of a turbocharged hydrogen engine. *Energy Convers. Manag.* **2018**, *162*, 1–12. [CrossRef]
22. Shi, C.; Chai, S.; Wang, H.; Ji, C.; Ge, Y.; Di, L. An insight into direct water injection applied on the hydrogen-enriched rotary engine. *Fuel* **2023**, *339*, 127352. [CrossRef]
23. Sun, B.; Luo, Q.; Bao, L. Development and trends of direct injection hydrogen internal combustion engine technology. *J. Automot. Saf. Energy* **2021**, *12*, 265–278. [CrossRef]

24. Wang, H.; Ji, C.; Shi, C.; Yang, J.; Ge, Y.; Wang, S.; Meng, H.; Chang, K.; Wang, X. Parametric modeling and optimization of the intake and exhaust phases of a hydrogen Wankel rotary engine using parallel computing optimization platform. *Fuel* **2022**, *324 Pt A*, 124381. [CrossRef]
25. Wang, H.; Ji, C.; Shi, C.; Ge, Y.; Meng, H.; Yang, J.; Chang, K.; Yang, Z.; Wang, S.; Wang, X. Modeling and parametric study of the performance-emissions trade-off of a hydrogen Wankel rotary engine. *Fuel* **2022**, *318*, 123662. [CrossRef]
26. Wang, H.; Ji, C.; Su, T.; Shi, C.; Ge, Y.; Yang, J.; Wang, S. Comparison and implementation of machine learning models for predicting the combustion phases of hydrogen-enriched Wankel rotary engines. *Fuel* **2022**, *310 Pt B*, 122371. [CrossRef]
27. Luo, Q.; Hu, J.; Sun, B.; Liu, F.; Wang, X.; Li, C.; Bao, L. Experimental investigation of combustion characteristics and NO$_x$ emission of a turbocharged hydrogen internal combustion engine. *Int. J. Hydrog. Energy* **2019**, *44*, 5573–5584. [CrossRef]
28. Takagi, Y.; Mori, H.; Mihara, Y.; Kawahara, N.; Tomita, E. Improvement of thermal efficiency and reduction of NO$_x$ emissions by burning a controlled jet plume in high-pressure direct-injection hydrogen engines. *Int. J. Hydrog. Energy* **2017**, *42*, 26114–26122. [CrossRef]
29. Takagi, Y.; Oikawa, M.; Sato, R.; Kojiya, Y.; Mihara, Y. Near-zero emissions with high thermal efficiency realized by optimizing jet plume location relative to combustion chamber wall, jet geometry and injection timing in a direct-injection hydrogen engine. *Int. J. Hydrog. Energy* **2019**, *44*, 9456–9465. [CrossRef]
30. Wang, H.; Ji, C.; Shi, C.; Yang, J.; Wang, S.; Ge, Y.; Chang, K.; Meng, H.; Wang, X. Multi-objective optimization of a hydrogen-fueled Wankel rotary engine based on machine learning and genetic algorithm. *Energy* **2023**, *263*, 125961. [CrossRef]
31. Ma, D.; Sun, Z. Progress on the studies about NOx emission in PFI-H2ICE. *Int. J. Hydrog. Energy* **2020**, *45*, 10580–10591. [CrossRef]
32. Xu, P.; Ji, C.; Wang, S.; Cong, X.; Ma, Z.; Tang, C.; Meng, H.; Shi, C. Effects of direct water injection on engine performance in a hydrogen (H$_2$)-fueled engine at varied amounts of injected water and water injection timing. *Int. J. Hydrog. Energy* **2020**, *45*, 13523–13534. [CrossRef]
33. Szwaja, S. Dilution of fresh charge for reducing combustion knock in the internal combustion engine fueled with hydrogen rich gases. *Int. J. Hydrog. Energy* **2019**, *44*, 19017–19025. [CrossRef]
34. Szwaja, S.; Naber, J.D. Dual nature of hydrogen combustion knock. *Int. J. Hydrog. Energy* **2013**, *38*, 12489–12496. [CrossRef]
35. Evans, A.; Wang, Y.; Wehrfritz, A.; Srna, A.; Hawkes, E.; Liu, X.; Kook, S.; Chan, Q.N. Mechanisms of NOx production and heat loss in a dual-fuel hydrogen compression ignition engine. In *Proceedings of the SAE Technical Paper*; No. 2021-01-0527; SAE International: Warrendale, PA, USA, 2021.
36. Babayev, R.; Andersson, A.; Dalmau, A.S.; Im, H.G.; Johansson, B. Computational characterization of hydrogen direct injection and nonpremixed combustion in a compression-ignition engine. *Int. J. Hydrog. Energy* **2021**, *46*, 18678–18696. [CrossRef]
37. Wang, Y.; Evans, A.; Srna, A.; Wehrfritz, A.; Hawkes, E.; Liu, X.; Kook, S.; Chan, Q.N. A numerical investigation of mixture formation and combustion characteristics of a hydrogen-diesel dual direct injection engine. In *Proceedings of the SAE Technical Paper*; No. 2021-01-0526; SAE International: Warrendale, PA, USA, 2021.
38. Liu, X.; Srna, A.; Yip, H.L.; Kook, S.; Chan, Q.N.; Hawkes, E.R. Performance and emissions of hydrogen-diesel dual direct injection (H2DDI) in a single-cylinder compression-ignition engine. *Int. J. Hydrog. Energy* **2021**, *46*, 1302–1314. [CrossRef]
39. Chen, Z.; He, J.; Chen, H.; Geng, L.; Zhang, P. Comparative study on the combustion and emissions of dual-fuel common rail engines fueled with diesel/methanol, diesel/ethanol, and diesel/n-butanol. *Fuel* **2021**, *304*, 121360. [CrossRef]
40. Mavrelos, C.; Theotokatos, G. Numerical investigation of a premixed combustion large marine two-stroke dual fuel engine for optimising engine settings via parametric runs. *Energy Convers. Manag.* **2018**, *160*, 48–59. [CrossRef]
41. Stoumpos, S.; Theotokatos, G.; Boulougouris, E.; Vassalos, D.; Lazakis, I.; Livanos, G. Marine dual fuel engine modelling and parametric investigation of engine settings effect on performance-emissions trade-offs. *Ocean. Eng.* **2018**, *157*, 376–386. [CrossRef]
42. Tripathi, G.; Sharma, P.; Dhar, A.; Sadiki, A. Computational investigation of diesel injection strategies in hydrogen-diesel dual fuel engine. *Sustain. Energy Technol. Assess.* **2019**, *36*, 100543. [CrossRef]
43. Sharma, P.; Dhar, A. Compression ratio influence on combustion and emissions characteristic of hydrogen diesel dual fuel CI engine: Numerical Study. *Fuel* **2018**, *222*, 852–858. [CrossRef]
44. Köse, H.; Ciniviz, M. An experimental investigation of effect on diesel engine performance and exhaust emissions of addition at dual fuel mode of hydrogen. *Fuel Process. Technol.* **2013**, *114*, 26–34. [CrossRef]
45. Ramsay, C.J.; Dinesh, K.K.J.R.; Fairney, W.; Vaughan, N. A numerical study on the effects of constant volume combustion phase on performance and emissions characteristics of a diesel-hydrogen dual-fuel engine. *Int. J. Hydrog. Energy* **2020**, *45*, 32598–32618. [CrossRef]
46. Taghavifar, H.; Nemati, A.; Salvador, F.J.; De la Morena, J. 1D energy, exergy, and performance assessment of turbocharged diesel/hydrogen RCCI engine at different levels of diesel, hydrogen, compressor pressure ratio, and combustion duration. *Int. J. Hydrog. Energy* **2021**, *46*, 22180–22194. [CrossRef]
47. Wu, H.-W.; Wu, Z.-Y. Combustion characteristics and optimal factors determination with Taguchi method for diesel engines port-injecting hydrogen. *Energy* **2012**, *47*, 411–420. [CrossRef]
48. Wang, H.; Ji, C.; Yang, J.; Wang, S.; Ge, Y. Towards a comprehensive optimization of the intake characteristics for side ported Wankel rotary engines by coupling machine learning with genetic algorithm. *Energy* **2022**, *261*, 125334. [CrossRef]
49. Shi, C.; Chai, S.; Di, L.; Ji, C.; Ge, Y.; Wang, H. Combined experimental-numerical analysis of hydrogen as a combustion enhancer applied to wankel engine. *Energy* **2023**, *263*, 125896. [CrossRef]

50. Wang, H.; Gan, H.; Theotokatos, G. Parametric investigation of pre-injection on the combustion, knocking and emissions behaviour of a large marine four-stroke dual-fuel engine. *Fuel* **2020**, *281*, 118744. [CrossRef]
51. Richards, K.J.; Senecal, P.K.; Pomraning, E. *CONVERGE 3.0 Manual*; Convergent Science: Madison, WI, USA, 2020.
52. Rahimi, A.; Fatehifar, E.; Saray, R.K. Development of an optimized chemical kinetic mechanism for homogeneous charge compression ignition combustion of a fuel blend of n-heptane and natural gas using a genetic algorithm. *Proc. Inst. Mech. Eng. Part D J. Automob. Eng.* **2010**, *224*, 1141–1159. [CrossRef]
53. Liu, J.; Wang, J.; Zhao, H. Optimization of the injection parameters and combustion chamber geometries of a diesel/natural gas RCCI engine. *Energy* **2018**, *164*, 837–852. [CrossRef]
54. Huang, H.; Lv, D.; Zhu, J.; Zhu, Z.; Chen, Y.; Pan, Y.; Pan, M. Development of a new reduced diesel/natural gas mechanism for dual-fuel engine combustion and emission prediction. *Fuel* **2019**, *236*, 30–42. [CrossRef]
55. Wang, H.; Ji, C.; Shi, C.; Wang, S.; Yang, J.; Ge, Y. Investigation of the gas injection rate shape on combustion, knock and emissions behavior of a rotary engine with hydrogen direct-injection enrichment. *Int. J. Hydrog. Energy* **2021**, *46*, 14790–14804. [CrossRef]
56. Bao, J.; Qu, P.; Wang, H.; Zhou, C.; Zhang, L.; Shi, C. Implementation of various bowl designs in an HPDI natural gas engine focused on performance and pollutant emissions. *Chemosphere* **2022**, *303*, 135275. [CrossRef]
57. Huang, H.; Zhu, Z.; Chen, Y.; Chen, Y.; Lv, D.; Zhu, J.; Ouyang, T. Experimental and numerical study of multiple injection effects on combustion and emission characteristics of natural gas–diesel dual-fuel engine. *Energy Convers. Manag.* **2019**, *183*, 84–96. [CrossRef]
58. Chen, Z.; Wang, L.; Wang, X.; Chen, H.; Geng, L.; Gao, N. Experimental study on the effect of water port injection on the combustion and emission characteristics of diesel/methane dual-fuel engines. *Fuel* **2022**, *312*, 122950. [CrossRef]

Article

Assessment of an Optimal Design Method for a High-Energy Ultrasonic Igniter Based on Multi-Objective Robustness Optimization

Liming Di [1,2,*], Zhuogang Sun [2], Fuxiang Zhi [2], Tao Wan [2] and Qixin Yang [2]

[1] Hebei Key Laboratory of Special Delivery Equipment, Qinhuangdao 066004, China
[2] School of Vehicle and Energy, Yanshan University, Qinhuangdao 066004, China
*   Correspondence: diliming@ysu.edu.cn

Abstract: The current deterministic optimization design method ignores uncertainties in the material properties and potential machining error which could lead to unreliable or unstable designs. To improve the design efficiency and anti-jamming ability of a high-energy ultrasonic igniter, a Six Sigma multi-objective robustness optimization design method based on the response surface model and the design of the experiment has been proposed. In this paper, the initial structural dimensions of a high-energy ultrasonic igniter have been obtained by employing one-dimensional longitudinal vibration theory. The finite element simulation method of COMSOL Multiphysics software has been verified by the finite element simulation results of ANSYS Workbench software. The optimal igniter design has been achieved by using the proposed method, which is based on the finite element method, the Optimal Latin Hypercube Design method, Grey Relational Analysis, the response surface model, the non-dominated sorting genetic algorithm, and the mean value method. Considering the influence of manufacturing errors on the igniter's performance, the Six Sigma method was used to optimize the robustness of the igniter. The Eigenfrequency analysis and the vibration velocity ratio calculation were conducted to verify the design's effectiveness. The results reveal that the longitudinal resonant frequency of the deterministic optimization scheme and the robustness optimization scheme are closer to the design's target frequency. The relative error is less than 0.1%. Compared with the deterministic optimization scheme, the vibration velocity ratio of the robustness optimization scheme is 2.8, which is about 15.7% higher than that of the deterministic optimization scheme, and the quality level of the design targets is raised to above Six Sigma. The proposed method can provide an efficient and accurate optimal design for developing a new special piezoelectric transducer.

Keywords: high-energy ultrasonic igniter; piezoelectric transducer; response surface model; Six Sigma multi-objective robustness optimization

Citation: Di, L.; Sun, Z.; Zhi, F.; Wan, T.; Yang, Q. Assessment of an Optimal Design Method for a High-Energy Ultrasonic Igniter Based on Multi-Objective Robustness Optimization. *Sustainability* **2023**, *15*, 1841. https://doi.org/10.3390/su15031841

Academic Editor: Antonio Galvagno

Received: 13 December 2022
Revised: 10 January 2023
Accepted: 16 January 2023
Published: 18 January 2023

1. Introduction

Traditional gasoline engines generally employ spark plug single-point ignition, which requires certain application conditions, including lean burn, in-cylinder direct injection, exhaust gas turbocharging, and exhaust gas recirculation to resolve problems such as ignition difficulty and poor flame stability [1,2]. Improving the spark plug ignition energy is an effective technical solution that achieves high efficiency and energy saving for the engine, but it will affect the lifetime of the spark plug to a certain extent [3,4]. Transient nonthermal plasma generated by laser [5], nanosecond pulse discharge [6], and dielectric barrier discharge [7] can ignite the lean fuel mixture and improve the ignition and combustion performance of the gas-liquid two-phase combustible working medium to a certain extent [8]. It cannot be widely used because of the high cost, narrow working range, and complexity of the required devices, and for other reasons [9]. Microwaves can induce nonequilibrium plasma and improve the ignition and combustion performance [10], which has the potential to achieve large-area multipoint ignition. However, the increment of the combustion chamber pressure will seriously affect the ignition performance [11]. Therefore, research and development of a high-energy igniter which can stably achieve the multipoint ignition of the fuel mixture in the combustion chamber is an important research direction for energy savings and emission reduction in internal combustion engines.

An ultrasonic wave is a high-frequency mechanical longitudinal wave greater than 20 kHz that displays large energy, strong penetration, weak diffraction, and good directionality. Its influence on chemical reactions mainly comes from ultrasonic mechanical action and the ultrasonic cavitation effect. High-frequency ultrasonic vibration and radiation pressure can form directional agitation and jet effects in the air and in liquid media. Due to the absorption of ultrasonic energy and the internal friction loss phenomenon of the media, a thermal effect temperature rise in the sound field area of the media under continuous ultrasonic action can be generated. In addition, at the moment of collapse, ultrasonic cavitation bubbles will produce extreme environments such as nanoscale transient high temperature, high pressure, and a high electric field, which can easily lead to complex physical, chemical, and biological effects [12–14]. According to international safety regulations, ultrasound is considered an ignition source [15]. The experiment and numerical simulation work by Ion et al. assessed the combustion characteristics of gas in an ultrasonic field. It was found that NO_x and CO emissions dropped, and combustion efficiency increased [16]. Di et al. used spherically focused ultrasound to carry out a spatially localized noncontact ignition study on a gas-liquid two-phase combustible working medium, and the results showed that the temperature of working medium reached the ignition threshold of traditional fuels such as gasoline and diesel when the sound source frequency was 300 kHz [17,18]. Nevertheless, there is a lack of studies on power ultrasound intervention in the ignition and combustion process, and its actual mechanisms of action and acoustic chemical effect are still unclear. Meanwhile, the design methods and performance parameters of a high-energy ultrasonic igniter attached to an internal combustion engine have not been studied systematically.

The optimal design of a high-energy ultrasonic igniter is carried out based on piezoelectric transducers. The non-dominated sorting genetic algorithm (NSGA-II) is widely used in engineering applications [19,20]. Based on the electromechanical equivalence method, Li et al. accomplished the optimized design of an ultrasonic scalpel by using a response surface model and a multi-objective genetic algorithm [21]. Ji et al. completed structure optimization of the ultrasonic horn based on finite element simulation and NSGA-II [22,23]. Karl et al. designed an amplifier through shape optimization using genetic algorithms and verified the effectiveness of the methodology [24], but this method needs multiple finite element simulations, which increase the computing cost. The above methods ignore the influence of uncertainties in the material properties and of machining error, which usually results in unreliable designs and increases the risk of design failure.

This paper focuses on how to design and optimize a high-energy ultrasonic igniter which is applied to the field of internal combustion engines and special burners. Owning to the limitation of the installation space and the traditional design theory of the igniter, the design target frequency of the ultrasonic igniter is set as 35 kHz in this paper. Based on the design theory of the conventional piezoelectric transducer and the traditional spark plug scale of the internal combustion engine, the initial structural dimensions of a high-energy ultrasonic igniter are determined according to the one-dimensional longitudinal vibration theory. The grid size is determined through grid independence verification. Moreover, this paper proposes a new optimization method to develop the optimal igniter, which maximizes the ratio of the front-end vibration velocity to the back-end vibration velocity and minimizes the difference between the igniter's longitudinal resonant frequency and the design's target frequency. The approach is based on finite element (FE) analysis, the Optimal Latin Hypercube Design (OLHD) method, the response surface model, NSGA-II, and the Six Sigma robustness optimization method. The optimal structure dimensions of the igniter are obtained based on the proposed optimization method. Furthermore, the dynamic characteristics of the optimal igniter have been verified using finite element method (FEM)-based simulation.

2. Structure and Theoretical Analysis of High-Energy Ultrasonic Igniter

The high-energy ultrasonic igniter and the constant-volume combustion bomb are rigidly connected through a flange structure, as shown in Figure 1. The mounting flange of the spark plug and the mounting flange of the igniter have interchangeability, and the design of the experiment can be carried out by changing the installation location and numbers of igniters. Figure 2 shows that the igniter consists of a prestressed bolt, back mass, electrode slices, ring-shaped piezoceramics slices, and stepped-type horn, and that they are connected in a coaxial series.

Figure 1. Structural composition of constant-volume combustion bomb system.

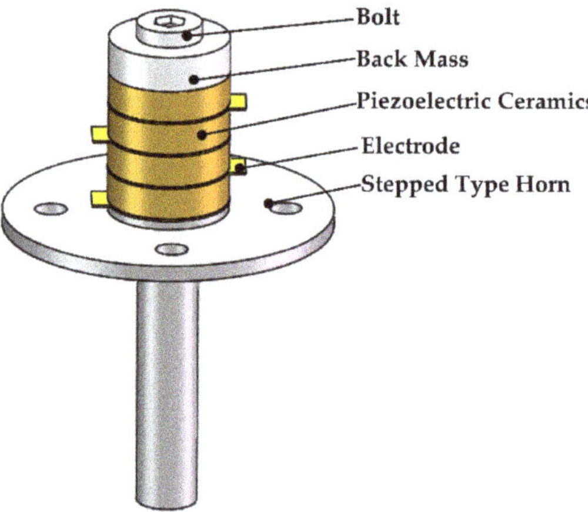

Figure 2. Schematic diagram of high-energy ultrasonic igniter.

In order to enhance the transmitting efficiencies of the igniter, according to the law of conservation of momentum, the back mass and stepped-type horn are made of 45Cr and 7075-T6. Their material properties are summarized in Table 1.

Table 1. Material parameters for the ignition.

Components	Back Mass	Stepped-Type Horn
Material	40Cr	7075-T6
Density (kg/m^3)	7850	2810
Poisson's ratio	0.29	0.33
Young's modulus (GPa)	206	71.7

PZT-4 is selected as the actuator, because it has a large mechanical quality factor, a high Curie temperature of 350 °C, and a low dissipation coefficient of 0.004, which is suitable for the field of high output ultrasonic intensity. The related parameters of PZT-4 are given as follows:

1. Relative Permittivity Matrix

$$[\varepsilon] = \begin{bmatrix} 762.5 & 0 & 0 \\ 0 & 762.5 & 0 \\ 0 & 0 & 663.2 \end{bmatrix}$$

2. Elasticity Constant Matrix

$$\left[C^{E}\right] = \begin{bmatrix} 139 & 77.8 & 74.3 & 0 & 0 & 0 \\ 77.8 & 139 & 74.3 & 0 & 0 & 0 \\ 74.3 & 74.3 & 115 & 0 & 0 & 0 \\ 0 & 0 & 0 & 25.6 & 0 & 0 \\ 0 & 0 & 0 & 0 & 25.6 & 0 \\ 0 & 0 & 0 & 0 & 0 & 25.6 \end{bmatrix} \text{GPa}$$

3. Piezoelectric Stress Matrix

$$[e] = \begin{bmatrix} 0 & 0 & 0 & 0 & 12.7 & 0 \\ 0 & 0 & 0 & 12.7 & 0 & 0 \\ -5.2 & -5.2 & 15.1 & 0 & 0 & 0 \end{bmatrix} \text{C/m}^2$$

The back mass and the stepped-type horn of the high-energy ultrasonic igniter are composed of equal-section cylinders. The longitudinal vibration model of the equal-section cylinder is shown in Figure 3. The transverse dimension of the equal-section cylinder is far less than the one-fourth wavelength of the material corresponding to its working frequency. Therefore, the vibration of the equal-section cylinder can be regarded as one-dimensional longitudinal vibration [25], and the one-dimensional longitudinal vibration wave equation is shown in Equation (1).

$$\frac{\partial^2 \xi}{\partial x^2} + k^2 \xi = 0 \tag{1}$$

where ξ is the particle displacement function, $k = \omega/c$ is the circular wave number, c is the longitudinal vibration speed in the cylindrical rod, and ω is the vibration frequency.

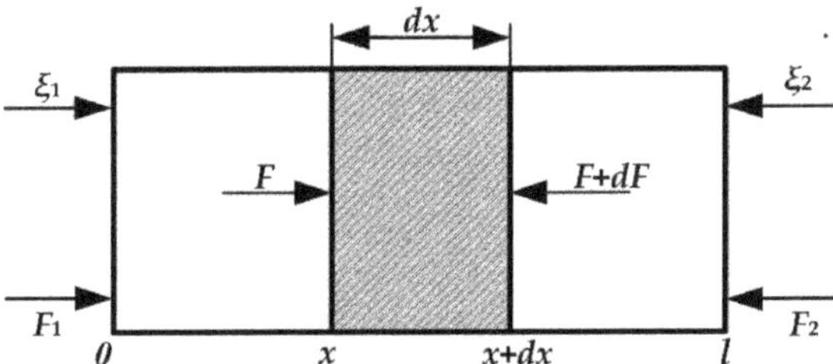

Figure 3. Longitudinal vibration model of uniform section rod.

Figure 4 is a schematic diagram of the ultrasonic igniter structure with the node-plane located in the front mass, where L_1 is the length of the front mass, L_2 is the thickness of the piezoceramics stack, L_3 is the length of the rear mass, and L_4 and L_5 are the lengths of the thick section and thin section of the stepped-type horn. Generally, the front mass and the stepped-type horn are made as one, namely, $L_1 = 0$. Because the nodal plane location displacement is close to 0 mm, the flange of the igniter is set at the nodal plane location to make the thin section of the stepped-type horn enter the combustion chamber. Therefore, the flange is set at the interface between the piezoceramics stack and the thick section of the stepped-type horn. According to the quarter-wavelength theory, the igniter is divided into a quarter-wavelength vibrator and a quarter-wavelength horn. Ignoring the influence of prestressed bolt and electrode slices, the frequency equations of a quarter-wavelength vibrator and a quarter-wavelength horn are shown in Equations (2) and (3). Through theoretical calculation and analysis, the igniter is designed in Table 2.

$$(Z_3/Z_2)\tan k_2 L_2 \tan k_3 L_3 + (Z_3/Z_1)\tan k_1 L_1 \tan k_3 L_3 + (Z_2/Z_1)\tan k_1 L_1 \tan k_2 L_2 = 1 \tag{2}$$

$$\tan k_4 L_4 \tan k_5 L_5 = Z_4/Z_5 \tag{3}$$

where Z_1, Z_2, Z_3, Z_4, and Z_5 are, respectively, the equivalent impedances of the front mass, the piezoceramics stack, the rear mass, the thick section, and the thin section of the stepped-type horn, and $k_i(i = 1, 2, 3, 4, 5)$ are respectively the circular wave number of the front mass, the piezoceramics stack, the rear mass, the thick section, and the thin section of the stepped-type horn.

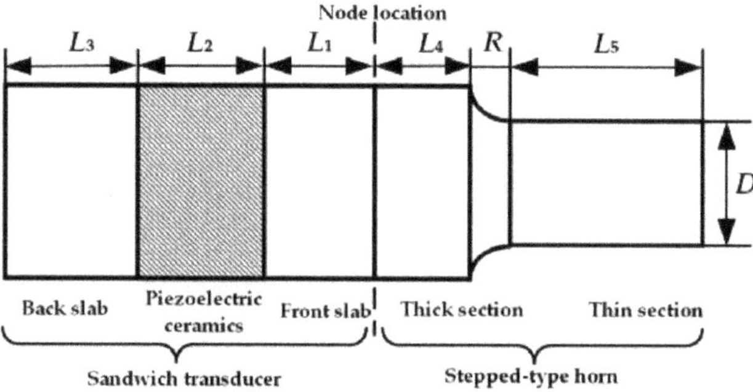

Figure 4. Geometric dimensions of the igniter.

Table 2. The initial dimensions of igniter.

Parameter	Value
F (kHz)	35
L_3 (mm)	4.54
L_4 (mm)	3.13
L_5 (mm)	39.39
R (mm)	1.66
D (mm)	7.44

3. Verification of Finite Element Model

The high-energy ultrasonic igniter is designed based on a piezoelectric transducer. The FEM of the latter is also suitable for the igniter. ANSYS has been regarded as the authoritative software for the finite element simulation design and analysis of piezoelectric transducers [26,27]. Because COMSOL Multiphysics has unique advantages in the multi-field coupling and structure parameters optimization of piezoelectric transducers, the longitudinal resonant frequency obtained from ANSYS modal analysis has been used as the basis for evaluation. To evaluate the accuracy of the COMSOL Multiphysics finite element simulation calculation, Figure 5 shows the thin end face vibration amplitude (TEVA)-frequency curve of the stepped-type horn obtained by the simulations of the two pieces of software, and the frequency corresponding to the maximum amplitude is the longitudinal resonant frequency of the igniter.

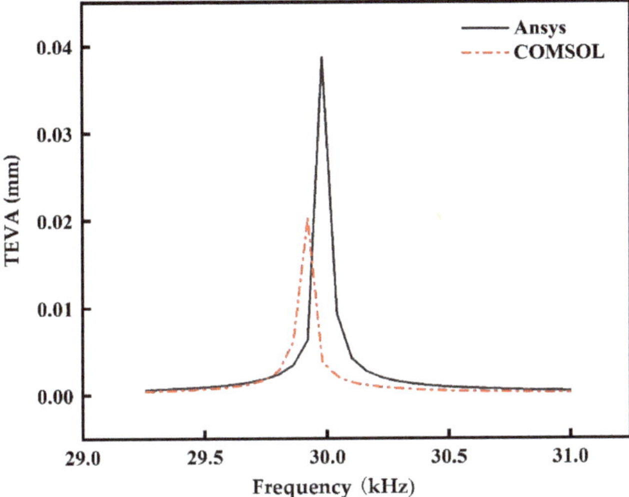

Figure 5. TEVA-frequency curve from ANSYS and COMSOL.

The longitudinal vibrational modal shapes of the igniter at a frequency of 30 kHz are calculated by using the simulation method of reference [28]. As shown in Figure 6, its longitudinal resonant frequency is 29.98 kHz. Figure 7 shows the longitudinal vibrational modal shapes obtained by simulation with COMSOL Multiphysics for an igniter with the same structural dimensions and grid size: its longitudinal resonant frequency is 29.906 kHz. The longitudinal resonant frequency relative error of the two pieces of software is less than 1%, and both of them coincide well with the design frequency of 30 kHz. Therefore, COMSOL Multiphysics can ensure the scientificity and accuracy of the FE analysis.

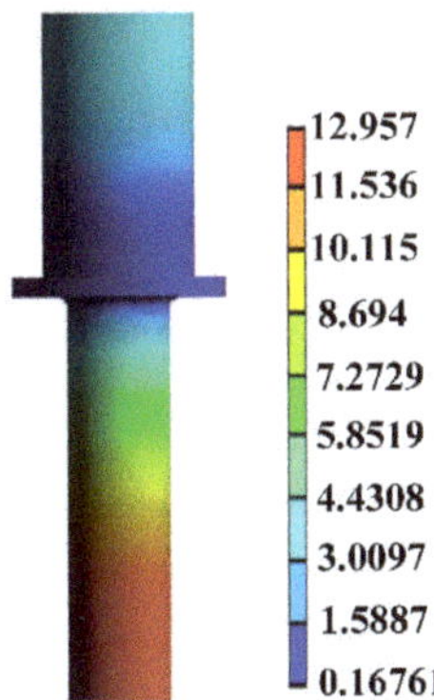

Figure 6. Longitudinal modal shape of for ANSYS.

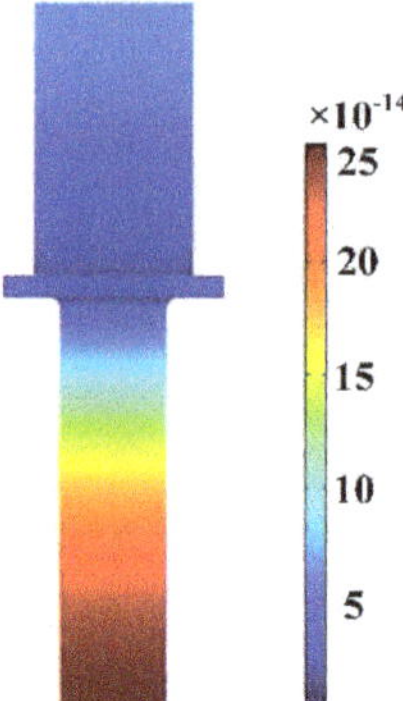

Figure 7. Longitudinal modal shape of for COMSOL.

4. Vibration Characteristic Analysis

"Solid Mechanics" physics and "Electrostatic" physics in COMSOL Multiphysics are chosen to analyze the dynamic characteristics of the igniter. "Piezoelectric Effect" is used to couple the "Solid Mechanics" and "Electrostatics" equations solved in the PZT-4 domains via the linear constitutive equations that model the piezoelectric effect by coupling stresses and strains with the electric field and electric displacement. When setting the properties of the piezoelectric material in the solution domain, the piezoelectric parameters based on the stress-charge material are selected, namely the E-type piezoelectric equation shown in Equations (4) and (5). The electrostatic interface solves the PZT-4 according to Equations (6) and (7). The piezoceramics are stacked alternately, the ground potential is equal to 0 V, and the terminal potential is set to 30 V.

$$T = c_E \cdot S - e^t \cdot E \tag{4}$$

$$D = e \cdot S + \varepsilon_s \cdot E \tag{5}$$

$$\nabla \cdot D = p_v \tag{6}$$

$$E = -\nabla V \tag{7}$$

where T is the stress vector and D denotes the electric flux density vector, S expresses the strain vector, and E is the electric field intensity vector; c_E represents the elasticity matrix for constant electric field, e is the piezoelectric stress matrix, and ε_S is the dielectric matrix; and $\nabla \cdot D$ is the electric charge density, p_v is the electric charge concentration, and E is the electric field due to the electric potential V.

Due to the significant impact of bolt pretension on the performance of the igniter, 12.9-grade alloy steel bolts have been selected in the simulation model. The range of piezoceramics pretension is generally 3000–3500 N/cm². A pretension force of about 3600 N has been applied. As a result of the prestress in the igniter, the harmonic variation of stress and other physical quantities during vibration takes place on top of the static bias stress. Hence, we need to solve this model using a two-step approach, where the first step involves solving for the static stress distribution using a "Stationary Study" step. The solution from this step is then used as a linearization point for solving the vibration problem in the "Frequency Domain Perturbation Study" step and the "Eigenfrequency Study" step [29]. With the COMSOL Multiphysics software, grid independence verification has been performed by contrasting the longitudinal resonant frequency under different grid scales. Base grid sizes from 0.8 mm to 1.5 mm are selected. The results of the grid independence verification are shown in Figure 8. The maximum relative error of the longitudinal resonant frequency is less than 1%, which indicates the grid independence is well verified. Considering the calculation accuracy and time cost, 1.3 mm is used as the grid size of the finite element model of igniter in this study. This model has 38,623 elements in total, and the minimum element quality of the mesh is 0.24. Figure 9 depicts the mesh model of the igniter.

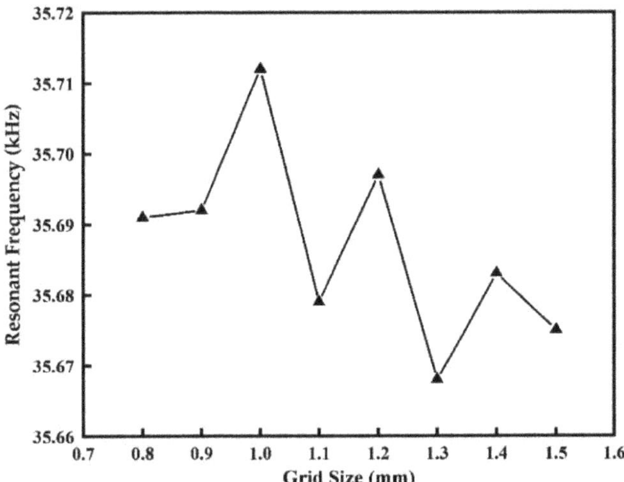

Figure 8. Grid independence verification.

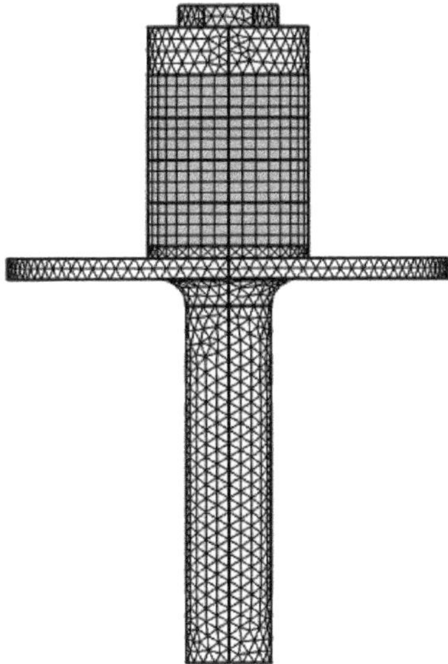

Figure 9. Mesh model of the igniter.

The longitudinal vibration modal and TEVA-frequency curve can be obtained by Eigenfrequency analysis and Frequency Domain Perturbation analysis using COMSOL Multiphysics, as shown in Figures 10 and 11. It can be seen that the longitudinal resonant frequency of the igniter is 35.668 kHz, slightly higher than the theoretical design frequency of 35 kHz. Therefore, it needs to be modified to achieve optimal performance.

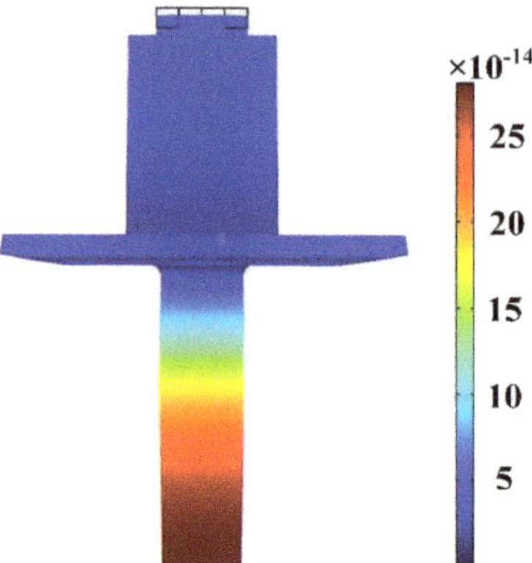

Figure 10. Longitudinal modal shape of igniter.

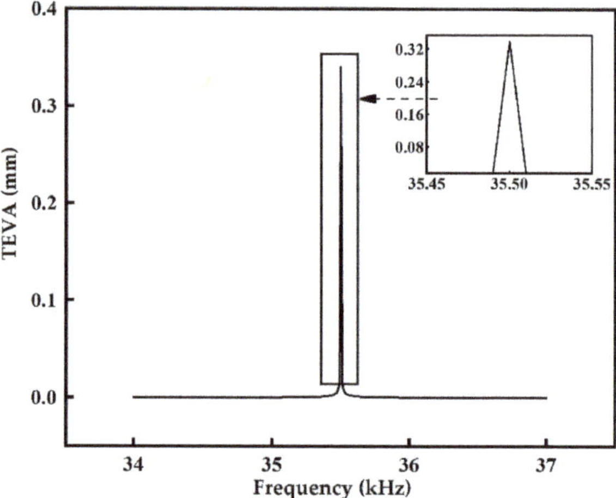

Figure 11. TEVA-frequency curve of igniter.

5. Establishment of Approximate Model

5.1. Calculation Method of Vibration Velocity Ratio

The practice has proved that the simplified theoretical analysis model can reflect the working state of the igniter to a large extent. To simplify the structure of the igniter and reduce the energy loss, the stepped-type horn and front mass are designed as a whole. Based on one-dimensional longitudinal vibration theory and the longitudinal boundary conditions and the influence of the prestressed bolt, electrodes slices, rounded corners, and flange, the electromechanical equivalent circuit of the igniter can be obtained, as shown in Figure 12 [30].

For simplicity, the forces on the front end and back end of the igniter can be neglected, $F_f = F_b = 0$. In Figure 12, C and N are the clamped capacitance and electromechanical transformation coefficient of the piezoceramics stack, as shown in Equations (8) and (9).

$$C = p^2 S_2 \varepsilon_{33}^T (1 - k_{33}^2) / L_2 \tag{8}$$

$$N = p d_{33} S_2 / (s_{33}^E L_2) \tag{9}$$

where $p = 4$ is the number of piezoceramics rings; S_2 and L_2 are the cross-sectional area and length of the piezoceramics stack; d_{33}, ε_{33}^T, s_{33}^E, and k_{33} are piezoelectric constant, free dielectric constant, elastic compliance constant, and electromechanical coupling coefficient; V is the voltage applied to the piezoceramics stack; V_b is the longitudinal vibration speed at the outer surface of the back mass; V_f is the longitudinal vibration velocity of the outer surface of the thin section of the stepped-type horn; V_1 is the longitudinal vibration velocity at the interface between the back mass and the piezoceramics stack; V_2 is the longitudinal vibration velocity of the interface between the thick section of stepped-type horn and the piezoceramics stack; V_3 is the longitudinal vibration velocity of the interface between the thick section and the thin section of the stepped-type horn; Z_{p1}, Z_{p2}, and Z_{p3} are the equivalent impedances of the piezoceramics stack; Z_{fl} and Z_{bl} are the load impedances of the igniter, namely $Z_{fl} = Z_{bl} = 0$; Z_{i1}, Z_{i2}, and Z_{i3} ($i = 3, 4, 5$) are the back mass, thick section, and thin section of the stepped-type horn; and the equivalent impedances of the thick section and the thin section of the stepped-type horn are calculated as shown in Equations (10) and (11).

$$Z_{i1} = Z_{i2} = j\rho_i c_i S_i \tan(k_i L_i/2) \tag{10}$$

$$Z_{i3} = -j\rho_i c_i S_i \sin k_i L_i \tag{11}$$

where $\rho_i (i = 2, 3, 4, 5)$ are the densities of the piezoceramics stack, back mass, thick section, and thin section of the stepped-type horn. $c_i (i = 2, 3, 4, 5)$ are the sound speeds of the piezoceramics stack, back mass, thick section, and thin section of the stepped-type horn. $S_i (i = 2, 3, 4, 5)$ are the cross-sectional areas of the piezoceramics stack, back mass, thick section, and thin section of the stepped-type horn.

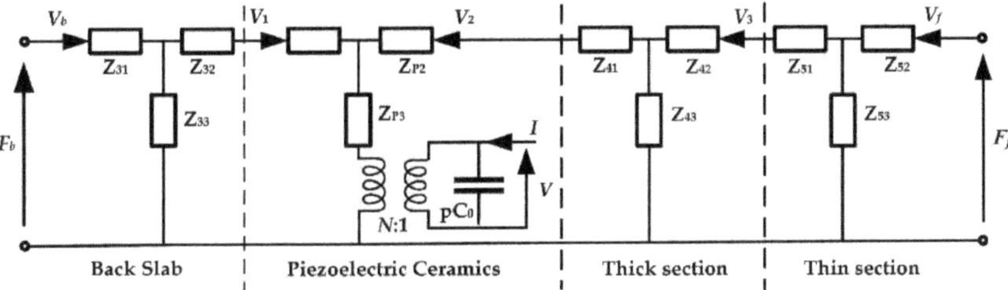

Figure 12. Electromechanical equivalent circuit of igniter.

The impedances of each igniter component are expressed in Equations (12)–(14). The calculation of the longitudinal vibration velocity is shown in Equations (15)–(18), and the calculation of the vibration velocity ratio of the front and back mass of the igniter is shown in Equation (19).

$$Z_3 = \frac{Z_{31} Z_{33}}{Z_{31} + Z_{33}} + Z_{32} \tag{12}$$

$$Z_4 = \frac{Z_{43}(Z_5 + Z_{42})}{Z_{43} + Z_5 + Z_{42}} + Z_{41} \tag{13}$$

$$V_3 \cdot (Z_{42} + Z_5) = (V_2 - V_3) \cdot Z_{53} \tag{14}$$

$$V_f \cdot Z_{52} = (V_3 - V_f) \cdot Z_{53} \tag{15}$$

$$V_3 \cdot (Z_{42} + Z_5) = (V_2 - V_3) \cdot Z_{53} \tag{16}$$

$$V_1 \cdot (Z_{p1} + Z_3) = V_2 \cdot (Z_{p2} + Z_4) \tag{17}$$

$$V_b \cdot Z_{31} = (V_1 - V_b) \cdot Z_{33} \tag{18}$$

$$M = \left| \frac{V_f}{V_b} \right| = \left| \frac{Z_{53} \cdot Z_{43} \cdot (Z_{p1} + Z_3) \cdot Z_3}{(Z_{52} + Z_{53}) \cdot (Z_{42} + Z_5 + Z_{43}) \cdot (Z_{p1} + Z_4) \cdot (Z_3 + Z_{33})} \right| \tag{19}$$

5.2. Selection of Design Variables

The optimization of the igniter has been preceded by a sensitivity analysis. The major purpose of the analysis is to identify the correlation between the values of the variable and the objective variable. The closer the correlation coefficient value is to 1, the stronger the correlation degree is with the objective variables. The sensitivity values are based on the Grey Relational Analysis method [21]. The dimensions of the igniter have an important influence on its performance [31,32]. In this paper, L_3, L_4, L_5, R, and D are selected as the design variables, with 30 simulations and calculations, and the results of the sensitivity analysis are presented in Figure 13. Therefore, it can be concluded that L_3, L_4, L_5, R, and D are the major variables. In Figure 13, M is the ratio of the igniter's front-end vibration velocity to its back-end vibration velocity, and f denotes the longitudinal resonant frequency.

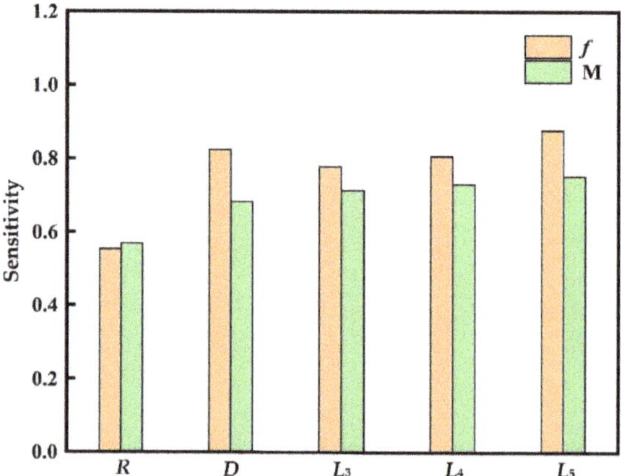

Figure 13. The results of the sensitivity analysis.

5.3. Objective Functions and Constraints

The longitudinal resonant frequency f is an important parameter of the igniter. Only when igniter is driven at its longitudinal resonant frequency can the maximum amplitude of the output end achieve the maximum values. The vibration velocity ratio M reflects the transmitted sound power of the igniter, and the transmitted sound power has a positive ratio with the vibration velocity ratio. In order to improve the comprehensive performance of the igniter, f and M are taken as the optimization objectives. The optimization of igniter can be viewed as a multi-objective optimization problem and shown in Equation (20).

$$\text{Min}\left[\frac{1}{|M|}, |f - 35000|\right]^{\text{T}}$$

$$\text{s.t.} \begin{cases} 4.05 \le L_3 \le 4.95 \\ 3.03 \le L_4 \le 3.25 \\ 38.32 \le L_5 \le 39.88 \\ 7.2 \le D \le 8.8 \\ 1 \le R \le 3 \end{cases} \tag{20}$$

where M is the ratio of the front-end vibration velocity to the back-end vibration velocity of the igniter; f denotes the longitudinal resonant frequency; R is the radius of the rounded corners between the stepped-type horn's thick section and thin section. D is the diameter of the stepped-type horn's thin section.

5.4. Approximate Model Validation

The design of the experiment for design points is implemented based on the OLHD method. Thirty groups of design points were set in COMSOL Multiphysics [32]. In addition, for the convenience of expression, the diameter of the stepped-type horn thin section, the length of the back mass, the length of stepped-type horn thick and thin section, and the radius of rounded corners were replaced by x_1, x_2, x_3, x_4, and x_5. The approximate second-order response surface model was established by calculating their corresponding longitudinal resonant frequencies and vibration velocity ratios, giving the following expressions:

$$
\begin{aligned}
f = & -227.19736 + 1.8921x_1 - 1.4866x_2 + 13.2063x_3 + 12.5087x_4 + 12.4241x_5 \\
& -0.1368x_1^2 - 0.4803x_2^2 - 0.29406x_3^2 - 8.13305x_4^2 - 0.16808x_5^2 - 0.003628x_1x_2 \\
& -0.09705x_1x_3 + 0.01087x_1x_4 - 0.012x_1x_5 - 0.215x_2x_3 + 0.6418x_2x_4 + 0.1345x_2x_5 \\
& -1.564x_3x_4 - 0.205x_3x_5 + 0.2898x_4x_5
\end{aligned}
$$

$$
\begin{aligned}
M = & -1131.63 + 3.8615x_1 + 20.8285x_2 + 18.977x_3 + 83.4454x_4 + 51.1719x_5 \\
& -0.4153x_1^2 - 2.0981x_2^2 - 2.005x_3^2 - 29.6261x_4^2 - 0.6379x_5^2 - 0.0279x_1x_2 \\
& -0.04614x_1x_3 - 0.7383x_1x_4 - 0.02687x_1x_5 + 0.06317x_2x_3 + 1.0165x_2x_4 \\
& -0.16617x_2x_5 - 0.4283x_3x_4 - 0.0152x_3x_5 - 0.42154x_4x_5
\end{aligned}
$$

Cross-validation was used to test the prediction accuracy of the second-order response surface model [33]. Figure 14a,b shows the simulated value of the 10 verification points versus the predicted values from the developed response surface model, where f_{pre} is the predicted value of longitudinal resonant frequency and f_{act} is the simulated value of the longitudinal resonant frequency. M_{pre} is the predicted value of the vibration velocity ratio, and M_{act} is the simulated value of the vibration velocity ratio. The closer the verification points are to the diagonal line, the better the response surface model fits the points. The accuracy of the response surface model was evaluated using the coefficient of determination (R^2), average error, maximum error, and root mean square error. The error analysis results are listed in Table 3. Thus, it can be concluded that the response surface model has sufficient accuracy, demonstrating that all variations can be explained with the developed response model.

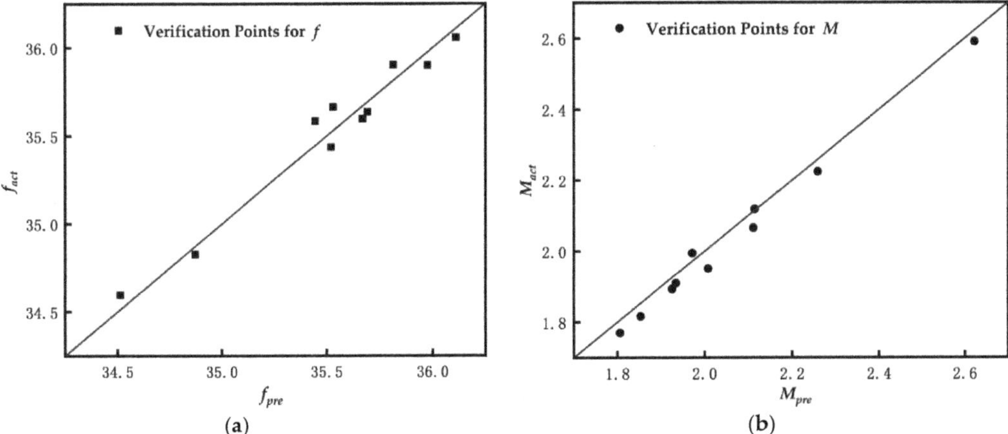

(a) (b)

Figure 14. (a) Goodness of fit of the longitudinal resonant frequency; (b) Goodness of fit of the vibration velocity ratio.

Table 3. Error analysis results of the response surface model.

	Average Error	Maximum Error	Root Mean Square Error	Determination Coefficient R^2
f	0.05693	0.09602	0.06052	0.9602
M	0.03989	0.0701	0.04311	0.9757
Acceptance level	<0.2	<0.3	<0.2	>0.9

6. Multi-Objective Deterministic Optimization Analysis

The NSGA-II algorithm has advantages in finding Pareto optimal solutions to multi-objective optimization problems [34,35]. The multi-objective optimization processes of the igniter have been illustrated in Figure 15. The initial population number of the NSGA-II algorithm was set as 12, and the multi-objective Pareto optimal solution was obtained after 20 iterations. The red dots represent the Pareto front of the deterministic optimization in Figure 16. A comprehensive optimum is selected from the Pareto front. A group of the optimal dimensions of the igniter is listed in Table 4.

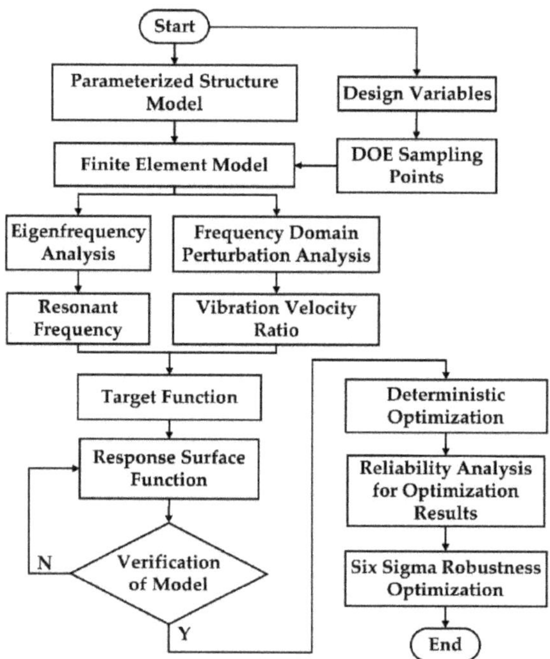

Figure 15. Optimization methodology for the ignition.

An Eigenfrequency analysis of the optimized igniter has been conducted to obtain the longitudinal vibration frequency and the corresponding vibration modal. The results show that the longitudinal resonant frequency is 35.01 kHz, which coincides well with the objective frequency of 35 kHz. Frequency Domain Perturbation analysis has been conducted to study the dynamic characteristics of the igniter under a sinusoidal excitation voltage of 30 V. Figures 17–19 indicate the longitudinal vibration modal, the displacement distribution, and the TEVA-frequency curve of the igniter, respectively. It can be included from Figure 18 that the wave node is close to the flange location, which can be used to fix without affecting the working of the igniter. Using Matlab to calculate the vibration velocity ratio yields a ratio of 2.42, which is about 1.2% higher than that before optimization, ensuring that most of the ultrasonic energy is transmitted to the combustion chamber.

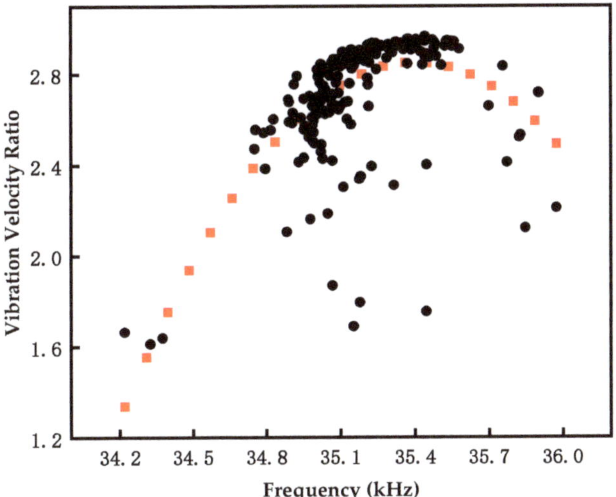

Figure 16. Pareto front of the deterministic optimization.

Table 4. The optimal dimensions of the igniter.

	Initial Value	Optimal Value
L_3 (mm)	4.54	4.47
L_4 (mm)	1.13	1.13
L_5 (mm)	39.39	39.57
R (mm)	1.66	1.5
D (mm)	7.44	7.42

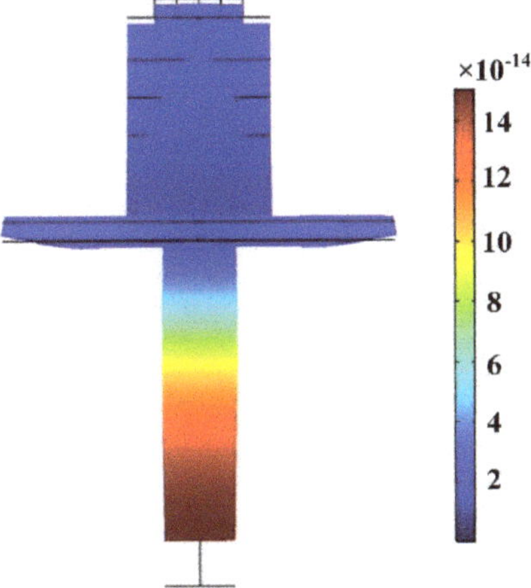

Figure 17. Longitudinal modal shape of deterministic optimization.

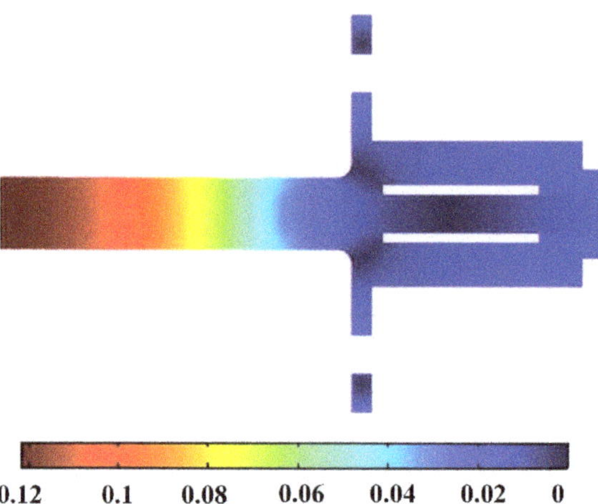

Figure 18. Displacement distribution of deterministic optimization.

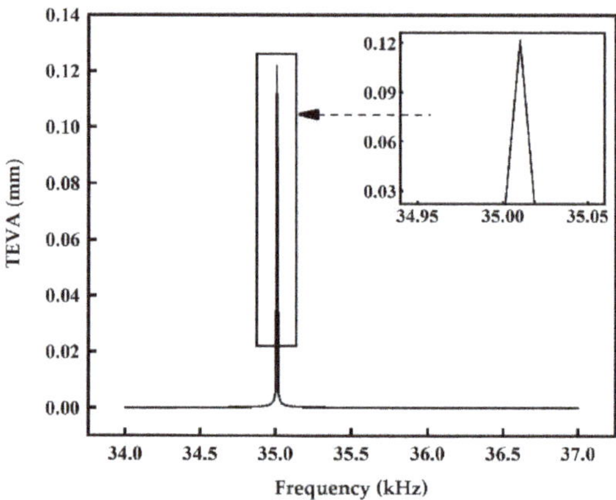

Figure 19. TEVA-frequency curve of the deterministic optimization.

7. Multi-Objective Robustness Optimization Analysis

Owning to the limitations of the levels of processing and manufacturing, the sizes of the igniter's structures will fluctuate, leading to deviations in its performance. However, deterministic optimization ignores these uncertain factors. It is necessary to conduct a robustness analysis. The Isight software integrated response surface approximation model was used, and the deterministic optimization results were used as input conditions for the Mean Value Method Reliability analysis. The probability distribution of the design variables and the objective variables was set as the normal distribution, and the coefficient of variation was set as 0.01. The results show that the Sigma level of the longitudinal resonant frequency is 4.015, as shown in Figure 20. The failure risk of the product is high, and a robustness optimization design is needed.

The NSGA-II algorithm was used to optimize the Six Sigma robustness model, and the optimized design scheme is shown in Table 5. Figures 21–23 show the results for the robustness optimization design. Figure 24 shows the Sigma level of the longitudinal resonant frequency in the robustness optimization. Compared with the deterministic optimization design results, the relative difference of the longitudinal resonant frequency has little variation, which is better than the optimization results of literature [21,36], the vibration velocity ratio is raised by 15.7%, and the Sigma levels of both are above Six Sigma. Therefore, the results indicate that the proposed approach could effectively solve igniter optimization problems.

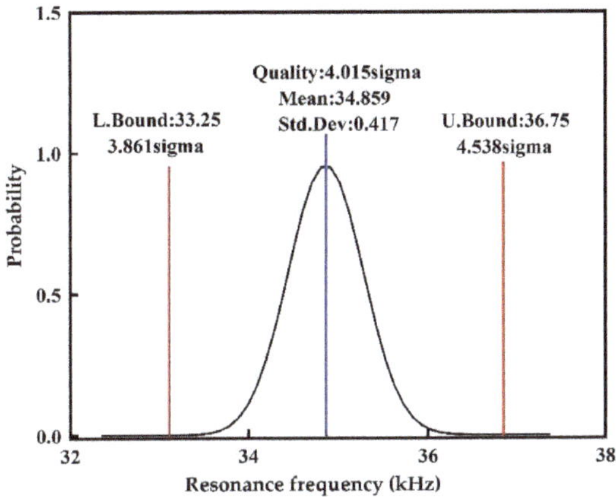

Figure 20. Six Sigma analysis result of the deterministic optimization.

Table 5. Six Sigma analysis results.

Parameter	Initial Value	Deterministic Optimization		Robust Optimization	
		Result	Sigma Level	Result	Sigma Level
L_3 (mm)	4.54	4.47	1.23	4.25	3.76
L_4 (mm)	3.13	3.13	8	3.14	8
L_5 (mm)	39.39	39.57	8	38.93	8
R (mm)	1.66	1.5	8	1	8
D (mm)	7.44	7.42	8	7.88	8
f (kHz)	35.668	35.01	4.015	34.976	6.088
M	1.935	2.084	6.58	2.086	6.188

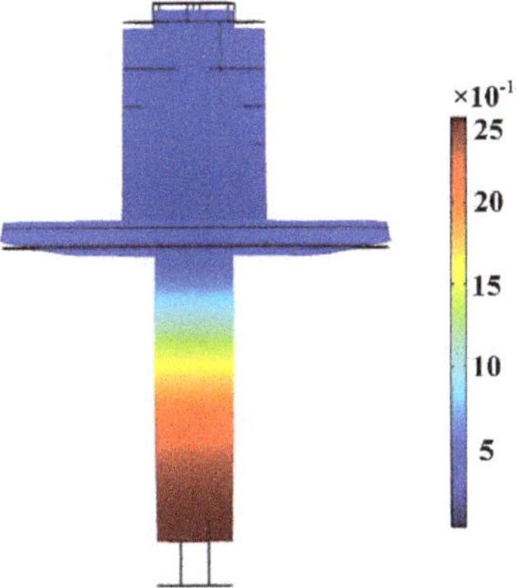

Figure 21. Longitudinal modal shape of robustness optimization.

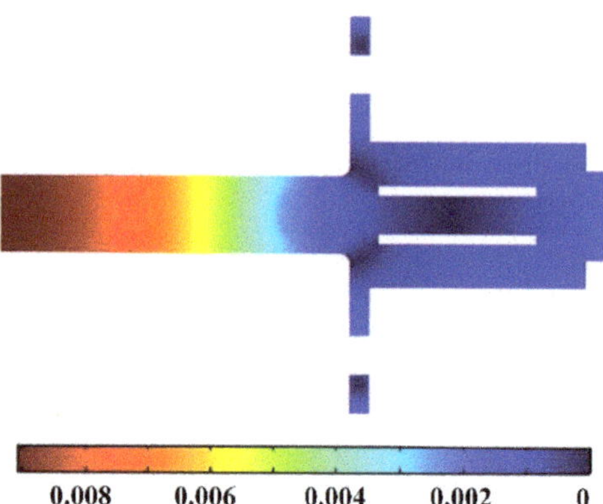

Figure 22. Displacement distribution of robustness optimization.

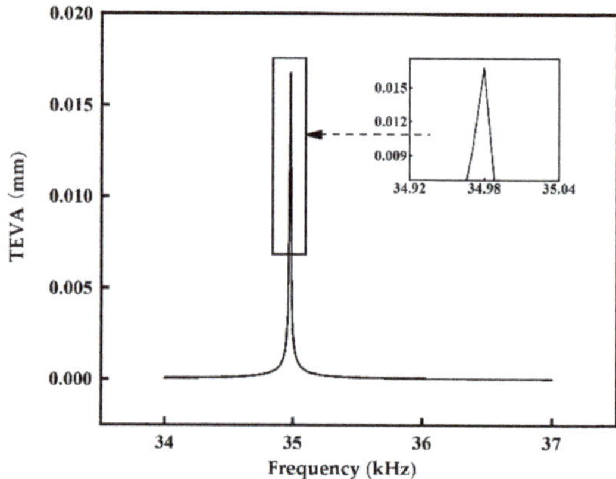

Figure 23. TEVA-frequency curve of the robustness optimization.

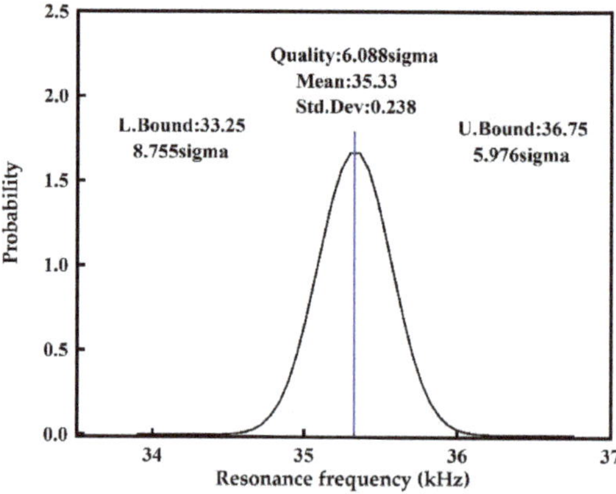

Figure 24. Six Sigma analysis result of the robustness optimization.

8. Conclusions

Using a Six Sigma multi-objective reliability-based optimization design method, a novel high-energy ultrasonic igniter has been designed. The proposed method can provide the necessary guidance for the optimization design of the high-energy ultrasonic igniter. Simultaneously, it has an important reference value for a new special piezoelectric transducer to improve research and development efficiency.

The initial dimensions of different components of the high-energy ultrasonic igniter have been determined using the one-dimensional longitudinal vibration theory combined with the boundary conditions. The multi-field coupling prestressed modal analysis of the igniter was carried out using COMSOL Multiphysics. The longitudinal resonant frequency of the igniter is 35.668 kHz, and the relative error with the design target frequency is 1.4%. The length of the thin section of the stepped-type horn has the biggest impact on the resonant frequency and vibration velocity ratio by means of Grey Relational Analysis. An approach based on the OLHD method, FEM, the response surface model, and NSGA-II is used to complete the deterministic optimization design and the Six Sigma robustness optimization design. The robustness optimization results show that the dynamic characteristics of the igniter have been greatly improved. The relative difference of the longitudinal resonant frequency is just 0.07%. At the same time, the vibration velocity ratio is increased by 15.7%. The optimal igniter has better reliability and quality levels, a much smaller resonant frequency shift, and a higher vibration velocity ratio.

Future work will be focused on developing the direct ultrasonic ignition and auxiliary catalytic ignition control methods in which ultrasound actively feeds into the internal engines and the constant-volume combustion bomb.

Author Contributions: Conceptualization, methodology, funding acquisition, L.D.; software, validation, formal analysis, investigation, resources, data curation, writing—original draft preparation, writing—review and editing, Z.S.; visualization, supervision, F.Z.; project administration, T.W. and Q.Y. All authors have read and agreed to the published version of the manuscript.

Funding: This research was supported by the National Natural Science Foundation of China, grant number 51406174; the Hebei Natural Science Foundation of China, grant number E2020203127 and the Cultivation Project for Basic Research and Innovation of Yanshan University, grant number 2021LGZD014.

Data Availability Statement: The data presented in this study are available on request from the corresponding author. The data are not publicly available due to privacy restriction.

Conflicts of Interest: The authors declare no conflict of interest.

References

1. Lu, H.; Wang, Y.; Liang, J.; Li, C.; Li, L. Effect of innovative two-stage ignition system on lean burn efficiency. *Trans. CSICE* **2017**, *35*, 530–537.
2. Shi, C.; Chai, S.; Di, L.; Ji, C.; Ge, Y.; Wang, H. Combined experimental-numerical analysis of hydrogen as a combustion enhancer applied to Wankel engine. *Energy* **2023**, *263*, 125896. [CrossRef]
3. Bao, J.; Qu, P.; Wang, H.; Zhou, C.; Zhang, L.; Shi, C. Implementation of various bowl designs in an HPDI natural gas engine focused on performance and pollutant emissions. *Chemosphere* **2022**, *303*, 135275. [CrossRef] [PubMed]
4. Shi, C.; Chai, S.; Wang, H.; Ji, C.; Ge, Y.; Di, L. An insight into direct water injection applied on the hydrogen-enriched rotary engine. *Fuel* **2023**, *339*, 127352. [CrossRef]
5. Wang, C.; Huang, S.; Hu, E.; Gao, Q.; Huang, Z. Laser-induced plasma ignition characteristics methane/hydrogen/air mixture. *J. Cent. South Univ. Sci. Technol.* **2022**, *53*, 2111–2121.
6. Ban, Y.; Zhang, F.; Zhong, S.; Zhu, J. Nanosecond pulsed plasma assisted ignition of methane/air under lean-burn conditions of engines. *J. Combust. Sci. Technol.* **2022**, *28*, 573–582.
7. Cheng, L.; Shen, J.; Jiang, Y.; Wu, J.; Chen, J.; Fang, S. Experimental Study on Premixed Ammonia/Air Swirl Combustion Assisted by Dielectric Barrier Discharge. Clean Coal Technology. 1–8. Available online: https://kns.cnki.net/kcms/detail/11.3676.TD.20 221021.1553.004.html (accessed on 4 January 2023).
8. Fei, L.; He, L.; Chen, Y.; Deng, J.; Lei, J.; Zhao, B. Numerical study on plasma-assisted combustion in One type of aero-engine combustion chamber. *J. Combust. Sci. Technol.* **2019**, *25*, 451–459.

9. He, L.; Liu, X.; Zhao, B.; Jin, T.; Yu, J.; Zeng, H. Current investigation progress of plasma-assisted ignition and combustion. *J. Aerosp. Power* **2016**, *31*, 1537–1551.
10. Wu, H.; Wang, Z.; Zhang, X.; Liu, C.; Cheng, X.; Wang, Z. Experimental study on discharge and flame kernel appearance under different microwave parameters. *Trans. CSICE* **2022**, *40*, 20–28.
11. Zhang, X.; Wang, Z.; Wu, H.; Zhou, D.; Huang, S.; Cheng, X.; Chen, J. Experimental study of microwave assisted spark ignition on expanding C_2H_2-Air spherical flames. *Combust. Flame* **2020**, *222*, 111–122. [CrossRef]
12. Miller, D.; Lu, X.; Dou, C.; Zhu, Y.; Fuller, R.; Fields, K.; Fabiilli, M.; Owens, G.; Gordon, D.; Kripfgans, O. Ultrasonic cavitation-enabled treatment for therapy of hypertrophic cardiomyopathy: Proof of principle. *J. Acoust. Soc. Am.* **2018**, *44*, 1439–1450. [CrossRef] [PubMed]
13. Sagin, S.V.; Solodovnikov, V.G. Cavitation treatment of high-viscosity marine fuels for medium-speed diesel engines. *Mod. Appl. Sci.* **2015**, *9*, 269–278. [CrossRef]
14. Nasi. Explosive Atmospheres-Explosion Prevention and Protection-Part 2: Basic Concepts and Methodology for Mining. 2014. Available online: https://standards.iteh.ai/catalog/standards/sist/09b2efb1-977f-434c-a7aa-022e12b861e9/sist-en-1127-2-2014 (accessed on 5 August 2014).
15. *EN B. S. 1127-1-2011*; Explosive Atmospheres-Explosion Prevention and Protection: Basic Concepts and Methodology. Standards Policy and Strategy Committee: London, UK, 2011.
16. Ion, I.V.; Dimofte, E.; Popescu, F.; Akhmetova, I.G. Investion of flame acoustic excitation of a gas burner. *Energy Rep.* **2022**, *8*, 263–269. [CrossRef]
17. Di, L.; Sun, W.; Sun, T.; Zhang, S.; Liu, Z. Study on spatial location ignition of flammable mixtures based on spherical focused ultrasound. *J. Combust. Sci. Technol.* **2021**, *27*, 493–500.
18. Di, L.; Zhang, S.; Shi, C.; Sun, Z.; Ouyang, Q.; Zhi, F.; Yang, Q. Effect of ultrasonic-fed time on combustion and emissions performance in a single-cylinder engine. *Chemosphere* **2022**, *302*, 134924. [CrossRef] [PubMed]
19. Wang, H.; Ji, C.; Shi, C.; Yang, J.; Wang, S.; Ge, Y.; Chang, K.; Meng, H.; Wang, X. Multi-objective optimization of a hydrogen-fueled Wankel rotary engine based on machine learning and genetic algorithm. *Energy* **2023**, *263*, 125961. [CrossRef]
20. Shen, Y.; Yang, H. Multi-objective optimization of integrated solar-driven CO_2 capture system for an industrial building. *Sustainability* **2023**, *15*, 526. [CrossRef]
21. Li, J.; Liu, H.; Li, J.; Yang, Y.; Wang, S. Piezoelectric transducer design for an ultrasonic scalpel with enhanced dexterity for minimally invasive surgical robots. *Proc. Inst. Mech. Eng. Part C J. Mech. Eng. Sci.* **2020**, *234*, 1271–1285. [CrossRef]
22. Ji, H.; Zhao, S.; Hu, X. Optimal design method for bezier ultrasonic horn based on multi-objective genetic algorithm. *China Mech. Eng.* **2016**, *27*, 1716.
23. Wang, F.; Ge, X.; Li, Y.; Zheng, J.; Zheng, W. Optimising the distribution of multi-cycle emergency supplies after a disaster. *Sustainability* **2023**, *15*, 902. [CrossRef]
24. Deibel, K.R.; Wegener, K. Methodology for shape optimization of ultrasonic amplifier using genetic algorithms and simplex method. *J. Manuf. Syst.* **2013**, *32*, 523–528. [CrossRef]
25. Lin, S. *The Theory and Design of Ultrasonic Transducers*; Science Press: Beijing, China, 2004; pp. 103–111.
26. Liu, S.; Feng, P.; Cha, H.; Feng, F. Optimized design for a piezoelectric ultrasonic transducer based on the six-terminal network. *Chin. J. Eng.* **2022**, *44*, 933–939.
27. Jia, Y.; Shen, J. Finite element analysis of stepped ultrasonic horn with different transitional sections. *Tech. Acoust.* **2006**, *25*, 242–245.
28. Mo, X. Simulation and analysis of acoustics transducers using the ANSYS software. *Tech. Acoust.* **2007**, *26*, 1279–1290.
29. Available online: https://cn.comsol.com/model/piezoelectric-tonpilz-transducer-with-a-prestressed-bolt-14535 (accessed on 4 October 2022).
30. Li, G.; Qu, J.; Xu, L.; Zhang, X.; Gao, X. Study on multi-frequency characteristics of a longitudinal ultrasonic transducer with stepped horn. *Ultrasonics* **2022**, *121*, 106683. [CrossRef]
31. Lin, S. Optimization of the performance of the sandwich piezoelectric ultrasonic transducer. *J. Acoust. Soc. Am.* **2004**, *115*, 182–186.
32. Xu, L.; Chen, Y.; Wu, G.; Zhang, H.; Li, F. Optimization of the sandwich piezoelectric ultrasonic welding transducer. *J. Shaanxi Norm. Univ. Nat. Sci. Ed.* **2020**, *48*, 55–59.
33. Wang, H.; Zhang, G.; Zhou, S.; Ouyang, L. Implementation of a novel Six Sigma multi-objective robustness optimization method based on the improved response surface model for bumper system design. *Thin-Walled Struct.* **2021**, *167*, 108–257. [CrossRef]
34. Lai, Y. *Isight Parameter Optimization Theory and Examples*; Beijing University Press: Beijing, China, 2012; pp. 193–194.
35. Jing, Z.; Zhang, C.; Cai, P.; Li, Y.; Chen, Z.; Li, S.; Lu, A. Multiple-objective optimization of a methanol/diesel reactivity controlled compression ignition engine based on non-dominated sorting genetic algorithm-II. *Fuel* **2021**, *300*, 120953. [CrossRef]
36. Shen, G.; Cai, W.; Yu, D. Design and optimization of two-section ultrasonic stepped horn. *J. Vib. Shock.* **2015**, *34*, 104–108.

 *sustainability*

Article

Investigating a New Method-Based Internal Joint Operation Law for Optimizing the Performance of a Turbocharger Compressor

Rong Huang *, Jimin Ni, Houchuan Fan, Xiuyong Shi and Qiwei Wang

School of Automotive Studies, Tongji University, Shanghai 201804, China
* Correspondence: hr1209@tongji.edu.cn; Tel.: +86-21-695-899-80

Abstract: A well-matched relationship between the compressor and turbine plays an important role in improving turbocharger and engine performance. However, in the matching of turbocharger and engine, the internal operation relationship between compressor and turbine is not considered comprehensively. In order to fill this gap, this paper proposed the internal joint operation law (IJOL) method based on the internal operating characteristics of the compressor and turbine using a combination of experimental and simulation methods. On this basis, the optimization method of the compressor was proposed. Firstly, according to the basic conditions of turbocharger, the compressor power consumption and the turbine effective power at a fixed speed were solved. Secondly, the power consumption curve of the compressor and the effective power curve of the turbine were coupled to obtain the power balance point of the turbocharger. Then, the internal joint operating point was solved and coupled to obtain the IJOL method. Finally, the IJOL method was used to optimize the blade number and the blade tip profile of the compressor. The simulation results showed that for the blade number, the 8-blade compressor had the best overall performance. For the blade tip profile, compared with the original compressor, the surge performance of the impeller inlet diameter reduced by 3.12% was better than that of the original compressor. In addition, in order to compare this to engine performance with different compressor structures, a 1D engine model was constructed using GT-Power. The simulation results showed that the maximum torque of the engine corresponding to the impeller designed by the IJOL method was 4.2% higher than that of the original engine, and the minimum brake specific fuel consumption was 3.1% lower. Therefore, compared with the traditional method, the IJOL method was reasonable and practical.

Keywords: engine; turbocharger compressor; internal joint operation law; optimization method; collaborative design

Citation: Huang, R.; Ni, J.; Fan, H.; Shi, X.; Wang, Q. Investigating a New Method-Based Internal Joint Operation Law for Optimizing the Performance of a Turbocharger Compressor. *Sustainability* **2023**, *15*, 990. https://doi.org/10.3390/su15020990

Academic Editor: Luca Cioccolanti

Received: 5 November 2022
Revised: 28 December 2022
Accepted: 3 January 2023
Published: 5 January 2023

1. Introduction

Turbocharging technology is widely used to increase the power output per swept displacement of an internal combustion engine (ICE). As a result, the engine size can be greatly decreased, and fuel consumption and gas emissions may be reduced [1–5]. The centrifugal compressor is a key component of the turbocharger, and its aerodynamic performance has a significant impact on the operation of turbocharged ICEs. An accurate understanding of the matching characteristics of the compressor and turbine is beneficial to improve the performance of the turbochargers and the ICEs [6–10]. In addition, it is necessary to carry out design and optimization investigations based on matching operating characteristics between the compressor and turbine in the design and optimization of the turbocharger compressors. Hatami et al. [11] and Hosseinimaab et al. [12] concluded that the matching operating characteristics in the compressor and turbine had an important role in turbocharger performance. However, most of the existing studies have focused on the design and optimization of the turbocharger based only on the improvement of the performance of the compressor itself, but not on the comprehensive performance of the turbocharger itself.

At present, scholars have carried out research on the optimization of the aerodynamic performance and design of the structures of the compressors. Ekradi et al. [13] presented an integrating three-dimensional (3D) blade parameterization method to optimize the centrifugal compressor impeller with splitter blades. The results showed that at the design point, the isentropic efficiency increased by 0.97%, and the mass flow rate and total pressure ratio increased by 0.65% and 0.74%, respectively. Ma et al. [14] compared four optimization algorithms to improve the operating stability of a centrifugal compressor and found that, compared to the reference design, the stall margin of the centrifugal compressor was improved by 1.87% with the optimization using the particle swarm optimization algorithm. Shaaban et al. [15] proposed a new radial vaneless diffuser design method to improve the centrifugal compressors' performance. The results indicated that under swirl flow and jet-wake conditions, the diffuser pressure coefficient increased 6.6% and the diffuser loss coefficient decreased 4.7%. Tüchler et al. [16] used a coupled approach of computational fluid dynamics and genetic algorithm to optimize an automotive compressor. The results showed that the shorter splitter and varied pitch fraction both increase near surge and peak efficiency. Zamiri et al. [17] investigated the influences of blade squealer tips on the aerodynamic performance of a centrifugal compressor. The results showed that at the design point, the use of squealer tips increased the compressor efficiency by 0.32%. Aparna et al. [18] and Moussavi et al. [19] concluded the same results. However, the compressor is affected by the fluctuation of the ICEs' admission, which make the flow rate and pressure of the compressor periodically change with the ICEs' operating process. This could further affect the engine performance. For example, the compressed air supercharging system could improve the driving force during the phase of the engine's increasing crankshaft rotational speed [20]. Therefore, it is noteworthy to pay attention to the operating characteristics between the compressor and the engine, which are important for improving the overall performance of the supercharger and the ICEs [21,22].

Currently, scholars have conducted research on the matching of the compressor with the ICEs. Mousavi et al. [23] suggested a new algorithm for turbocharger matching. Chen et al. [24] proposed a novel pseudo-Mean Average Precision (MAP) optimization method to achieve full-operation-range performance optimization of a compressor. The results indicated that by using the optimization method, the choke flow rate increased by 20% and the maximum efficiency increased to 80%. In order to improve the accuracy of matching, Huang et al. [25] established a lumped model to calculate compressor adiabatic efficiency and the heat transfer properties of a turbocharger. Wu et al. [26] also proposed a method to match a two-stage turbocharging system, and they found that with adopting the method, the engine torque increased more than 10% and the engine low fuel consumption area was broadened. In order to improve engine performance, Hosseinimaab et al. [12] employed a hybrid optimization approach (including modern and numerical optimizers) to modify the compressor geometry. However, they did not consider the effect of the turbine on the supercharger and engine performance. Li et al. [27] presented a new method to predict the performance MAPs of automotive turbocharger compressors. The results indicated that the method could be used for the preliminary design of turbocharger compressors with both vaneless and vaned diffusers. However, most of the existing studies only match the compressor with the ICEs. Actually, the performance at both ends of the turbocharger is limited by the structural properties of another end, and a complete performance map is usually unavailable [28]. Therefore, the research on the internal matching characteristics of the compressor and turbine play an important role in improving the performance of turbochargers and ICEs.

Turbochargers can effectively improve the power and fuel economy of ICEs, reduce emissions, and miniaturize ICEs. For a turbocharger, the internal matching characteristics between the compressor and turbine determine its operating performance, which in turn determines the matching performance of the turbocharger to the engine. There is no in-depth and systematic research on the internal matching between the compressor and turbine, and the important role of the internal matching of their joint points, in the design of turbochargers, is not fully reflected. To fill this gap, this study proposed an optimization method for the compressor based on the internal joint operating characteristics between the compressor and turbine. The rest of this study is organized as follows: Section 2 is devoted to the design and validation of the compressor and turbine models, introducing the test turbocharger performance test bench. Section 3 proposes an optimization method for the compressor based on the IJOL method of the turbocharger, including the calculation of turbine and compressor powers, and the coupling to obtain the internal joint operation curve of the turbocharger. Section 4 presents the optimized design of the blade number and the blade tip profile of the compressor using the IJOL method and compares the optimization results with the traditional method. Section 5 summarizes the main results of this study.

2. Materials and Methods

To clearly demonstrate the research idea of this study, an investigation procedure of the study is shown in Figure 1. Firstly, based on the basic conditions of turbocharger operation, since there was energy loss between the output power at the turbine end to the power consumption at the compressor end, the power transfer coefficient was proposed to characterize the bearing friction loss of the intermediate body. Then, based on the simulation and self-cycling experiments, the powers of turbine and compressor were calculated, and the coupling obtained the internal joint operation curve. Secondly, a compressor optimization method based on the internal joint operation law (IJOL) was proposed, and the blade number and blade tip profile were optimized. The optimization simulation results were compared with the traditional method. Finally, experiments were conducted to verify the results. Compared with the traditional method, the IJOL method was more reasonable and practical.

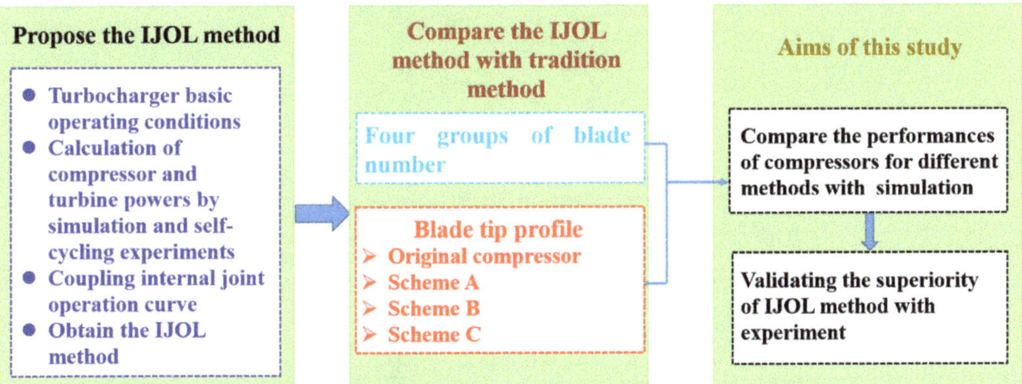

Figure 1. Investigation procedure.

2.1. Research Objects

A 4-cylinder automotive engine turbocharger was used as the research object. Figure 2 shows the three-dimensional (3D) structure of the physical prototype of the turbocharger. Table 1 lists the parameters of turbocharger.

2.2. Setting Model Parameters

In the compressor numerical model, the impeller of the compressor was the equal-length blade and the impeller outlet with angle bending. The diffuser adopted a bladeless diffuser. The principle geometric structure and original turbocharger compressor fluid domain model are shown in Figures 3 and 4.

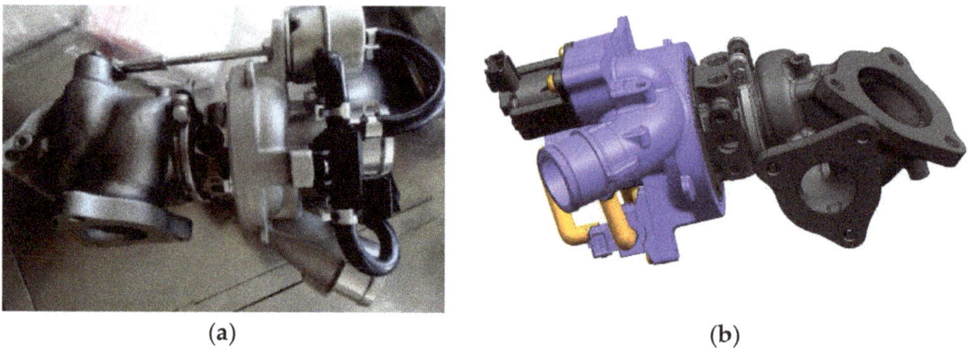

(a)   (b)

Figure 2. (**a**) Physical prototype; (**b**) three-dimensional structure model.

Table 1. Turbocharger parameters.

Item	Value	Item	Value
Outlet diameter D_{1t} of compressor impeller (mm)	44	Inlet diameter of turbine impeller (mm)	37.6
Inlet diameter of compressor impeller (mm)	32.1	Outlet diameter of turbine impeller (mm)	33.1
Blade number of compressor	8	Blade number of turbine	11
Diffuser height (mm)	2.5	Turbine impeller inlet blade angle (°)	0
Design pressure ratio	2.2	Turbine impeller inlet blade height (mm)	5.1
Rated speed (r/min)	220,000	Turbine impeller axial length (mm)	18.9
Flow range (kg/s)	0.02–0.13	Turbine impeller exit mean blade angle (°)	56.4
Displacement of gasoline engine (L)	1.5	Type of cooling	oil cooling + water cooling

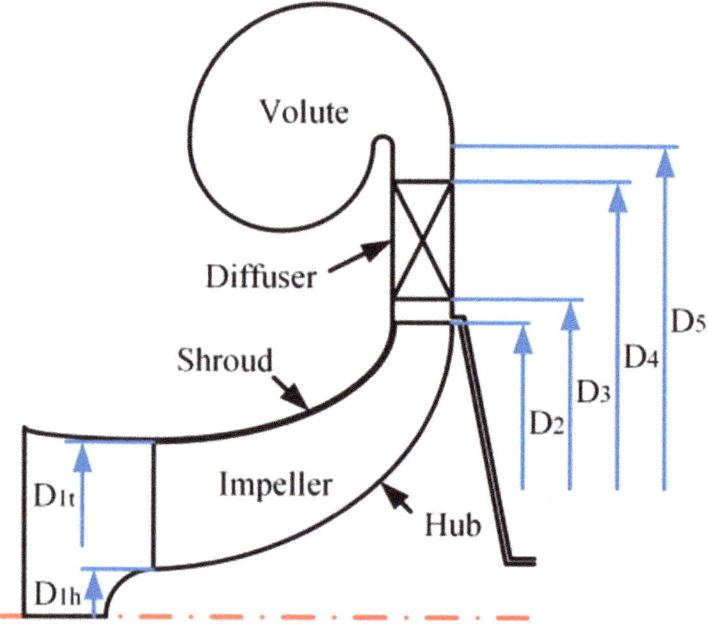

Figure 3. Compressor geometric structure [29].

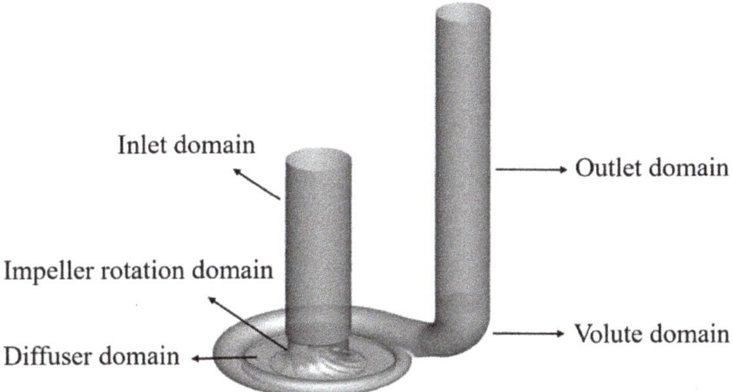

Figure 4. The original compressor fluid domain.

The compressor model consisted of five parts, which were the inlet domain, impeller rotation domain, diffuser domain, volute domain and outlet domain. The turbulence model used the Shear Stress Transport (SST) two-equation model, and the mathematical model used the Reynolds-averaged Navier–Stokes system of equations [30]. The model walls were all set as smooth, non-slip adiabatic walls [31], and wall temperature in the simulation was set as the compressor or turbine wall temperature under fixed boundary conditions (speed and flow rate) of the experiments. The fluid was defined as an ideal gas, and the fluid viscosity was set as a function of temperature. The total inlet pressure of the compressor was set to 101.325 kPa, the total inlet temperature was 293.15 K, and outlet boundary set to the mass flow rate (kg/s) corresponding to the turbocharger speed. For turbine model, the specific parameter settings can be seen in our previous paper [32].

In order to validate the accuracy of numerical model, a mesh independence analysis was conducted. The total pressure, efficiency and power consumption of the compressor with different grid numbers are shown in Figure 5. It can be seen from the figure that the performance curve of the compressor was smoothly distributed and almost constant when the grid number reached above 3,358,000, so it could be considered that the grid number met the simulation requirements. Therefore, the grid number of 3,358,000 was chosen to use in the calculation.

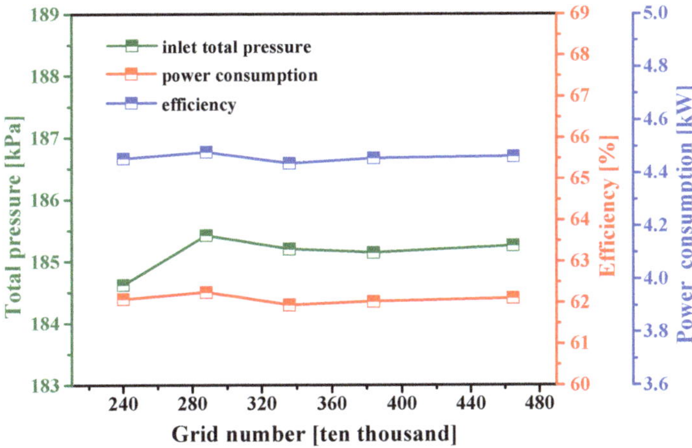

Figure 5. The effect of grid number on compressor performance at speed of 150,000 r/min.

2.3. Test Preparation and Model Validation

The compressor and turbine performance experiments were conducted on a Kratzer turbocharger test bench. Figure 6 shows the physical illustration and schematic diagram of the bench arrangement. The test bench was capable of measuring performance parameters of both the compressor and turbine, with automated data acquisition and processing functions. The bench could realize accurate control of turbocharger speed, compressor and turbine inlet and outlet parameters. The compressor inlet was equipped with a pressure and temperature regulator, which could be automatically adjusted in real time to ensure a smooth inlet temperature and pressure. Sensors were arranged in the turbine inlet and outlet pipes to improve the accuracy of temperature and pressure acquisition. The fuel for the experiment was liquefied petroleum gas, which could automatically correct the air-fuel ratio of the gas according to the turbine inlet temperature.

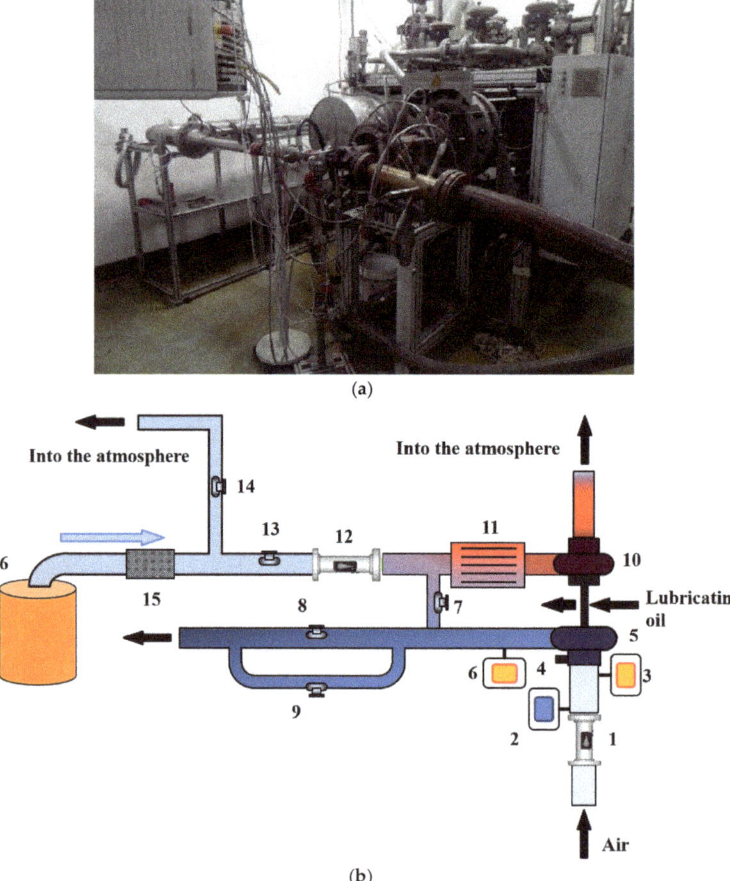

1. win-twist flow meter	2. Pressure differential sensor of flowmeter	3. Compressor inlet temperature and pressure sensor	4. Speed sensor
5. Compressor	6. Compressor outlet temperature sensor	7. Automatic circulating valve	8. Electric exhaust control valve
9. Electric trimming valve	10. Turbine	11. Burner	12. Turbine inlet flowmeter
13. Turbine inlet control valve	14. Air source vent valve	15. Filter	16. Air source

Figure 6. (**a**) Kratzer test bench; (**b**) schematic diagram of test bench arrangement.

During the experiment of compressor performance, at the set turbocharger speed, the surge point and choke point of the compressor were to be found firstly. Then, experimental points evenly between the surge point and choke point were set. Finally, data were collected in turn to form the pressure ratio-flow performance curve at the same speed.

The validation experiments of the compressor and turbine were carried out on the Kratzer turbocharger test bench, and the following three speed conditions of 110,000 r/min, 150,000 r/min and 190,000 r/min were selected to represent the low, medium and high speeds of the turbocharger operation, respectively. In this study, the validation method was used in previous papers [31,33].

Figure 7 shows the turbocharger performances between the experiments and simulation. For pressure ratio of compressor, it can be seen that the maximum error was 4.71% at the high pressure ratio condition at 190,000 r/min, which was within the acceptable range. For expansion ratio of turbine, the experiment and simulation of the swallowing capacity were in good agreement with each other and had a maximum error value of 2.62% at the speed of 150,000 r/min, which was within the acceptable range. This discrepancy was attributed to the simplification of the secondary feature in the geometry, the optimal application scope of the SST model and manufacturing error [34,35]. Therefore, the turbocharger model could meet the simulation requirements and can be used for further study.

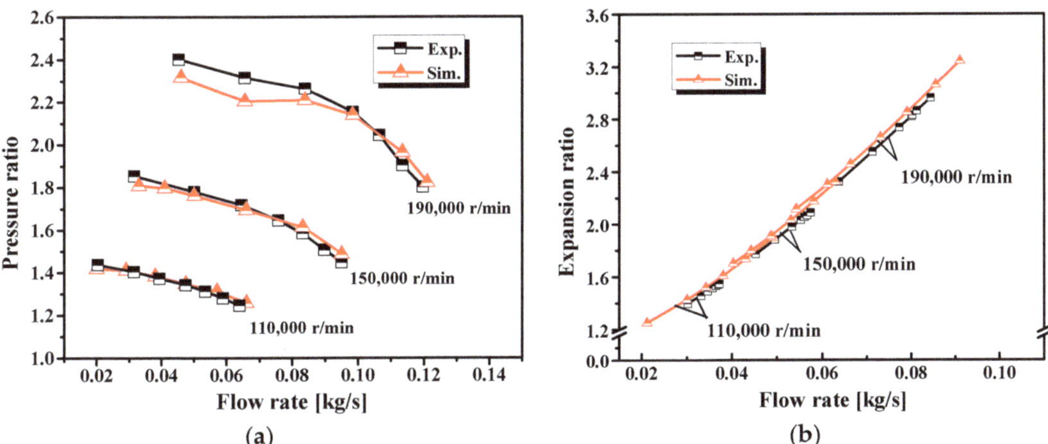

Figure 7. Turbocharger model test verification. (**a**) Compressor model; (**b**) Turbine model.

3. Internal Joint Operation Law Analysis

3.1. Coupling of Internal Joint Operation Curves

The compressor and turbine are the two working parts of the turbocharger; both are rigidly connected by the rotor shaft, and the compressor and turbine have the same speed, and the working condition of the compressor and turbine is related to the engine working process. In this study, the numerical simulations were conducted based on the assumption of static state operation. There were three basic conditions for the operation of the turbocharger, including speed balance, flow rate balance and power balance.

Speed balance: The compressor, turbine and rotor shaft have the same speed.

$$n_C = n_T = n_R \qquad (1)$$

where, n_C is the compressor speed (r/min), n_T is the turbine speed (r/min), and n_R is the rotor shaft speed (r/min).

Flow rate balance: The leakage mass flow rate is negligible because it is very small [36]. The compressor operating mass flow rate plus the engine fuel mass flow rate equals the turbine operating mass flow rate.

$$m_T = m_C + m_F \tag{2}$$

where m_T is the turbine operating flow rate (kg/s), m_C is the compressor operating flow rate (kg/s), and m_F is the fuel flow rate (kg/s).

Power balance: Figure 8 shows the energy transfer process in the turbocharger. Power transfer from the turbine end to the compressor end of the process would lose part of the energy, which is the bearing friction loss. The turbine output power minus the losses of intermediate body is equal to the compressor power consumption. Based on the law of conservation of energy, the calculation formula is as follows:

$$P_T = P_C + P_{Loss} \tag{3}$$

where P_T is the turbine output power (W), P_C is the compressor power consumption (W), and P_{Loss} is the power loss of intermediate body (W).

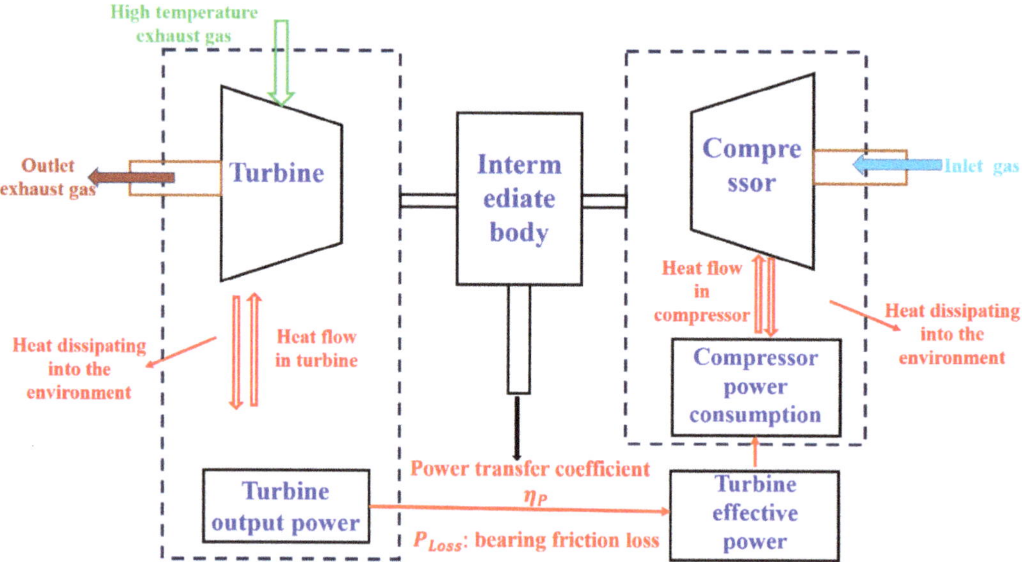

Figure 8. Energy transfer process in the turbocharger.

The heat losses at a turbocharger were heat dissipating into the environment, heat flows within the turbocharger as well as active cooling through cooling water and oil, all due to high temperature gradients. In the simulation, the heat loss dissipating into the environment was not considered, which was neglected. Due to the wall temperature in the simulation that was set as the compressor or turbine wall temperature under fixed boundary conditions (speed and flow rate) of the experiments, the heat loss from heat flow in the turbocharger was considered, and the heat loss was included in the turbine output power and compressor consumption power. This was because the turbocharger bench test was a hot blow experiment, and it can be seen from Figure 9 that the simulated values of the compressor and turbine efficiencies basically match the experimental values, indicating that the heat flows in the turbocharger had been considered in the simulation calculation.

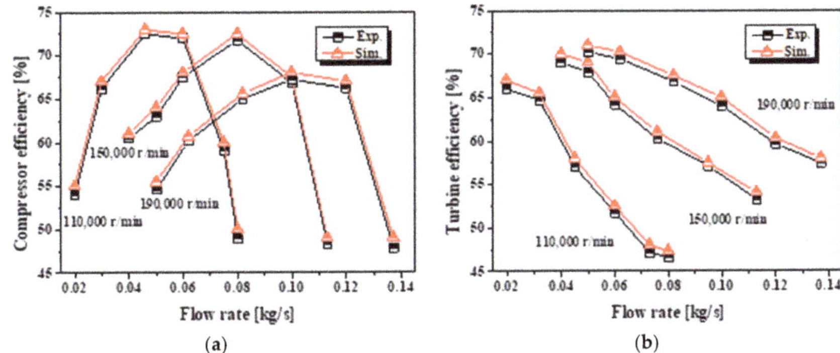

Figure 9. Comparison of simulated and tested turbocharger efficiencies. (**a**) Compressor; (**b**) Turbine.

Turbine output power refers to the power generated by the exhaust gas impinging on the rotation of all blades. Compressor power consumption refers to the work consumed to drive all the blades to rotate for compressing the fluid. The power calculation formula is as follows:

$$P = T_{all\ blades} \times \omega \tag{4}$$

where, P is the turbine output power or the compressor power consumption (W), $T_{all\ blades}$ is the total torque of all blades (N·m), ω is the angular velocity (rad/s).

The compressor power consumption is equal to the turbine effective power, and there is power transfer loss (bearing friction loss) between the turbine output power and turbine effective power in intermediate body. Therefore, the power transfer coefficient η_P is used to define the ratio of the compressor power consumption to the turbine output power.

$$\eta_P = P_C / P_T \tag{5}$$

The power transfer coefficient η_P could relate the compressor power consumption to the turbine output power and thus to couple the joint operation curve inside the turbocharger.

At the fixed speed, the turbine output power and the compressor power consumption were calculated by the simulation and turbocharger self-cycling experiments. Figure 10 shows the self-cycling experiment values of the compressor outlet pressure and turbine inlet pressure compared with the simulated values. As it can be seen that the self-cycling experiment values and simulation values were in good agreement, the error was in the allowable range.

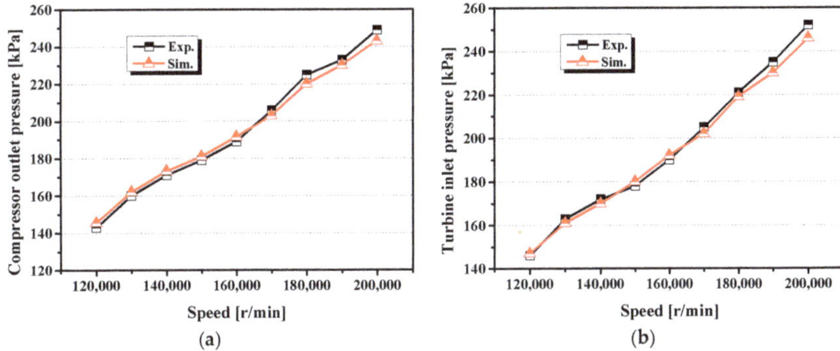

Figure 10. Comparison of self-cycling experiment values and simulated values. (**a**) Compressor model; (**b**) Turbine model.

Based on the turbine output power and compressor power consumption calculated under the self-cycling experiment conditions, the power transfer coefficient is calculated by Equation (5), as shown in Figure 11.

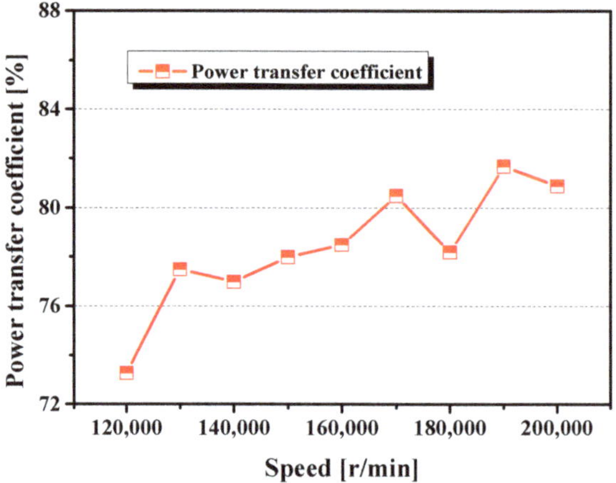

Figure 11. Distribution of power transfer coefficient.

It should be noted that the power transfer coefficients obtained by the self-cycling experiments could not cover all the flow rate points on the speed line, and there was a certain error. However, the focus of this study was the performance difference caused by the change of structures, and the power transfer coefficients could meet the needs of performance evaluation and comparison between different compressor structures.

The turbine effective power is the maximum power that can be obtained at the compressor end. The turbine effective power is obtained by transforming the turbine output power through Equation (5), as shown in Figure 12.

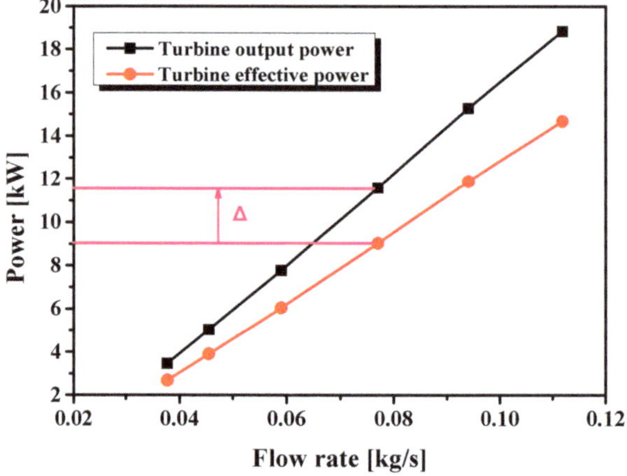

Figure 12. Conversion relationship between the output power and effective power of the turbine at 150,000 r/min.

In the fixed-speed line, the compressor power consumption curve and turbine effective power curve were superimposed and coupled. The horizontal coordinate of intersection point of both the compressor and turbine at the time of coupling were the compressor working flow rate, and the intersection point of the power balance was obtained, as shown in Figure 13. The significance of the proposed joint operation point was to obtain a joint curve reflecting the turbocharger operation on the basis of considering the internal matching of the compressor and turbine.

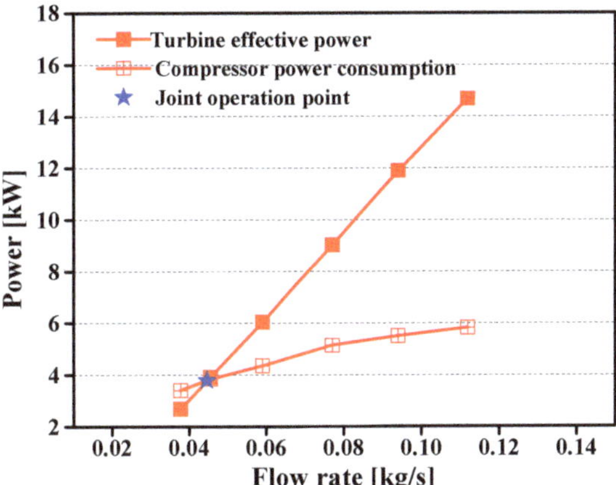

Figure 13. The power balance at both ends of the turbocharger at 150,000 r/min.

By analogy, the compressor power consumption curves and turbine effective power curves for all speeds were superimposed and coupled to obtain the joint operation points of the turbocharger at each speed, as shown in Figure 14.

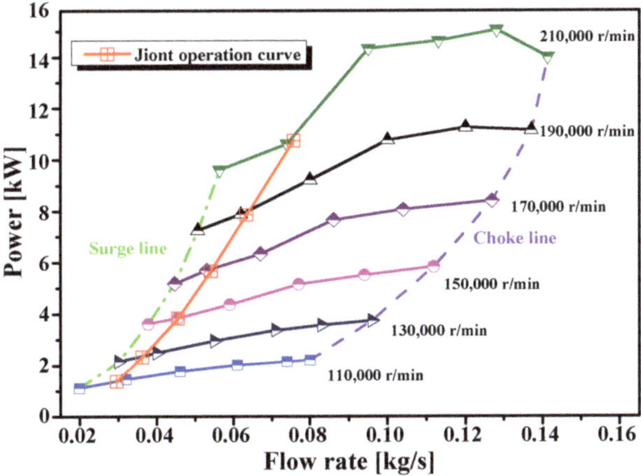

Figure 14. Distribution of the joint operation curve on the compressor power consumption MAP.

3.2. Propose the Total Efficiency Calculation Method of the Compressor

According to the internal joint operation curve, the total efficiency at the joint point is calculated by the following equation:

$$\eta_{tot} = \eta_T \times \eta_C \times \eta_P \tag{6}$$

where η_{tot} is the total efficiency of the turbocharger.

The traditional method of calculating the total efficiency of a turbocharger is to multiply the maximum efficiency of the turbine, the maximum efficiency of the compressor and the mechanical efficiency, which is not objective enough. This was because, on the one hand, the highest efficiency point of the turbine and the highest efficiency point of the compressor are often not in the same operating point. On the other hand, the work of the supercharger is a dynamic process; the actual efficiency of both the turbine and the compressor are changing, making different operating conditions have different total efficiency of the supercharger [37]. Therefore, the concept of joint point was proposed in this study to calculate the turbocharger total efficiency, compressor and turbine performances.

Figure 15 shows the distribution of compressor efficiency for the joint operating conditions on the compressor efficiency MAP. It can be seen from the figure that the efficiency of the joint point operating condition did not go through the maximum efficiency point of the compressor. Therefore, the efficiency of the compressor calculated based on the joint operating point was related to the speed and compressor working flow rate, synchronized with the turbine operating condition, and closed to the actual situation of the turbocharger, which was practical.

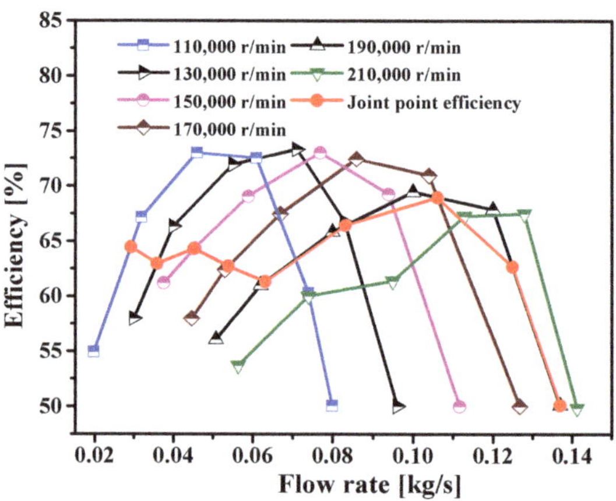

Figure 15. Efficiency distribution for joint operating conditions on the compressor efficiency MAP.

3.3. Internal Joint Point Operation Law (IJOL) Method

In the process of compressor design and optimization, pressure ratio, surge margin, choke flow rate and efficiency are important performance parameters [38]. In this study, the compressor performance index included pressure ratio, surge margin, choke flow rate, joint point efficiency and maximum efficiency. The formula for calculating the surge margin is shown below [39]:

$$SM = \left(\frac{\frac{\pi_s}{m_{scor}}}{\frac{\pi_w}{m_{wcor}}} - 1 \right) \times 100\% \tag{7}$$

where π_s is the pressure ratio at surge, m_{scor} is corrected flow rate at surge, π_w is the pressure ratio of working point, m_{wcor} is corrected flow rate at the working point. The larger the surge margin, the better the surge performance of the compressor.

The formula for calculating the compressor performance index is shown below:

$$\phi_c = A_1 \times \frac{S_{R_{surge}}}{S_{O_{surge}}} + A_2 \times \frac{\eta_{R_{joint\ point}}}{\eta_{O_{joint\ point}}} + A_3 \times \frac{\pi_R}{\pi_O} + A_4 \times \frac{m_R}{m_O} + A_5 \times \frac{\eta_{R_{max}}}{\eta_{O_{max}}} \tag{8}$$

where Φ_C is the calculated compressor performance index. $S_{R_{surge}}$ and $S_{O_{surge}}$ were the surge margin of remodel and original compressors. $\eta_{R_{joint\ point}}$ and $\eta_{O_{joint\ point}}$ were the efficiency of joint point. π_R and π_O were the pressure ratio. m_R and m_O were the mass flow rate. $\eta_{R_{max}}$ and $\eta_{O_{max}}$ were the maximum efficiency. The parameters of A_1, A_2, A_3, A_4 and A_5 were the weighting factors of surge margin, joint point efficiency, pressure ratio, choke flow rate and maximum efficiency. For the weight distribution of each parameter, turbocharger compressor work was increasingly close to the surge line, operating stability was critical for the compressor [40] and needed to broaden the compressor stable working regions, so the joint point of the surge performance weight distribution was the largest. The efficiency and pressure ratio of the joint point of the compressor directly reflected the compressor performance, which in turn determined the engine intake performance, so the efficiency and pressure ratio of the joint point of the compressor weight distribution were second only to that of surge performance. The choke flow of the compressor only needed to be greater than the maximum working flow of the engine, so the choke flow and maximum efficiency weights were the smallest. The weighting coefficients are shown in Table 2.

Table 2. Weight assignment for each performance parameter.

	Surge Margin	Joint Point Efficiency	Pressure Rate	Choke Flow Rate	Maximum Efficiency
Weight distribution	30	25	25	10	10

4. Results and Discussion

4.1. Analysis of Compressor Blade Number Based on IJOL Method

The compressor impeller is the most important component that affects the compressor performance, so the compressor impeller was selected for optimization.

4.1.1. Optimization of the Blade Number

Four groups of compressors with blade numbers of 7, 8, 9 and 10 were designed, and the 3D modeling of each group was kept consistent with that of the original compressor. The same topology and meshing method were used for the impeller models, and the optimization study was carried out at three speeds of 110,000 r/min, 150,000 r/min and 190,000 r/min, respectively.

Figure 16 shows the effect of the blade number on the pressure ratio at different speeds. It can be seen that the pressure ratio increased with an increase in the blade number. When the blade number was greater than 8, the increased amplitude of the pressure ratio was limited. At 190,000 r/min, the difference between the outlet pressures of a 7-blade compressor and an 8-blade compressor were more than 4.5 kPa. Further observation found that the overall surge line of the four groups of blades was relatively close, but the choke line moved to the left with an increase in the blade number. With an increase in the blade number, the working flow range of the compressor became narrower. Specifically, at low and medium speeds, with the blade number increased, the surge flow slightly increased and the joint point surge margin gradually decreased. At high speed, there was no significant difference in the surge flow and surge margin for different blade numbers (as shown in Table 3).

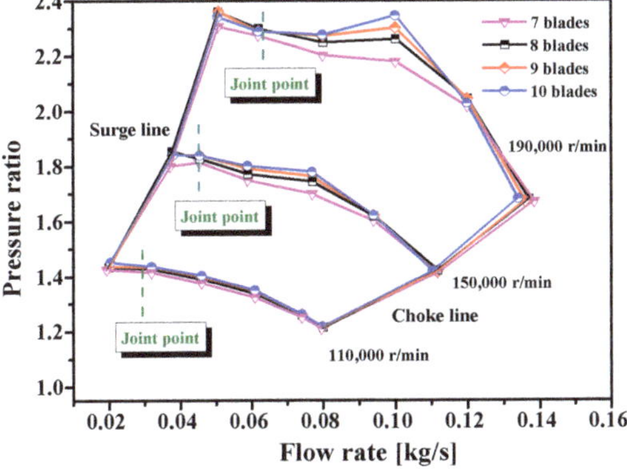

Figure 16. Pressure ratio for joint operating conditions.

Table 3. Comparison of surge performance of compressors with different blade numbers.

	7		8		9		10	
Speed (r/min)	Surge Flow Rate (kg/s)	Surge Margin (%)	Surge Flow Rate (kg/s)	Surge Margin (%)	Surge Flow Rate (kg/s)	Surge Margin (%)	Surge Flow Rate (kg/s)	Surge Margin (%)
110,000	0.0193	10	0.0198	9.5	0.0201	9.2	0.0204	8.9
150,000	0.0373	8	0.0377	7.6	0.0380	7.3	0.0381	7.2
190,000	0.0508	12.4	0.0508	12.4	0.0509	12.3	0.0506	12.6

Figure 17 shows the effect of the blade number on the efficiency. It can be seen that near the joint point, the more the blade number, the higher was the efficiency. However, the difference was not obvious; the design point efficiency of the 10-blade compressor was about 1% higher than the design point efficiency of the 8-blade compressor. The 7-blade compressor's joint point efficiency decreased significantly compared with other blade numbers, especially in high speed. As can be seen in Table 4, for the compressor performance index, an 8-blade compressor was the highest, while the 7-blade compressor performance index decreased more, indicating that the 8-blade compressor had the best performance and high comprehensive performance at the joint point.

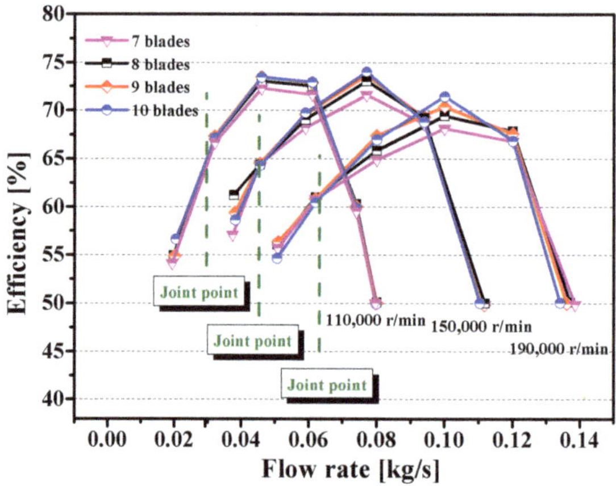

Figure 17. Effect of the blade number on the efficiency distribution.

Table 4. Comparison of performance indexes of compressors with different blade numbers.

	7	8	9	10
Mean surge margin (%)	10.13	9.83	9.60	9.57
Mean joint point efficiency (%)	64.5	65.1	65.8	65.9
Mean joint point pressure ratio	1.871	1.953	1.967	1.978
Maximum efficiency (%)	72.296	73.029	73.675	74.07
Choke flow rate (kg/s)	0.1385	0.137	0.136	0.134
Performance index	99.6	100.0	99.8	99.8

4.1.2. Blade Tip Profile Optimization

The compressor's impeller inlet diameter, outlet diameter and diffuser height have an important impact on the performance of the compressor. The change of blade tip profile could change the compressor inlet diameter, diffuser height and compressor transition arc profile, and therefore affected the performance of the compressor.

There were three schemes for the optimization of the blade tip profile. Scheme A: Compared with the original compressor, the impeller inlet diameter was reduced by 3.12%, and the diffuser height remained unchanged. Scheme B: Compared with the original compressor, the impeller inlet diameter remained unchanged, and the diffuser height was reduced by 20%. Scheme C: Compared with the original compressor, the impeller inlet diameter and diffuser height were reduced by 3.12% and 20%, respectively. The percentage of impeller inlet diameter and diffuser height for three structures as compared with original compressor is shown in Figure 18.

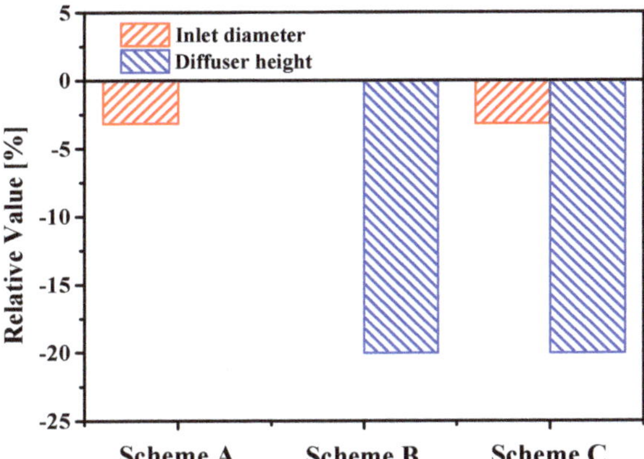

Figure 18. The Percentage of impeller inlet diameter and diffuser height for three structures as compared with original compressor.

Figure 19 shows the comparison of the pressure ratio of the four compressors at various speeds. It can be seen from the figure that there was no significant difference in the pressure ratio at the joint point of low and middle speeds. At the joint point of high speed, the pressure ratio of Scheme A was slightly smaller than that of the original compressor, but the difference was within 1.2%. Near the joint point of low and medium speeds, the pressure ratio of Scheme A and the original compressor had a wide overlap part. Overall, the pressure ratio of Scheme A was closest to those of the original compressor.

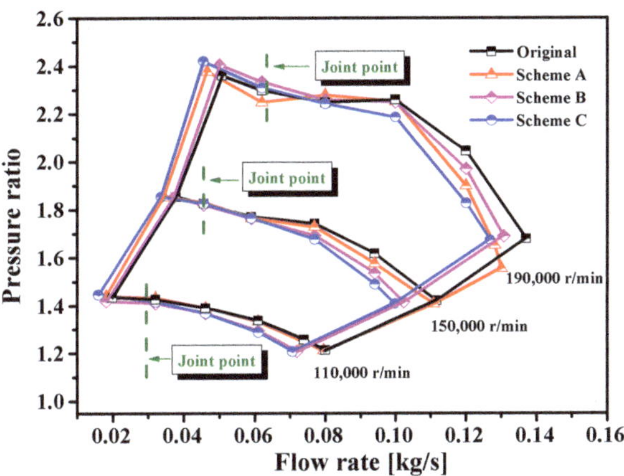

Figure 19. The effect of blade tip profile on pressure ratio.

For surge performance, it can be seen from the Table 5 that all three schemes improved the surge flow rate as compared with the original compressor. The surge margin of Schemes A and C was above 10%, which met the requirements of the surge margin, while the surge margin of the original compressor was below 10% at low and medium speeds.

Table 5. Comparison surge performance of compressors with different blade tip profile.

Speed (r/min)	Original		Scheme A		Scheme B		Scheme C	
	Surge Flow Rate (kg/s)	Surge Margin (%)	Surge Flow Rate (kg/s)	Surge Margin (%)	Surge Flow Rate (kg/s)	Surge Margin (%)	Surge Flow Rate (kg/s)	Surge Margin (%)
110,000	0.0198	9.5	0.0180	11.3	0.0180	11.3	0.0158	13.5
150,000	0.0377	7.6	0.0345	10.8	0.0370	8.3	0.0335	11.8
190,000	0.0508	12.4	0.0465	16.7	0.0500	13.2	0.0454	17.8

The effect of blade tip profile on the efficiency of the compressors is shown in Figure 20. It can be seen that the efficiencies of the three schemes were improved as compared with the original compressor near the joint point of each speed, and the joint point of Scheme A was 0.7–1.5% higher than that of the original compressor. In addition, the highest efficiency of Scheme A was 0.8% higher than that of the original compressor.

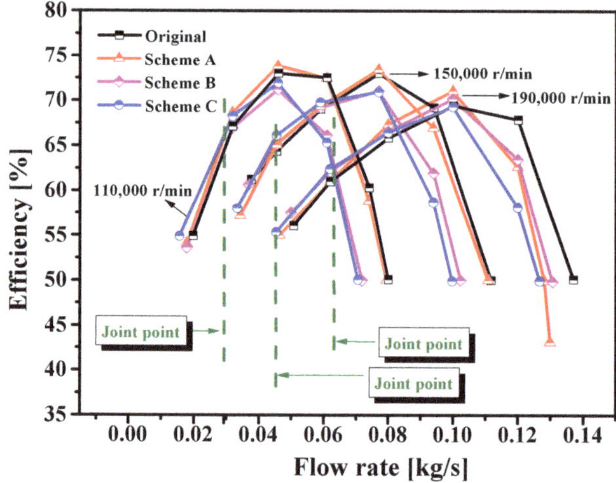

Figure 20. Effect of blade tip profile on compressor efficiency.

For compressor performance index, as can be seen in Table 6, because the choke flow rate of Scheme C did not meet the maximum working flow rate of the engine, it could not be used as an optimized structure. Scheme A had the highest compressor performance index and the best overall performance among the available options. Compared with the original compressor, Scheme A had obvious comprehensive advantages.

Table 6. Comparison of performance indices of different blade tip profiles.

	Original	Scheme A	Scheme B	Scheme C
Mean surge margin (%)	9.83	12.93	10.93	14.37
Mean joint point efficiency (%)	65.1	66.5	66.1	66.5
Mean joint point pressure ratio	1.953	1.943	1.954	1.936
Maximum efficiency (%)	73.029	73.845	71.225	71.965
Choke flow rate (kg/s)	0.137	0.130	0.131	0.127
Performance index	100.0	109.5	103.0	113.3

In summary, Scheme A was the best blade tip profile in the operating conditions range. Therefore, when the impeller inlet diameter was reduced by 3.12% and diffuser height remained unchanged, the overall performance of the compressor was higher than that of the original compressor, especially in terms of surge performance and joint point efficiency.

4.2. Comparison and Validation Analysis of the IJOL and Traditional Methods

For the traditional method, the optimization goal is to pursue the maximum efficiency of the compressor [41,42]. Therefore, the index for evaluating the optimization results of traditional methods is the maximum efficiency. In order to compare the IJOL method with the traditional method, the original impeller structure, the splitter-blade impeller designed by the traditional method and the structure determined by the IJOL method with the impeller inlet diameter reduced by 3.12% were selected for comparison. The inlet and outlet diameters of the compressor impeller optimized based on the traditional method were the same as those of the original compressor. The numbers of main blades and splitter blades were each 5. In addition, the main blade was 3.5 mm higher than the splitter blade at the inlet guide vane, and the radial parts of the main and splitter blades had the same 3D shape. The main blade shape remained the same as the original compressor blade.

A comparison of the pressure ratio of the three structures is shown in Figure 21. It can be seen that the surge line of the traditional method moved to the left compared with the original compressor, but the leftward shift of its surge line was not as large as that of the IJOL method. The specific values are shown in Table 7. The lower the speed, the smaller was the surge margin of the traditional method, while the surge line of the IJOL method basically moved to the left as a whole, among which the improvement of the surge margin at low speed was obvious. At the joint points of low and middle speeds, the pressure ratio of the traditional method was not significantly different from that of the original compressor. While at 190,000 r/min, its pressure ratio was lower than that of the original compressor, which was especially obvious at high speed and high flow rate.

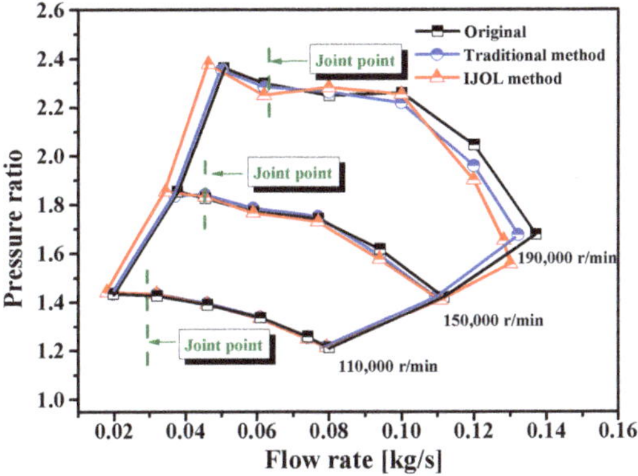

Figure 21. Comparison of pressure ratio of three structures.

Table 7. Comparison of surge performance of different compressor structures.

Speed (r/min)	Original		Traditional Method		IJOL Method	
	Surge Flow Rate (kg/s)	Surge Margin (%)	Surge Flow Rate (kg/s)	Surge Margin (%)	Surge Flow Rate (kg/s)	Surge Margin (%)
110,000	0.0198	9.5	0.0180	11.3	0.0196	9.7
150,000	0.0377	7.6	0.0345	10.8	0.0371	8.4
190,000	0.0508	12.4	0.0465	16.7	0.0502	13.0

The efficiency comparison of the three structures is shown in Figure 22. Near the joint point of 150,000 r/min, the efficiency of the traditional method was the same as the original compressor, while at other speed joint points the efficiency of the traditional method was lower than that of the original compressor. The efficiency of the IJOL method was higher than that of the original compressor at all joint points. As can be seen in Table 8, the performance index of the traditional method was lower than that of the original compressor, while the performance index of the IJOL method had obvious advantages and was the best performance structure in the operating conditions.

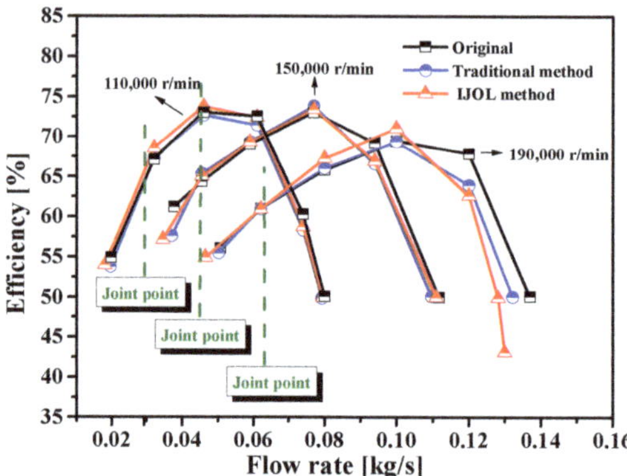

Figure 22. Comparison of efficiency of three structures.

Table 8. Comparison of performance indexes of different compressor structures.

	Original	Traditional Method	IJOL Method
Mean surge margin (%)	9.83	12.93	10.37
Mean joint point efficiency (%)	65.1	66.5	64.3
Mean joint point pressure ratio	1.953	1.943	1.861
Maximum efficiency (%)	73.029	73.845	73.901
Choke flow rate (kg/s)	0.137	0.13	0.132
Performance index	100.0	99.9	109.5

4.3. Comparative Analysis of Engine Performance with Optimized Structures of Two Methods

The optimal compressor structures based on the traditional method and the IJOL method were the splitter-blade impeller and the impeller inlet diameter reduced by 3.12% structures. A 1D engine model in Figure 23 was constructed using GT-Power for engine performance simulation and analysis. The basic assumptions [43] in this simulation are as follows:

1. The working fluid was a uniform state, and the air entering the cylinder and the residual exhaust gas were completely mixed instantaneously;
2. Air and mixed gas were considered ideal gases, and their thermodynamic parameters were affected by the temperature and composition of the gas;
3. A steady flow process was regarded for the process of working fluid;
4. The import and export kinetic energy of the working fluid was negligible, and there was no leakage during the combustion process;
5. The combustion heat release process was regarded as a thermodynamic process in which the external heats the working fluid inside the system in accordance with the established heat release law.

In the simulation, except for the change of the turbocharger structure, the rest of the engine structural components and settings were kept unchanged.

As can be seen in Figure 24, the IJOL method had the highest overall torque and the lowest brake-specific fuel consumption (BSFC) at low and medium engine speeds, while the traditional method was the opposite. Compared with the original engine, the maximum torque of the IJOL method was 4.2% higher and the minimum BSFC was 3.1% lower, while the maximum torque of the traditional method was 2.4% lower and the minimum BSFC was 1.2% higher than that of the original engine. The reason was that the performance index of the IJOL method was highest in the three compressor structures, while the performance index of the traditional method was smallest. The better overall performance of the compressor could further increase engine torque and reduce BSFC [44]. Therefore, the impeller inlet diameter that was reduced by 3.12% designed by the IJOL method instead of the original impeller could improve the power and fuel economy of the engine at low and medium speeds.

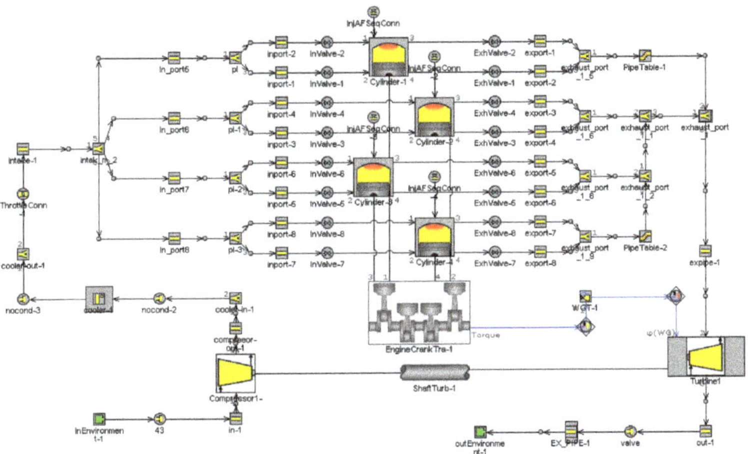

Figure 23. Engine 1D model.

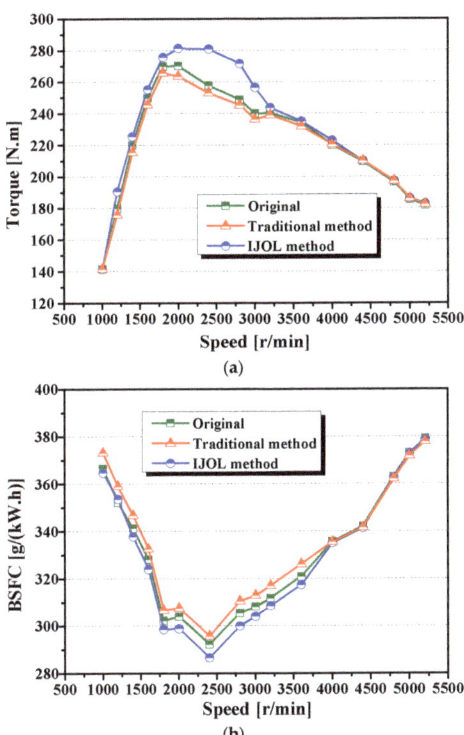

Figure 24. Effect of three structures on engine performance. (**a**) Torque; (**b**) BSFC.

4.4. Experimental Verification and Comparison of Compressor Optimization

Experimental analysis was performed on three structures of the compressor optimized based on the IJOL method and the traditional method. Figure 25 shows the pressure ratio for the three structures. It can be seen from the figure that the pressure ratios of the IJOL method were all equal to or slightly higher than that of the original compressor, and the pressure ratio was 1.49% higher than that of the original compressor near the joint point at high speed. The pressure ratio of the traditional method was basically the same as the original compressor at the joint point of low and medium speeds, and was lower than the original compressor at high speed. It was further observed that both the IJOL and the traditional methods had smaller surge flow and higher surge margin than that of the original compressor. Compared with original compressor, the IJOL method had a maximum reduction of 13.91% in the surge flow rate and a maximum increase of 6.07% in the surge margin, which was a significant improvement.

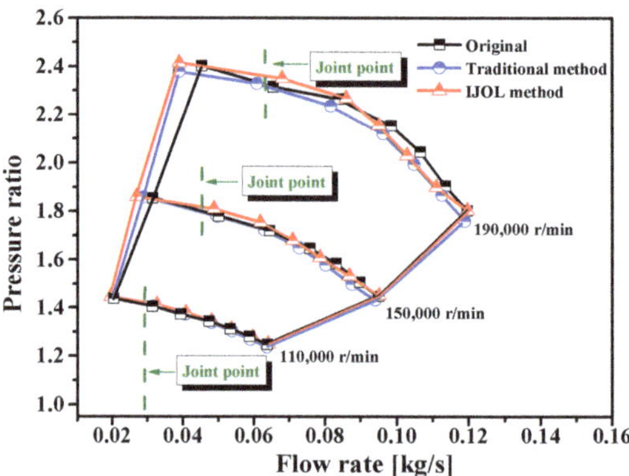

Figure 25. Experimental comparison of pressure ratio for three structures.

Figure 26 shows the efficiency for the three structures. The maximum efficiency of the original compressor was set to the reference value of 100%. Near each joint point, the efficiency of the IJOL method was higher than that of the original compressor, with higher values of 0.8% and above.

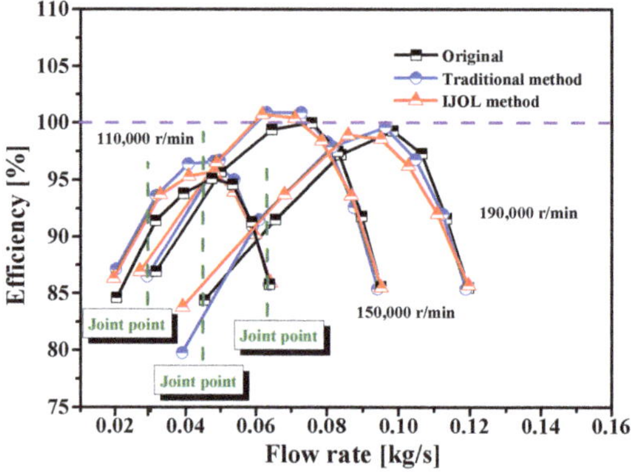

Figure 26. Experimental comparison of efficiency for three structures.

From the analysis, the experimental results of compressor performance and simulation conclusions were consistent. Therefore, both optimized compressor structures could reduce the surge flow of the compressor, and the impeller optimized by the IJOL method with the impeller inlet diameter reduced by 3.12% had pressure ratio and efficiency advantages near the joint points, which could improve the power and fuel economy of the engine in the low and medium speeds.

5. Conclusions

In this study, according to the basic conditions of turbocharger internal operating characteristics, the turbocharger joint points were determined, and the internal joint operation law (IJOL) was obtained. A compressor optimization method was proposed based on the IJOL, and the blade number and blade tip profile were selected for optimal application design. The following conclusions were obtained within the scope of this study:

1. Based on the joint operating characteristics of the two ends of the turbocharger compressor and turbine, the IJOL method of the turbocharger was coupled using the performance distribution of the compressor and turbine, and the calculation method for the total efficiency of the turbocharger was improved. The efficiency of the compressor obtained using the IJOL method was synchronized with the working of the two ends, which was closer to the actual situation, and more practical.
2. Based on the IJOL method, the effects of the blade number and the blade tip profile on the performance were analyzed. For the blade number, the 8-blade compressor had the best overall performance. For the blade tip profile, the compressor with the impeller inlet diameter reduced by 3.12% as compared with the original compressor had better surge performance.
3. Compared with the traditional method, the maximum efficiency of the IJOL method was slightly lower, but its joint point performance was higher than that of the traditional method.
4. Compared with the performance of the original engine, the power and fuel economy of the engine designed based on the traditional method were worse than those of the original engine. The maximum torque of the engine based on the IJOL method was 4.2% higher than that of the original engine, and the minimum BSFC was 3.1% lower. Compared with the traditional method, the IJOL had obvious advantages.

Author Contributions: Methodology, J.N.; Formal analysis, H.F. and X.S.; Investigation, R.H.; Resources, J.N.; Data curation, H.F.; Writing—original draft, R.H. and Q.W.; Writing—review & editing, R.H.; Supervision, R.H.; Project administration, J.N.; Funding acquisition, X.S. and Q.W. All authors have read and agreed to the published version of the manuscript.

Funding: This research was funded by National Natural Science Foundation of China Youth Science Foundation project (grant number 2210020419), State Key Laboratory of Internal Combustion Engine Reliability Open Subject Foundation of China (grant number skler-202114) and Natural Science Foundation of Shanghai (grant number 16ZR1438500).

Institutional Review Board Statement: Not applicable.

Informed Consent Statement: Not applicable.

Data Availability Statement: The data presented in this study are available in the main text of the article.

Acknowledgments: The team of authors acknowledges anonymous reviewers for their feedback, which certainly improved the clarity and quality of this paper.

Conflicts of Interest: The authors declare no conflict of interest.

References

1. Figari, M.; Theotokatos, G.; Coraddu, A.; Stoumpos, S.; Mondella, T. Parametric investigation and optimal selection of the hybrid turbocharger system for a large marine four-stroke dual-fuel engine. *Appl. Therm. Eng.* **2022**, *208*, 117991. [CrossRef]
2. Huang, R.; Ni, J.; Cheng, Z.; Wang, Q.; Shi, X.; Yao, X. Assessing the effects of ethanol additive and driving behaviors on fuel economy, particle number, and gaseous emissions of a GDI vehicle under real driving conditions. *Fuel* **2021**, *306*, 121642. [CrossRef]
3. Wang, D.; Shi, L.; Zhang, H.; Li, X.; Qian, Y.; Deng, K. Research on influence and demand of Miller cycle based on the coupling of marine low-speed engine and turbocharger. *Appl. Therm. Eng.* **2021**, *200*, 117624. [CrossRef]

4. Feneley, A.J.; Pesiridis, A.; Andwari, A.M. Variable Geometry Turbocharger Technologies for Exhaust Energy Recovery and Boosting-A Review. *Renew. Sustain. Energy Rev.* **2017**, *71*, 959–975. [CrossRef]
5. Bao, J.; Qu, P.; Wang, H.; Zhou, C.; Zhang, L.; Shi, C. Implementation of various bowl designs in an HPDI natural gas engine focused on performance and pollutant emissions. *Chemosphere* **2022**, *303*, 135275. [CrossRef]
6. Reihani, A.; Hoard, J.; Klinkert, S.; Kuan, C.-K.; Styles, D.; McConville, G. Experimental response surface study of the effects of low-pressure exhaust gas recirculation mixing on turbocharger compressor performance. *Appl. Energy* **2019**, *261*, 114349. [CrossRef]
7. Tanda, G.; Marelli, S.; Marmorato, G.; Capobianco, M. An experimental investigation of internal heat transfer in an automotive turbocharger compressor. *Appl. Energy* **2017**, *193*, 531–539. [CrossRef]
8. Broatch, A.; Ruiz, S.; García-Tíscar, J.; Roig, F. On the influence of inlet elbow radius on recirculating backflow, whoosh noise and efficiency in turbocharger compressors. *Exp. Therm. Fluid Sci.* **2018**, *96*, 224–233. [CrossRef]
9. Tancrez, M.; Galindo, J.; Guardiola, C.; Fajardo, P.; Varnier, O. Turbine adapted maps for turbocharger engine matching. *Exp. Therm. Fluid Sci.* **2011**, *35*, 146–153. [CrossRef]
10. Huang, R.; Ni, J.; Zheng, T.; Wang, Q.; Shi, X.; Cheng, Z. Characterizing and assessing the fuel economy, particle number and gaseous emissions performance of hybrid electric and conventional vehicles under different driving modes. *Atmos. Pollut. Res.* **2022**, *13*, 101597. [CrossRef]
11. Hatami, M.; Cuijpers, M.; Boot, M. Experimental optimization of the vanes geometry for a variable geometry turbocharger (VGT) using a Design of Experiment (DoE) approach. *Energy Convers. Manag.* **2015**, *106*, 1057–1070. [CrossRef]
12. Hosseinimaab, S.; Tousi, A. Optimizing the performance of a single-shaft micro gas turbine engine by modifying its centrifugal compressor design. *Energy Convers. Manag.* **2022**, *271*, 116245. [CrossRef]
13. Ekradi, K.; Madadi, A. Performance improvement of a transonic centrifugal compressor impeller with splitter blade by three-dimensional optimization. *Energy* **2020**, *201*, 117582. [CrossRef]
14. Ma, S.-B.; Afzal, A.; Kim, K.-Y. Optimization of ring cavity in a centrifugal compressor based on comparative analysis of optimization algorithms. *Appl. Therm. Eng.* **2018**, *138*, 633–647. [CrossRef]
15. Shaaban, S. Design optimization of a centrifugal compressor vaneless diffuser. *Int. J. Refrig.* **2015**, *60*, 142–154. [CrossRef]
16. Tüchler, S.; Chen, Z.; Copeland, C.D. Multipoint shape optimisation of an automotive radial compressor using a coupled computational fluid dynamics and genetic algorithm approach. *Energy* **2018**, *165*, 543–561. [CrossRef]
17. Zamiri, A.; Choi, M.; Chung, J.T. Effect of blade squealer tips on aerodynamic performance and stall margin in a transonic centrifugal compressor with vaned diffuser. *Aerosp. Sci. Technol.* **2022**, *123*, 107504. [CrossRef]
18. Aparna, K.; Sharath, R. Development of efficient compressor wheel for turbocharger by varying number of blades. *Mater. Today Proc.* **2022**, *62*, 198–202. [CrossRef]
19. Moussavi, S.A.; Benisi, A.H.; Durali, M. Effect of splitter leading edge location on performance of an automotive turbocharger compressor. *Energy* **2017**, *123*, 511–520. [CrossRef]
20. Mamala, J.J.; Praznowski, K.; Kołodziej, S.; Ligus, G. The Use of Short-Term Compressed Air Supercharging in a Combustion Engine with Spark Ignition. *Int. J. Automot. Mech. Eng.* **2021**, *18*, 8704–8713. [CrossRef]
21. Lu, D.; Theotokatos, G.; Zhang, J.; Zeng, H.; Cui, K. Parametric investigation of a large marine two-stroke diesel engine equipped with exhaust gas recirculation and turbocharger cut out systems. *Appl. Therm. Eng.* **2021**, *200*, 117654. [CrossRef]
22. Yang, M.; Hu, C.; Bai, Y.; Deng, K.; Gu, Y.; Qian, Y.; Liu, B. Matching method of electric turbo compound for two-stroke low-speed marine diesel engine. *Appl. Therm. Eng.* **2019**, *158*, 113752. [CrossRef]
23. Mousavi, S.; Nejat, A.; Alaviyoun, S.; Nejat, M. An Integrated Turbocharger Matching Program for Internal Combustion Engines. *J. Appl. Fluid Mech.* **2021**, *14*, 1209–1222.
24. Chen, Q.; Ni, J.; Wang, Q.; Shi, X. Match-based pseudo-MAP full-operation-range optimization method for a turbocharger compressor. *Struct. Multidiscip. Optim.* **2019**, *60*, 1139–1153. [CrossRef]
25. Huang, L.; Ma, C.; Li, Y.; Gao, J.; Qi, M. Improved Heat Transfer Correction Method for Turbocharger Compressor Performance Measurements. *IOP Conf. Ser. Mater. Sci. Eng.* **2019**, *470*, 012002. [CrossRef]
26. Wu, B.; Han, Z.; Yu, X.; Zhang, S.; Nie, X.; Su, W. A Method for Matching Two-Stage Turbocharger System and Its Influence on Engine Performance. *J. Eng. Gas Turbines Power* **2019**, *141*, 5–11. [CrossRef]
27. Li, X.; Zhao, Y.; Liu, Z.; Yao, H.; Chen, H. A new method for performance map prediction of automotive turbocharger compressors with both vaneless and vaned diffusers. *Proc. Inst. Mech. Eng. Part D J. Automob. Eng.* **2020**, *235*, 1734–1747. [CrossRef]
28. Payri, F.; Serrano, J.; Fajardo, P.; Reyes-Belmonte, M.; Gozalbo-Belles, R. A physically based methodology to extrapolate performance maps of radial turbines. *Energy Convers. Manag.* **2012**, *55*, 149–163. [CrossRef]
29. Liu, C.; Cao, Y.; Zhang, W.; Ming, P.; Liu, Y. Numerical and experimental investigations of centrifugal compressor BPF noise. *Appl. Acoust.* **2019**, *150*, 290–301. [CrossRef]
30. Nakayama, Y.; Boucher, R. Computational fluid dynamics. *Introd. Fluid Mech.* **1998**, *31*, 249–273.
31. Fan, H.; Ni, J.; Shi, X.; Jiang, N.; Zheng, Y.; Zheng, Y. Unsteady Performance Simulation Analysis of a Waste-Gated Turbocharger Turbine under Different Valve Opening Conditions. *SAE Tech. Pap.* **2017**, *1*, 2417. [CrossRef]
32. Yin, S.; Ni, J.; Fan, H.; Shi, X.; Huang, R. A Study of Evaluation Method for Turbocharger Turbine Based on Joint Operation Curve. *Sustainability* **2022**, *14*, 9952. [CrossRef]

33. Fan, H.; Ni, J.; Shi, X.; Qu, D.; Zheng, Y.; Zheng, Y. Simulation of a Combined Nozzled and Nozzleless Twin-Entry Turbine for Improved Efficiency. *J. Eng. Gas Turbines Power* **2019**, *141*, 051019. [CrossRef]
34. Nemati, A.; Cai, O.; Honoré, W. CFD analysis of combustion and emission formation using URANS and LES under large two-stroke marine engine-like conditions. *Appl. Therm. Eng.* **2022**, *216*, 119037. [CrossRef]
35. Galindo, J.; Fajardo, P.; Navarro, R.; García-Cuevas, L. Characterization of a radial turbocharger turbine in pulsating flow by means of CFD and its application to engine modeling. *Appl. Energy* **2013**, *103*, 116–127. [CrossRef]
36. Serrano, J.R.; Navarro, R.; García-Cuevas, L.M.; Inhestern, L.B. Turbocharger turbine rotor tip leakage loss and mass flow model valid up to extreme off-design conditions with high blade to jet speed ratio. *Energy* **2018**, *147*, 1299–1310. [CrossRef]
37. Cheng, L.; Dimitriou, P.; Wang, W.; Peng, J.; Aitouche, A. A novel fuzzy logic variable geometry turbocharger and exhaust gas recirculation control scheme for optimizing the performance and emissions of a diesel engine. *Int. J. Engine Res.* **2018**, *21*, 1298–1313. [CrossRef]
38. Han, G.; Yang, C.; Wu, S.; Zhao, S.; Lu, X. The investigation of mechanisms on pipe diffuser leading edge vortex generation and development in centrifugal compressor. *Appl. Therm. Eng.* **2023**, *219*, 119606. [CrossRef]
39. Liu, X.; Zhao, L. Approximate Nonlinear Modeling of Aircraft Engine Surge Margin Based on Equilibrium Manifold Expansion. *Chin. J. Aeronaut.* **2012**, *25*, 663–674. [CrossRef]
40. Zhao, B.; Zhou, T.; Yang, C. Experimental investigations on effects of the self-circulation casing treatment on acoustic and surge characteristics in a centrifugal compressor. *Aerosp. Sci. Technol.* **2022**, *131*, 108002. [CrossRef]
41. Chen, L.; Luo, J.; Sun, F.; Wu, C. Optimized efficiency axial-flow compressor. *Appl. Energy* **2005**, *81*, 409–419. [CrossRef]
42. Li, X.; Liu, Z.; Zhao, M.; Zhao, Y.; He, Y. Stability improvement without efficiency penalty of a transonic centrifugal compressor by casing treatment and impeller/diffuser coupling optimization. *Aerosp. Sci. Technol.* **2022**, *127*, 107685. [CrossRef]
43. Xing, K.; Huang, H.; Guo, X.; Wang, Y.; Tu, Z.; Li, J. Thermodynamic analysis of improving fuel consumption of natural gas engine by combining Miller cycle with high geometric compression ratio. *Energy Convers. Manag.* **2022**, *254*, 115219. [CrossRef]
44. Fontanesi, S.; Paltrinieri, S.; Cantore, G. CFD Analysis of the Acoustic Behavior of a Centrifugal Compressor for High Performance Engine Application. *Energy Procedia* **2014**, *45*, 759–768. [CrossRef]

 *sustainability*

Article

Effect of a Taper Intake Port on the Combustion Characteristics of a Small-Scale Rotary Engine

Run Zou [1,*], Yi Zhang [1], Jinxiang Liu [2], Wei Yang [1], Yangang Zhang [1], Feng Li [1] and Cheng Shi [3,*]

[1]   School of Energy and Power Engineering, North University of China, Taiyuan 030051, China; zhangyi@nuc.edu.cn (Y.Z.); yangwei2184@nuc.edu.cn (W.Y.); 20102507@nuc.edu.cn (Y.Z.); fengli@nuc.edu.cn (F.L.)
[2]   School of Mechanical Engineering, Beijing Institute of Technology, Beijing 100081, China; liujx@bit.edu.cn
[3]   School of Vehicle and Energy, Yanshan University, Qinhuangdao 066004, China
*   Correspondence: zourun@nuc.edu.cn (R.Z.); shicheng@ysu.edu.cn (C.S.)

Abstract: Taper intake ports are effective in improving the charging efficiency of small-scale rotary engines (REs), but it is unclear how their structural parameters affect the in-cylinder flow field and combustion characteristics. For this reason, the effects of the diameter-length ratio (D/L) of an intake port on the in-cylinder flow field and combustion characteristics of a small-scale RE were numerically investigated by utilizing a three-dimensional computational fluid dynamics (CFD) model. The results showed that the in-cylinder pressure of the RE did not follow a simple single-directional trend with the D/L of the intake port, but it was divided into three levels, where the peak in-cylinder pressure was at its maximum at the D/L of 0.6 and at its minimum at the D/L of 0.8. The gas flows in the intake port with different values of the D/L were all unidirectional, and they made a difference in the vortexes formed on the leading side of the combustion chamber of the RE, which was the main factor affecting the in-cylinder combustion performance. The vortexes formed on the leading side of the combustion chamber with $D/L = 0.6$ were maintained for a long period of time, thus promoting the propagation of flame and advancing the center of gravity of combustion. So, the heat release rate and combustion efficiency in the cylinder were increased at the price of a larger increment in nitrogen oxide formation.

Keywords: small-scale rotary engine; taper intake port; in-cylinder flow field; combustion characteristics

Citation: Zou, R. ; Zhang, Y.; Liu, J.; Yang, W.; Zhang, Y.; Li, F.; Shi, C. Effect of a Taper Intake Port on the Combustion Characteristics of a Small-Scale Rotary Engine. *Sustainability* **2022**, *14*, 15809. https://doi.org/10.3390/su142315809

Academic Editors: Firoz Alam

Received: 28 October 2022
Accepted: 24 November 2022
Published: 28 November 2022

Publisher's Note: MDPI stays neutral with regard to jurisdictional claims in published maps and institutional affiliations.

1. Introduction

In comparison with reciprocating piston engines, rotary engines (REs) possess many advantages, such as a simple structure, low vibration, low noise, high speed, high power-to-weight ratio, small frontal area, etc. [1,2]. These advantages make REs promising for applications in small unmanned aerial vehicles, electric vehicle range extenders, armored military vehicles, small ships, etc. [3,4]. However, REs also possess some drawbacks, such as poor sealing and high fuel consumption [5,6]. In particular, the elongated rotor chamber and the unidirectional flow field in the cylinder make it difficult for the flame to propagate to the trailing side of the combustion chamber, resulting in severely inadequate combustion [7]. To improve the combustion performance of REs, researchers have carried out some work from the perspectives of structural optimization, ignition scheme improvement, multi-fuel mixing, intake process optimization, etc. [8–10]. Unlike a traditional piston engine, an RE does not have intake valves and exhaust valves. Instead, the mixture enters the combustion chamber directly from the intake port and collides with the rotor recess wall. This collision affects the distribution of the in-cylinder flow field and the intensity of the turbulent kinetic energy (TKE), which, in turn, affects the efficiency of in-cylinder combustion. Therefore, the intake process of the RE is more significantly affected by the intake port structure.

The intake process of an RE is affected by a variety of factors of the intake port structure, mainly including the intake port's cross-sectional shape, its location, its structural size parameters (diameter, length, etc.), etc. Fan et al. [11] numerically investigated the effects of intake shape, including a round intake, rectangular intake, regular trapezoidal intake, and inverted trapezoidal intake, on the flow field in the cylinder of an RE, and they found that the inverted trapezoidal intake was the most conducive to improving the volume coefficient and holding a swirl structure. Ji et al. [12] studied the influences of a peripheral-ported intake, side-ported intake, and compound intake on the in-cylinder flow field, the flame propagation, and the emission formation of a gasoline RE. The results showed that the peripheral-ported intake had the fastest flame propagation and the highest combustion rate. At the moment of opening of the exhaust valve, the NOx emission was the lowest for the side-ported intake, and the CO emission was the lowest for the compound intake. Taskiran et al. [13] analyzed the effects of multiple side-ported intakes and exhaust ports on the flow field and combustion process in the combustion chamber of an RE and showed that the two side-ported intakes had the best charge coefficients and the fullest combustion of the mixtures. Jeng et al. [14] studied the influence of the length and diameter of an intake pipe and exhaust pipe with an equal cross-sectional area on the performance of an RE and concluded that the shorter the intake pipe and the longer the exhaust pipe, the higher the output power. These studies were carried out to assess the influence of intakes with an equal cross-section on the flow field and combustion performance, and this contributed to the improvement of the intake process of REs, but it did not involve studies of variable-section intake ports, such as a taper intake port. A taper intake port improved intake efficiency by increasing the differential pressure [15,16]. In general, a small-scale RE uses a taper intake port in order to achieve a high intake efficiency. The length and diameter of a taper intake port directly affect the intake resistance and intake pressure difference, which affects the mass flow rate and velocity of air entering the cylinder, and then influences the flow field distribution and combustion characteristics in the cylinder. However, due to the lack of studies on taper intake ports, it is unclear how the changes in intake charge and intake velocity caused by changes in the diameter and length of a taper intake port synthetically affect the in-cylinder flow field and combustion characteristics of a small-scaled RE.

In this paper, a three-dimensional computational fluid dynamics (CFD) model of a peripheral-intake RE was established in commercial software based on a reduced chemical reaction mechanism and the applicable turbulence model, and its accuracy was verified through a comparison with experimental results. Then, the influences of the diameter–length ratio of the taper intake port on the flow field, TKE, flame propagation, and combustion process inside the combustion chamber of the smal-scale RE were analyzed, which, in turn, provided theoretical guidelines for the design of the intake port of a small-scale RE.

2. Numerical Procedure

2.1. Engine Geometry and Computational Domain

The object of this study is a peripheral-ported-intake single-rotor engine with a dual spark plug ignition mode, whose model schematic and model parameters are shown in Figure 1 and Table 1, respectively.

Chamber III of the tested RE was selected as the calculation object in this study due to the identical working process of the three combustion chambers, which corresponded to a single rotor of the RE, and due to the consideration of the symmetrical similarity of the construction and working principle of the RE. The starting position of the calculation was 720°EA BTDC. In addition, to eliminate the influence of backflow on the exhaust flow, the exhaust port was extended in the calculation (as shown in the right diagram of Figure 1).

2.5. Model Validation

To ensure the simulation accuracy of the 3-D CFD simulation model of the RE established in this study, the parameter settings of the model need to be verified. Under operating conditions with an equivalent ratio of 1.25, a speed of 4500 rpm, and a wide-open throttle, a comparison was made between the simulation results and experimental results for the in-cylinder pressure, as shown in Figure 3. As can be seen from the figure, the in-cylinder pressure curve of the numerical simulation was slightly higher than the curve of the experimental results at the early stage of combustion (near the TDC). The main reasons for this were: (1) There were only two fuel components used in the simulation, which did not fully replace the multiple components of actual gasoline. (2) Leakage through the rotor tip was neglected in the simulation model. (3) The simplified chemical reaction mechanism of the alternative fuel that was adopted did not sufficiently consider the quenching phenomenon. Fortunately, the comparison showed a good agreement between the experiments and simulation with a deviation of less than 5%. The deviation values were within acceptable limits and on par with similar reference simulations from around the globe. Therefore, the 3-D CFD model established in this study can accurately predict the combustion process of a small-scale RE.

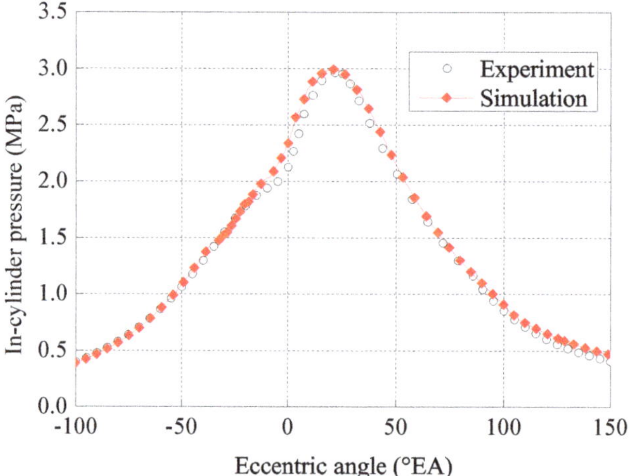

Figure 3. Comparison of the simulation results and experimental results.

3. Results and Discussion

3.1. Effect of D/L on In-Cylinder Pressure

Figure 4 shows the variation in the in-cylinder pressure according to the eccentric angle with different D/L parameters. It can be seen that when D/L was 0.6, the peak in-cylinder pressure was at its maximum, and when D/L was 0.8, the peak in-cylinder pressure was at its minimum. In addition, the in-cylinder pressure was divided into three levels in all calculated tests. The maximum in-cylinder pressure was found for D/L values of 0.2 and 0.6, the second highest was found for D/L values of 0.3, 0.5, 0.7, and 0.9, and the lowest was found for D/L values of 0.4 and 0.8. As illustrated in Figure 5, the peak in-cylinder pressure first decreased and then increased with the D/L parameter in a cyclic pattern, which was not a simple linear trend in a single direction. The eccentric angle corresponding to the peak pressure varied in proportion to the peak pressure. The above phenomenon was mainly caused by the influence of the change in the size of the taper intake port on the intake efficiency and the in-cylinder flow field of the small-scale RE.

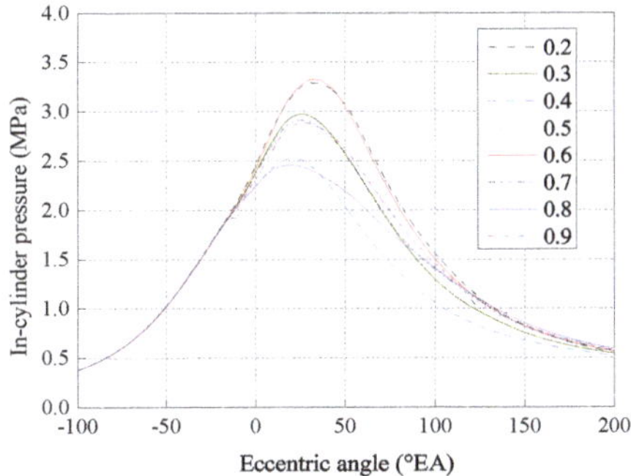

Figure 4. In-cylinder pressure versus eccentric angle with different *D/L* parameters.

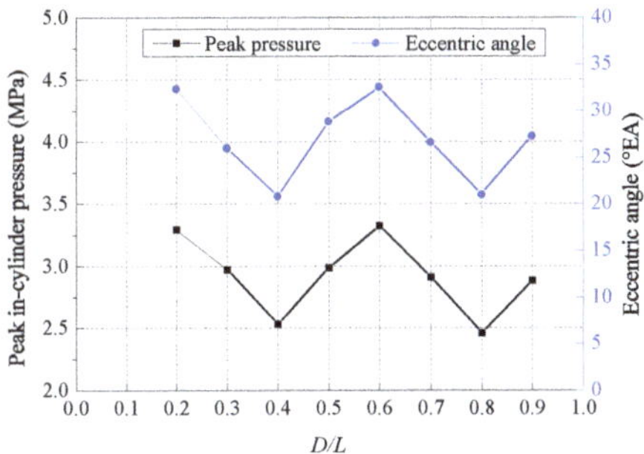

Figure 5. The peak in-cylinder pressure and relevant eccentric angle with different *D/L* parameters.

Figure 6 shows the results for the volumetric coefficient with different *D/L* parameters. From the figure, it can be seen that the volumetric coefficient showed a slightly increasing trend with the increase in *D/L*, except for the sudden increment in the volumetric coefficient for the intake port structure scheme with *D/L* = 0.4. This was mainly because as the *D/L* increased, the pressure difference between the constriction and expansion sections of the intake port increased, thus boosting the flow velocity of the mixtures in the constriction section and increasing the amount of air intake. However, the previous results showed that the peak in-cylinder pressure first decreased and then increases with the *D/L* parameter, rather than a simple change trend in a single direction, which indicated that the variation in the volumetric coefficient caused by the change in the *D/L* was not the main reason for the effect on the performance of the RE.

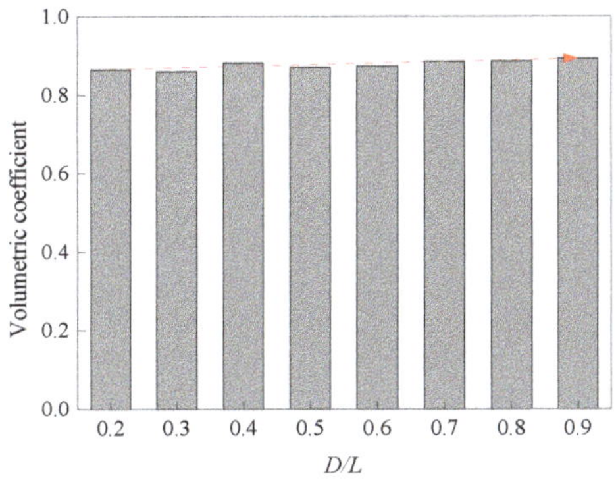

Figure 6. Volumetric coefficients with different *D/L* parameters.

To find out how the parameters of the taper intake port affected the engine performance, representative D/L parameters of 0.3, 0.6, and 0.8 at three pressure levels were selected for a subsequent analysis of the in-cylinder flow field and combustion characteristics of the small-scale RE.

3.2. Effect of D/L on the In-Cylinder Flow Field Characteristics

The change in the parameters of the taper intake port inevitably caused variations in the intake charge and intake flow velocity, which, in turn, affected the evolution of the in-cylinder flow field. Figure 7 displays the distribution of the velocity field and streamlines in the central plane of the combustion chamber at different moments for different values of D/L. It can be clearly seen that the flow field in the intake port was unidirectional at all times, without backflows or vortexes, which was the same as the results that were obtained by other researchers [16]. At the early stage of the intake stroke (500°EA BTDC), the mixtures from the intake port impinged on the rotor wall and moved in the direction of rotation, and there was no vortex phenomenon in the cylinder and a lower flow velocity in the intake port. This was because at the initial stage of the intake stroke, the smaller volume of the combustion chamber limited the development of vortexes, and the pressure difference between the combustion chamber and the intake port was minor, resulting in a low flow velocity in the intake port. At the median stage of the intake stroke (410°EA BTDC), one vortex was formed on the leading side and one on the trailing side of the combustion chamber. However, the vortex on the leading side of the combustion chamber for $D/L = 0.8$ was in the initial formation stage, and its intensity was lower than that of the vortex for D/L values of 0.3 and 0.6. In addition, as the D/L increased from 0.3 to 0.8, the velocity at the entrance of the combustion chamber was gradually enhanced. This was due to an increment in the contraction ratio at the throat of the taper intake port as the D/L increased. At the late stage of the intake stroke (360°EA BTDC), the vortex on the leading side of the combustion chamber with a D/L of 0.3 started to disappear, while the vortex on the leading side of the combustion chamber with D/L values of 0.6 and 0.8 showed an increasing trend in comparison with the median stage of the intake stroke. This was because the increase in the flow velocity at the combustion chamber's entry caused by the rise of the D/L boosted the effect of the in-cylinder mixtures on the rotor walls. This increased the intensity of the mixture flow after impacting the rotor walls and enhanced the intensity and duration of the vortex on the leading side of the combustion chamber. From the results given above, it can be seen that the flow velocity in the intake port with a D/L of 0.6 enabled the vortex on the leading side of the combustion chamber to form early and disappear late, indicating that its vortex duration was longer, which was conducive to the formation of homogeneous mixtures in the combustion chamber. In fact, similar results were obtained by other researchers for the above rules of vortex evolution in the intake stroke [11,13].

Figure 8 depicts the variation curves of the TKE according to the eccentric angle with D/L values of 0.3, 0.6, and 0.8. It can be concluded from Figure 8 that at the intake stroke (600°EA~250°EA BTDC), the TKE in the cylinder for the D/L values of 0.6 and 0.8 was basically the same, while the TKE for $D/L = 0.3$ was higher than that of both of the other D/L values. This was because the smaller flow velocity in the intake port with $D/L = 0.3$ enabled vortexes on the leading side of the combustion chamber to form early, resulting in higher TKE in the cylinder at the early stage. However, during the whole compression stroke (250°EA BTDC~0°EA), the TKE for $D/L = 0.8$ was lower than that of the other two intake ports. This was because the vortexes on the leading side of the combustion chamber for $D/L = 0.8$ formed late and were maintained for a short time, resulting in a lower TKE in the cylinder at the later stage. In addition, the TKE in the cylinder during the compression stroke was essentially the same for the D/L values of 0.3 and 0.6, but after the TDC, the TKE in the cylinder for $D/L = 0.3$ was lower than that for $D/L = 0.6$.

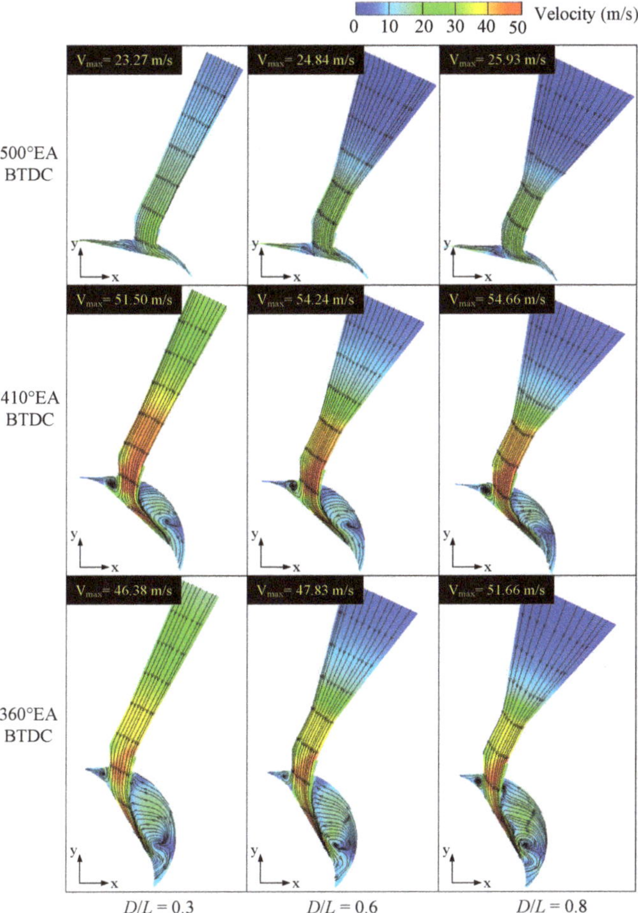

Figure 7. The distribution of the velocity field and streamlines in the intake port and the cylinder.

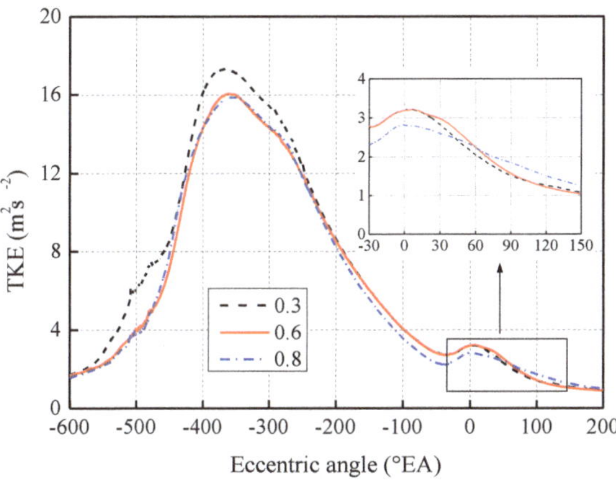

Figure 8. The variations in the TKE in the cylinder with the eccentric angle.

3.3. Effect of D/L on Combustion Characteristics

A difference in the intake port structure leads to changes in the in-cylinder flow field, which will certainly affect the development and propagation of the flame inside the combustion chamber of an RE. Figure 9 illustrates the influence of variations in the D/L on the flame propagation and velocity streamlines in the cylinder at 21°EA BTDC, TDC, 15°EA ATDC, and 48°EA ATDC. It can be seen that at the early stage of combustion, the flame propagation was significantly faster for D/L values of 0.3 and 0.6 than that for $D/L = 0.8$, and the flame spread area was the largest for $D/L = 0.6$. This was because the higher TKE in the cylinder for the D/L values of 0.3 and 0.6 promoted flame propagation at the early stage of combustion. After the TDC, the flame propagation area formed at the D/L of 0.6 was the widest, and the spread of the flame to the front and rear end caps partially disappeared at 48°EA ATDC. This showed that the intake port with $D/L = 0.6$ had the fastest flame propagation velocity and the highest combustion rate. This was attributed to the maximum in-cylinder TKE for $D/L = 0.6$ during the rapid combustion period. In addition, it could also be seen from Figure 9 that the flame was propagated from the spark plug position to the leading side of the combustion chamber, but it was difficult for it to propagate to the trailing side of the combustion chamber due to the unidirectional flow field, which was the main reason for the high fuel consumption and poor emission of the RE. This finding was also obtained by other researchers [8].

The flame propagation and combustion rate were influenced by the chemical kinetic reaction mechanism of the gasoline during the combustion process, as well as intermediate element reactions, i.e., $OH + H_2 \rightleftharpoons H + H_2O$ and $H + O_2 \rightleftharpoons O + OH$, which directly determined the combustion reaction rate [24]. Figure 10 shows the peak mass fractions of H, O, and OH under different intake port schemes. It can be observed from the figure that when $D/L = 0.6$, the concentration of active OH radicals produced in the combustion chamber was the highest, which is 53.6% higher than the level with 0.3 and 7.5% higher than the level with 0.8. The increase in OH concentration enabled the concentration of H and O to increase, and this then enhanced the fuel combustion rate. In addition, the heat release of the combustion reaction was mainly generated by the intermediate element reaction of $CO + OH \rightleftharpoons H + CO_2$ [25]. Therefore, the increment in the OH production also increased the rate and amount of heat release in the combustion reactions in the combustion chamber, which, in turn, boosted the in-cylinder temperature. This indicated that the intake port with $D/L = 0.6$ had the maximum heat release and the highest temperature.

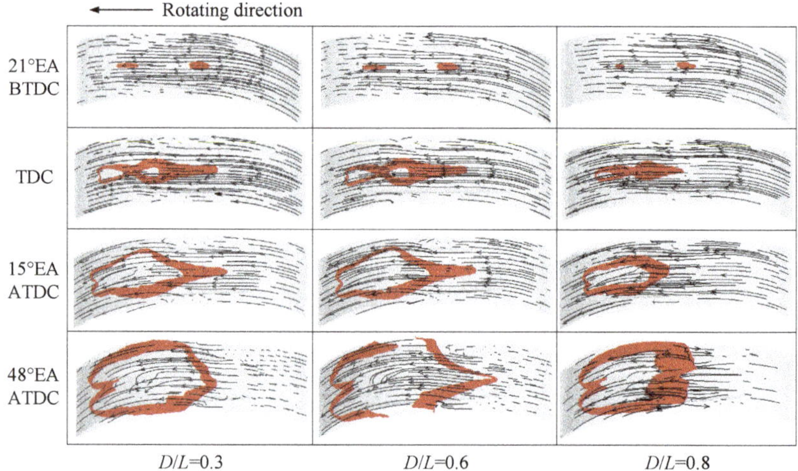

Figure 9. Flame position and velocity streamlines.

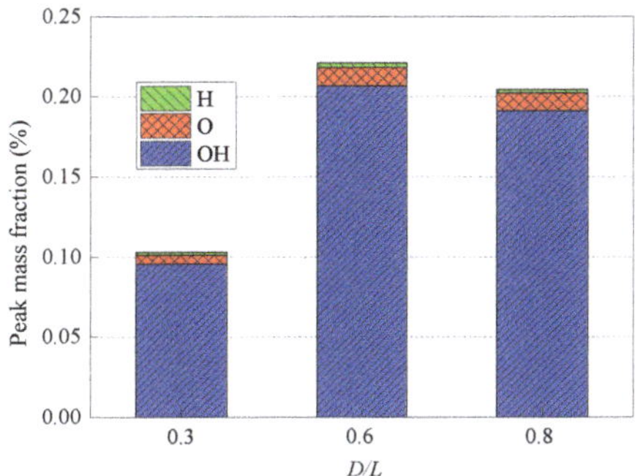

Figure 10. Peak mass fractions of H, O, and OH.

The variations in the burned mass fraction in the cylinder according to eccentric angle with different intake port structures are shown in Figure 11. According to the figure, it is easy to notice that the in-cylinder heat release was sequentially delayed for the D/L values of 0.6, 0.8, and 0.3. The intake port with the D/L of 0.6 had the earliest and fastest heat release in the cylinder. The eccentric angle for reaching 90% of the burned mass fraction for the intake port with the D/L of 0.6 was 78.85°EA ATDC, while the eccentric angles for achieving 90% of the burned mass fraction for the D/L values of 0.3 and 0.8 were 120°EA ATDC and 97.98°EA ATDC, respectively. This indicated that the intake port with $D/L = 0.6$ was the most conducive to promoting chemical reactions and increasing the rate of heat release. This is due to the fact that the intake port with $D/L = 0.6$ enhanced the local flame propagation rate and shortened the combustion duration.

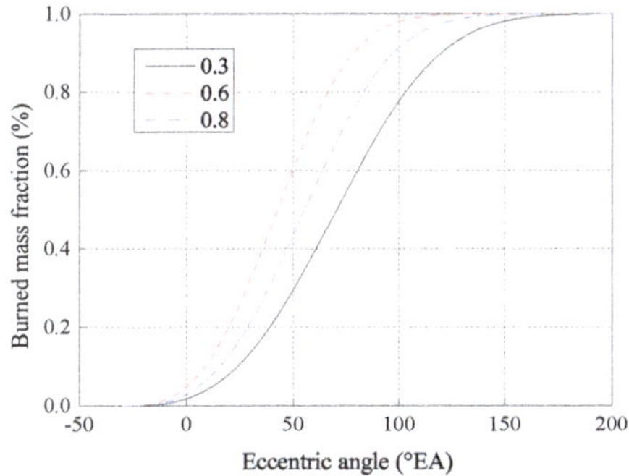

Figure 11. Variations in the burned mass fraction with the eccentric angle for different intake port structures.

A comparison of the combustion durations with different intake port structures is shown in Figure 12. EA0, EA50, and EA 90 indicate an eccentric angle corresponding to 0%, 50%, and 90% of the cumulative heat release from fuel combustion in the combustion chamber, respectively. Typically, EA0–10 and EA10–90 denote the flame development and flame propagation of burning mixtures, respectively, and EA50 denotes the combustion efficiency of the fuel. It can be observed from Figure 12 that the values of EA0–10 for the D/L values of 0.3 and 0.6 were significantly smaller than that with the D/L of 0.8, indicating that the flames for the D/L values of 0.3 and 0.6 developed more rapidly at the initial stage of combustion (as shown in Figure 9). Meanwhile, the value of EA10–90 for $D/L = 0.6$ was decreased by 33.49% and 13.04% in comparison with those for the D/L values of 0.3 and 0.8, respectively, indicating that the intake port with $D/L = 0.6$ was able to improve the flame propagation speed and promote the combustion reaction in the cylinder. This was mainly due to the fact that the intake port with $D/L = 0.6$ enabled the vortexes in the cylinder to form earlier and last longer, and the TKE was higher during the combustion stage, which, in turn, promoted mixture formation and flame propagation in the combustion chamber. Furthermore, the value of EA50 for D/L = 0.6 was significantly smaller than those for the D/L values of 0.3 and 0.8, indicating that the in-cylinder combustion's center of gravity was advanced for D/L = 0.6. This improved the combustion efficiency and reduced combustion losses, which is conducive to more useful work output.

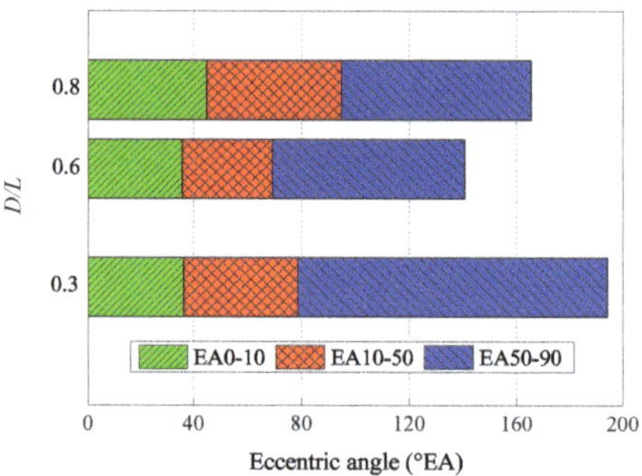

Figure 12. Comparison of the combustion durations with different intake port structures.

From the in-cylinder mean temperature in Figure 13 and the NO generation curve in Figure 14, it can be seen that the rate and amount of NO generation were consistent with the trend of variation in the in-cylinder mean temperature, which was attributed to the fact that the conditions for NO generation are a high temperature and oxygen-rich atmosphere. Near the TDC, due to the slow flame development and propagation speed (as shown in Figure 9), the temperature in the cylinder changed slowly, and a smaller amount of NO is generated. As the engine operated, the fuels burned rapidly, and the flame propagation speed and heat release increased rapidly, which caused the in-cylinder temperature to rise sharply and drove the rapid generation of NO. For the intake port with D/L = 0.6, the NO generation rate was the fastest, and its peak mass fraction of NO was increased by 197% and 104% in comparison with those for 0.3 and 0.8, respectively. This reason is that, on the one hand, the form of the movement of the mixtures entering the cylinder from the intake port with D/L = 0.6 accelerated the flame propagation (as shown in Figure 9) and improved the fuel combustion rate and heat release rate (as shown in Figure 11), which accelerated the NO generation rate. On the other hand, after the NO generation in the combustion chamber reached the maximum value, it kept a relatively stable value until the exhaust valve was opened.

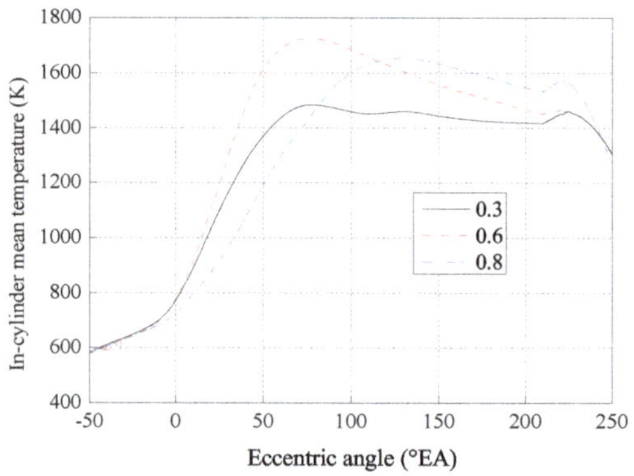

Figure 13. Comparison of the in-cylinder mean temperature for different intake port structures.

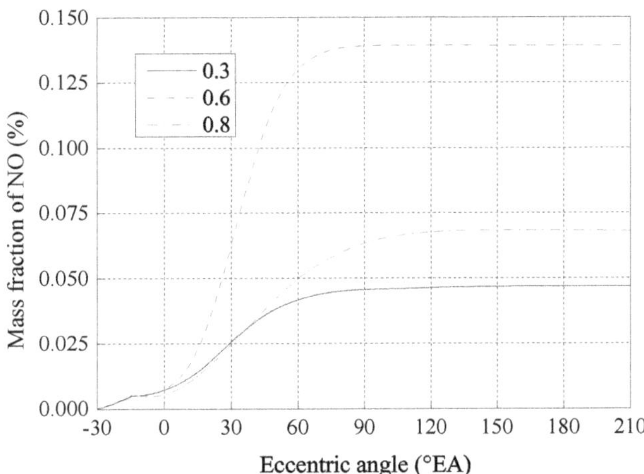

Figure 14. Comparison of the mass fraction of NO for different intake port structures.

It can also be concluded from Figure 14 that the amount of NO generated for the intake port with $D/L = 0.8$ was higher than that for $D/L = 0.3$. This was mainly due to the fact that when $D/L = 0.8$, the heat release of fuel combustion was faster (as shown in Figure 11), and the combustion duration was shorter (as shown in Figure 12), which led to a higher peak in-cylinder temperature and promoted the generation of NO. However, this problem can be improved through the post-treatment of exhaust and lean combustion.

4. Conclusions

In this work, a three-dimensional CFD simulation model of a small-scale RE based on a suitable turbulent model and a reduced gasoline chemical reaction mechanism was established. The effects of the taper intake port structure on the flow field distribution and combustion process in the combustion chamber of the small-scale RE were numerically analyzed. The main conclusions that were obtained are as follows:

(1) The in-cylinder pressure of the RE did not change linearly with the D/L of the intake port in a single direction, but was divided into three levels. The maximum in-cylinder pressure was at the D/L of 0.6, and the minimum in-cylinder pressure was at the D/L of 0.8. In addition, the eccentric angle corresponding to the peak in-cylinder pressure was consistent with the trend of the variation in the peak pressure.

(2) The gas flow in the intake port of the three representative pressure levels was unidirectional, which caused differences in the vortexes that were formed in the combustion chamber of the RE, and this was mainly reflected in the vortexes that were formed on the leading side of the combustion chamber. This was also the main factor affecting the combustion performance of the small-scale RE. When the D/L was 0.6, the vortexes on the leading side of the combustion chamber formed early and disappeared late, indicating that the vortexes in the combustion chamber lasted longer, and this was conducive to mixture formation.

(3) The intake port with the D/L of 0.6 promoted the in-cylinder combustion reaction and improved the combustion efficiency of the RE due to the higher flame propagation speed and the earlier center of gravity of combustion, which reduced the combustion losses and facilitated a more useful work output. However, owing to its higher peak in-cylinder temperature, it led to an increase in NO emissions. As for the high levels of emission of NO, these could be reduced through the post-treatment of exhaust and lean combustion.

Author Contributions: Conceptualization, R.Z. and J.L.; methodology, R.Z. and C.S.; software, J.L. and C.S.; validation, Y.Z. (Yi Zhang) and F.L.; formal analysis, Y.Z. (Yi Zhang); investigation, R.Z. and J.L.; resources, W.Y. and Y.Z. (Yangang Zhang); data curation, W.Y.; writing—original draft preparation, R.Z.; writing—review and editing, Y.Z. (Yi Zhang), J.L. and C.S.; visualization, R.Z. and F.L.; supervision, W.Y.; project administration, J.L.; funding acquisition, R.Z. and J.L. All authors have read and agreed to the published version of the manuscript.

Funding: This research was supported by the Applied Basic Research Programs of Shanxi Province in China (20210302123069; 20210302123072).

Data Availability Statement: Not applicable.

Conflicts of Interest: The authors declare no conflict of interest.

Abbreviations

CFD	Computational fluid dynamics
RE	Rotary engine
TKE	Turbulent kinetic energy
EA	Eccentric angle
TDC	Top dead center
BTDC	Before top dead center
ATDC	After top dead center
BDC	Bottom dead center
BBDC	Before bottom dead center
ABDC	After bottom dead center
RON	Research octane number
AMR	Adaptive mesh refinement
OH	Hydroxy
CO	Carbon monoxide
CO_2	Carbon dioxide
NO	Nitric oxide

References

1. Ribau, J.; Silva, C.; Brito, F.P.; Martins, J. Analysis of four-stroke, Wankel, and microturbine based range extenders for electric vehicles. *Energy Convers. Manag.* **2012**, *58*, 120–133. [CrossRef]
2. Ma, J.; Bowen, R.; Allen, R.; Gu, H. Mathematical simulation and study of the transient performance of a rotary engine. *SAE Trans.* **1993**, *102*, 2051–2061.
3. Siadkowska, K.; Wendeker, M.; Majczak, A.; Baranski, G.; Szlachetka, M. *The Influence of Some Synthetic Fuels on the Performance and Emissions in a Wankel Engine*; SAE Technical Paper; Society of Automotive Engineers (SAE): Warren, PA, USA, 2014. [CrossRef]
4. Fan, B.; Pan, J.; Yang, W.; Chen, W.; Bani, S. The influence of injection strategy on mixture formation and combustion process in a direct injection natural gas rotary engine. *Appl. Energy* **2017**, *187*, 663–674. [CrossRef]
5. Fan, B.; Zhang, Y.; Pan, J.; Wang, Y.; Otchere, P. Experimental and numerical study on the formation mechanism of flow field in a side-ported rotary engine considering apex seal leakage. *J. Energy Resour.* **2020**, *143*, 022303. [CrossRef]
6. Amrouche, F.; Erickson, P.A.; Varnhagen, S.; Park, J.W. An experimental study of a hydrogen-enriched ethanol fueled Wankel rotary engine at ultra Lean and full load conditions. *Energy Convers. Manag.* **2016**, *123*, 174–184. [CrossRef]
7. Shi, C.; Zhang, Z.; Ji, C.; Li, X.; Di, L.; Wu, Z. Potential improvement in combustion and pollutant emissions of a hydrogen-enriched rotary engine by using novel recess configuration. *Chemosphere* **2022**, *299*, 134491. [CrossRef] [PubMed]
8. Shi, C.; Zhang, P.; Ji, C.; Di, L.; Zhu, Z.; Wang, H. Understanding the role of turbulence-induced blade configuration in improving combustion process for hydrogen-enriched rotary engine. *Fuel* **2022**, *319*, 123807. [CrossRef]
9. Ji, C.; Meng, H.; Wang, S.; Wang, D.; Yang, J.; Shi, C.; Ma, Z. Realizing stratified mixtures distribution in a hydrogen-enriched gasoline Wankel engine by different compound intake methods. *Energy Convers. Manag.* **2020**, *203*, 112230. [CrossRef]
10. Finkelberg, L.; Kostuchenkov, A.; Zelentsov, A.; Minin, V. Improvement of combustion process of spark-ignited aviation Wankel engine. *Energies* **2019**, *12*, 2292. [CrossRef]
11. Fan, B.; Pan, J.; Yang, W.; An, H.; Tang, A.; Shao, X.; Xue, H. Effects of different parameters on the flow field of peripheral ported rotary engines. *Eng. Appl. Comp. Fluid Mech.* **2015**, *9*, 445–457. [CrossRef]
12. Ji, C.; Ma, Z.; Shi, C.; Wang, F.; Yang, J.; Meng, H. The influence of inlet structure on the combustion process of gasoline rotary engine. *J. Beijing Univ. Technol.* **2020**, *46*, 393–401.
13. Taskirana, O.; Calik, A.; Kutlarb, O. Comparison of flow field and combustion in single and double side ported rotary engine. *Fuel* **2019**, *254*, 115651. [CrossRef]

14. Jeng, D.-Z.; Hsieh, M.-J.; Lee, C.-C.; Han, Y. *The Intake and Exhaust Pipe Effect on Rotary Engine Performance*; SAE Technical Paper; Society of Automotive Engineers (SAE): Warren, PA, USA, 2013.

15. Smart, M.K. Design of three-dimensional hypersonic inlets with rectangular-to-elliptical shape transition. *J. Propul. Power* **1999**, *15*, 408–419. [CrossRef]

16. Yue, L.; Xiao, Y.; Chen, L.; Chang, X. Design of base flow for streamline-traced hypersonic inlet. In Proceedings of the 16th AIAA/DLR/DGLR International Space Planes and Hypersonic Systems and Technologies Conference, Bremen, Germany, 19–22 October 2009. [CrossRef]

17. Shi, C.; Chai, S.; Di, L.; Ji, C.; Ge, Y.; Wang, H. Combined experimental-numerical analysis of hydrogen as a combustion enhancer applied to Wankel engine. *Energy* **2023**, *263*, 125896. [CrossRef]

18. Ra, Y.; Reitz, R. A reduced chemical kinetic model for IC engine combustion simulations with primary reference fuels. *Combust. Flame* **2008**, *155*, 713–738. [CrossRef]

19. Zou, R.; Liu, J.; Wang, N.; Jiao, H. Combined effects of intake oxygen enrichment, intake pressure and temperature on combustion behavior and emission characteristics of a small-scaled rotary engine. *Appl. Therm. Eng.* **2022**, *207*, 118096. [CrossRef]

20. Wang, Z.; Wang, F.; Wang, J. *Study of Engine Knock in HCCI Combustion Using Large Eddy Simulation and Complex Chemical Kinetics*; SAE Technical Paper; Society of Automotive Engineers (SAE): Warren, PA, USA, 2014.

21. Senecal, P.K.; Pomraning, E.; Richards, K.J.; Briggs, T.E.; Choi, C.Y.; McDavid, R.M.; Patterson, M.A. *Multi-Dimensional Modeling of Direct-Injection Diesel Spray Liquid Length and Flame Lift-Off Length Using CFD and Parallel Detailed Chemistry*; SAE Technical Paper; Society of Automotive Engineers (SAE): Warren, PA, USA, 2003.

22. Spreitzer, J.; Zahradnik, F.; Geringer, B. *Implementation of a Rotary Engine (Wankel Engine) in a CFD Simulation Tool with Special Emphasis on Combustion and Flow Phenomena*; SAE Technical Paper; Society of Automotive Engineers (SAE): Warren, PA, USA, 2015.

23. Zou, R.; Liu, J.; Jiao, H.; Zhao, J.; Wang, N. Combined effect of intake angle and chamber structure on flow field and combustion process in a small-scaled rotary engine. *Appl. Therm. Eng.* **2022**, *203*, 117652. [CrossRef]

24. Pan, M.; Huang, R.; Liao, J.; Ouyang, T.; Zheng, Z.; Lv, D.; Huang, H. Effect of EGR dilution on combustion, performance and emission characteristics of a diesel engine fueled with n-pentanol and 2-ethylhexyl nitrate additive. *Energy Convers. Manag.* **2018**, *176*, 246–255. [CrossRef]

25. Lee, B.J.; Im, H.G. Dynamics of bluff-body-stabilized lean premixed syngas flames in a meso-scale channel. *Proc. Combust. Inst.* **2017**, *36*, 1569–1576. [CrossRef]

sustainability

MDPI

Article

Encoder–Decoder-Based Velocity Prediction Modelling for Passenger Vehicles Coupled with Driving Pattern Recognition

Diming Lou [1], Yinghua Zhao [1], Liang Fang [1,*], Yuanzhi Tang [1] and Caihua Zhuang [2]

[1]   College of Automotive Studies, Tongji University, Shanghai 201804, China
[2]   Propulsion Control and Software Engineering Department, SAIC MOTOR, Shanghai 201804, China
*   Correspondence: fangliang@tongji.edu.cn; Tel.: +86-021-69589207

Abstract: To improve the performance of predictive energy management strategies for hybrid passenger vehicles, this paper proposes an Encoder–Decoder (ED)-based velocity prediction modelling system coupled with driving pattern recognition. Firstly, the driving pattern recognition (DPR) model is established by a K-means clustering algorithm and validated on test data; the driving patterns can be identified as urban, suburban, and highway. Then, by introducing the encoder–decoder structure, a DPR-ED model is designed, which enables the simultaneous input of multiple temporal features to further improve the prediction accuracy and stability. The results show that the root mean square error ($RMSE$) of the DPR-ED model on the validation set is 1.028 m/s for the long-time sequence prediction, which is 6.6% better than that of the multilayer perceptron (MLP) model. When the two models are applied to the test dataset, the proportion with a low error of 0.1~0.3 m/s is improved by 4% and the large-error proportion is filtered by the DPR-ED model. The DPR-ED model performs 5.2% better than the MLP model with respect to the average prediction accuracy. Meanwhile, the variance is decreased by 15.6%. This novel framework enables the processing of long-time sequences with multiple input dimensions, which improves the prediction accuracy under complicated driving patterns and enhances the generalization-related performance and robustness of the model.

Keywords: passenger vehicle; velocity prediction; encoder–decoder; driving pattern recognition

Citation: Lou, D.; Zhao, Y.; Fang, L.; Tang, Y.; Zhuang, C. Encoder–Decoder-Based Velocity Prediction Modelling for Passenger Vehicles Coupled with Driving Pattern Recognition. *Sustainability* **2022**, *14*, 10629. https://doi.org/10.3390/su141710629

Academic Editors: Cheng Shi, Jinxin Yang, Jianbing Gao and Peng Zhang

Received: 26 July 2022
Accepted: 23 August 2022
Published: 26 August 2022

Publisher's Note: MDPI stays neutral with regard to jurisdictional claims in published maps and institutional affiliations.

1. Introduction

Energy management strategies are used to determine the power distribution between different energy sources at each moment to improve the performance of hybrid vehicles while satisfying torque requirements and are one of the key technologies of hybrid vehicles [1]. Up to now, scholars have conducted a great deal of research on energy management strategies, which are mainly divided into three major categories: rule-based, optimization-based, and learning-based [2–4]. Rule-based energy management strategies (RB-EMSs) are designed with suitable control rules for energy distribution according to the characteristics of the controlled object. The process of design depends mainly on the designers' knowledge and continuous exploration and experimentation. Moreover, the differences in the control rules between different models require repeated calibrations and cause difficulty with respect to obtaining optimal results. RB-EMSs are widely used due to their simple logic, easy implementation, and high stability [5]. As its research progressed, the addition of fuzzy logic improved the performance of RB-EMSs [6]; however, on the one hand, researchers continue to pursue higher requirements for all aspects of hybrid vehicle performance, and on the other hand, the disadvantages of RB-EMS will gradually appear due to its inherent poor portability, cumbersome tuning of parameters, and inability to achieve optimal control. Optimization-based energy management strategies use optimization algorithms for energy management, which are less empirically dependent than RB-EMSs and can achieve better results. Instantaneous optimization algorithms are those algorithms that do not need to know all the path-related information and can be optimized using only the current information, including the equivalent fuel consumption minimization strategy (ECMS) [7,8] and adaptive equivalent fuel consumption minimization strategy (A-ECMS) [9,10]. The optimization effect of both ECMS and A-ECMS are closely related to the equivalence factor and often have some distance from the global optimal solution. The global optimization algorithm represented by dynamic programming (DP) can obtain a global optimal solution, but at the same time, DP needs to anticipate the vehicle-state information of the entire future driving cycle in advance. The optimal strategy is obtained only for that particular cycle. Therefore, given the complex and variable driving conditions in the real world, such energy management strategies cannot be applied in practice and are often used as references for evaluating and optimizing other energy management strategies [11,12]. With the development of energy management, algorithms such as model predictive control (MPC) obtain locally optimal solutions by continuous roll-forward optimization within the prediction-sight distance, which are neither short-sighted nor sensitive compared to instantaneous optimization algorithms. An advanced knowledge of all the future driving information is not required compared to global optimization algorithms, so it has more potential for real-vehicle applications [13]. With the development of artificial intelligence technology, learning-based algorithms based on Reinforcement Learning (RL) have gained widespread attention, as they can learn energy management criteria from training samples [14], be applied to different driving conditions [15], adapt to different prediction sight distances [16], require no discretization of state nor control parameters [17], and provide the possibility of finding a theoretical global optimal solution [18]. In summary, short-term predictive energy management strategies that obtain locally optimal solutions by rolling the prediction sight distance are the focus of the current research. Among them, the short-term vehicle velocity prediction algorithm is the top priority of the predictive energy management strategies, which greatly affects the performance of these strategies [19,20], and the velocity prediction algorithm also plays a crucial role in the fields of traffic flow prediction and active vehicle safety control [21].

The mainstream of short-term vehicle velocity prediction methods is divided into stochastic and deterministic vehicle velocity prediction algorithms [22]. The Markov chain-based vehicle velocity prediction algorithm is the most representative stochastic vehicle velocity prediction algorithm [23,24]. Although the accuracy of vehicle velocity prediction can be improved by establishing a multi-level Markov chain model, it leads to an exponential expansion of the size of the probability matrix, which leads to an exponential rise in the computation time of the algorithm, and even so, it is still difficult to cover all possible Markov states [25]. Deterministic velocity prediction algorithms can be further divided into parametric and non-parametric velocity prediction algorithms. A parametric vehicle velocity prediction algorithm refers to the prediction by establishing a model with parameters, and the typical case is to establish an auto-regressive moving average model (ARMA) for a time series analysis, which is proven to have a wide range of engineering-control application cases [26]. However, for vehicles driving on roads, the randomness of the travel path and the lack of information sharing of individual vehicles leads to limitations on the ability of parametric vehicle velocity prediction models, causing larger prediction errors than non-parametric vehicle velocity prediction algorithms [27]. Instead, non-parametric velocity prediction algorithms use historical data to build a prediction model for future velocity prediction, so they are also called data-driven velocity prediction algorithms and currently represent a growing trend [28]. The neural network model is a typical data-driven algorithm due to its powerful nonlinear mapping capability and strong robustness. It has become one of the key technologies in the field of predictive modeling and optimal control of nonlinear systems [29] and is also widely applied to the field of vehicle velocity prediction. Scholars have conducted in-depth research on neural network-based vehicle velocity prediction methods and fruitful results have been achieved. Artificial neural networks (ANN) [16], long short-term memory neural networks (LSTM) [30], nonlinear autoregressive neural networks with additional inputs (NARX) [31], and RBFNN [32] are several neural networks that have been mostly studied in the field of vehicle velocity prediction. Among them, LSTM networks are widely used in time series prediction models and can maintain the internal input memory and compensate for the gradient vanishing and gradient explosion that exist in RNN during the training process [33]. In the field of vehicle velocity prediction, LSTM networks are currently used as the "Encoder" structure with a multi-step input and single-step output, and the "Decoder" structure with a single-step input and multi-step output. The former can achieve the effect of sequence prediction by combining the output as a new input, but the velocity of computation needs to be improved; the latter can achieve sequence prediction directly, but the structure lacks memory for the information of long history sequences.

Up to now, there are still few relevant studies on the sequence-to-sequence vehicle velocity prediction model established by the encoder-decoder framework. Nevertheless, with the continuous development of vehicle-networking technology, the information interaction data provided by vehicle networking can also be used jointly with neural networks in the future [34], and the relevant information used as the input of neural networks can also improve the vehicle velocity prediction accuracy and robustness. Therefore, it is necessary to provide a solution for the encoder–decoder-based time series prediction models with different input and output dimensions. In this paper, an innovative encoder–decoder-based velocity prediction modelling system for passenger vehicles coupled with driving pattern recognition is proposed. The model uses the encoder–decoder framework with LSTM units as neurons to establish a multi-input sequence-to-sequence model based on the characteristics of vehicle velocity sequences and the results of driving pattern recognition. Moreover, with the continuous development of the Internet of vehicles, an increasing amount of time-domain information can be accessed by vehicles. The multi-input feature of this model can improve the efficacy of velocity prediction by rearranging the input samples and retraining them as a "skeleton", which is very rare in the existing similar model. Section 2 introduces the establishment of the method of the driving-pattern-recognition model and recognition performance as the foundation for the following velocity prediction model. Section 3 proposes the velocity prediction based on the Encoder–decoder model, including the revised structure under the different dimensions of input and output sequences and the training method. The effects of the history window, prediction window, and number of neurons on the model performance are analyzed in Section 4, and the performance of the optimal model is compared with that of the traditional neural network. Finally, in Section 5, conclusions are presented.

2. Driving Pattern Recognition

This section will take the accuracy of recognition, the real-time calculation speed, and the compilation feasibility into consideration and find a method that can use the current and historical driving information for driving pattern recognition. In this section, the K-means clustering algorithm is used for driving pattern recognition. Firstly, the training matrix is obtained with standard driving cycles, and the K-means clustering algorithm model is trained. Then, the model is used to recognize the driving patterns in the test data and the results are shown.

2.1. Dataset and Characteristics

One of the keys to recognizing driving patterns is to find a set of features that can represent certain characteristics of driving pattern information, such as the average velocity of the past period of time, idle ratio, vehicle acceleration, etc., and they are used as the basis for recognizing the current driving patterns of the vehicle. In recent years, the selection of features to reflect the characteristics of driving patterns, which can then be applied to the process of driving pattern recognition, has been a hot spot of research, in which the difficulty lies in the appropriate number and type of features to reflect the characteristics of driving patterns. As far as the number of features is concerned, selecting too many features or combining them with the method of principal component analysis for driving pattern recognition may lead to an increase in the difficulty of real-time online analysis, whereas selecting too few features will not accurately reflect the characteristics of driving patterns and lead to the poor accuracy of driving pattern recognition [35]. Hu et al. selected four features—to name three: the average velocity, idle ratio, and cruising time ratio—for driving pattern recognition, which ensured the accuracy of the model and greatly improved the computational real time [36].

In view of the above research results, three features have been selected in this paper to balance the real-time properties, accuracy, and compilation feasibility of driving pattern recognition, including the average velocity (v_mean), maximum velocity (v_max), and idle ratio (T_idle).

After the selection of the features, typical driving cycles need to be selected as the training set; six standard driving cycles are selected in this paper. The features of each standard driving cycle are calculated separately and used to train the K-means clustering model, so that the model can realize the real-time recognition of any driving patterns in actual driving. Standard driving cycles are mostly designed for the driving pattern s in their respective regions. In order to enhance the adaptability of the traffic conditions in different regions, this paper takes into full consideration the representative driving cycles of various regions in the world when selecting the typical driving cycles so that the regional limitations of the recognition on driving patterns is reduced. As a result, this paper refers to six representative standard driving cycles, including the WLTC, which is the most commonly used in the world; the NEDC introduced by the European Union; the CLTC-P introduced by China; and the Artemis cycles introduced by the United States. The velocity versus time of the standard driving cycles selected in this paper are shown in Figure 1.

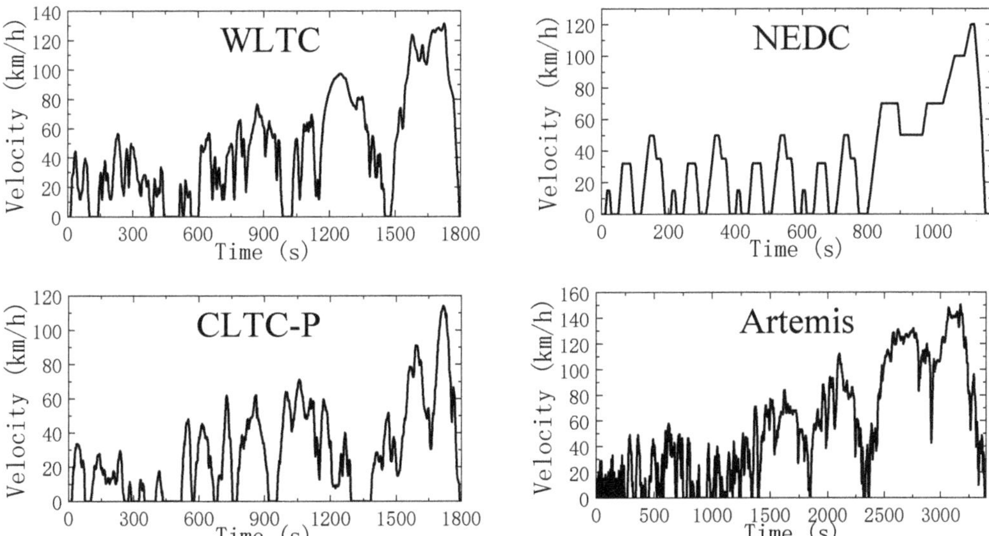

Figure 1. Velocity versus time of typical driving cycles for K-means clustering algorithm training.

After the typical driving cycles are selected, the division method of the driving cycle samples needs to be determined. The aim of the driving cycle division is to build a sample database, which is the premise of the driving pattern clustering analysis. In the existing work, there are two methods used to divide the driving cycle samples: one is the division method based on a certain time window, and the other is the division method based on micro-trip driving cycle samples. In this paper, we adopt the division method based on a certain time window, as shown in Figure 2. When the time window is determined (the length of the time window is selected as 60 s in this paper), the velocity sequence is intercepted every 60 s as a driving cycle sample from the starting point to the end.

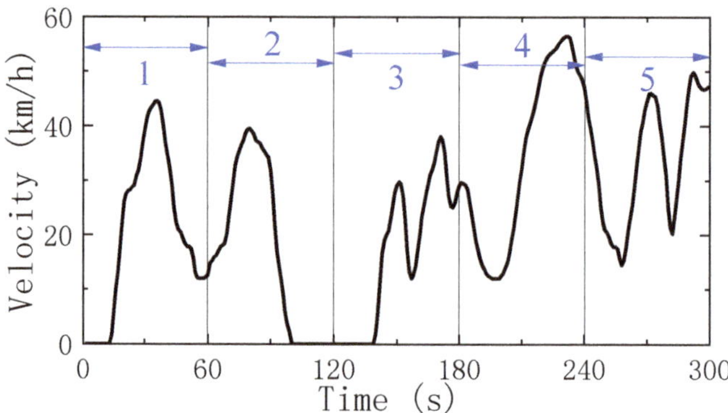

Figure 2. Sample division method based on 60 s time window.

2.2. K-Means Cluster Analysis

A K-means algorithm is applicable to large sample data for clustering and enabling a more reasonable classification. The optimization of K-Means entails minimizing the within-cluster sum of squared errors (*SSE*), as shown in Formula (1):

$$SSE = \sum_{m=1}^{k} \sum_{x \in C_i} dist(X, c_i)^2 = \sum_{m=1}^{k} \sum_{x \in C_i} \|X, c_i\| \tag{1}$$

where $dist(X, c_i)$ is the distance between object X and the clustering center c_i, and dist denotes the Euclidean distance between any two objects in D.

The *SSE* can be analyzed by the elbow rule. The *SSE* decreases with the increase in the number of clustering results, but for data with a certain degree of differentiation, the *SSE* improves greatly when a certain point is reached and decreases slowly afterwards, and this point can be considered as the point with the best clustering performance.

For a reasonable evaluation and analysis of the clustering performance, the Silhouette Coefficient (SC) evaluation method is also a commonly used evaluation method [37]. Assuming that the average distance of the objects in the clusters containing data x and x' is defined as $a(x)$, $b(x)$ is the minimum average distance from the cluster containing x to all the clusters that do not contain x'. $a(x)$, $b(x)$, and the Silhouette coefficient $S(i)$ are defined as follows:

$$a(x) = \frac{\sum_{x' \in C_i, x \neq x'} \|x - x'\|}{|C_i - 1|} \tag{2}$$

$$b(x) = \min_{C_j : 1 \leq i \leq K, x \neq x'} \frac{\sum_{x' \in C_j} \|x - x'\|}{|C_j|} \tag{3}$$

$$S(i) = \frac{b(i) - a(i)}{\max\{b(i) - a(i)\}} \tag{4}$$

where $a(i)$ denotes the maximum distance of the sample i from other samples in the same cluster, and $b(i)$ denotes the minimum distance of the sample i from all samples in the other clusters.

The optimal number of clustering centers can be determined based on $S(i)$. The closer $S(i)$ is to 1, the more reasonable the sample clustering result; the closer $S(i)$ is to -1, the less accurate the sample clustering result; if $S(i)$ is close to 0, the sample is at the boundary position between two clustering results.

Combining the SC evaluation method and the elbow rule of clustering, the *SSE* and $S(i)$ versus the number of driving pattern clustering centers are calculated, respectively, as shown in Figure 3.

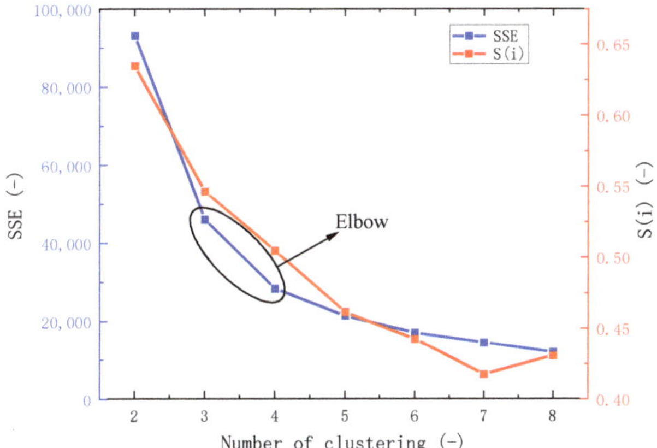

Figure 3. *SSE* and *S*(*i*) versus the number of driving pattern clustering.

It is evident from Figure 3 that the changing rate of *SSE* is greatly decreased when the number of clustering, k, is taken as 3 or 4, so k = 3 or k = 4 can be considered as the number of clustering; the Silhouette coefficient is 0.55 when k is taken as 3 and 0.50 when k is taken as 4. So, the clustering result is more reasonable when k is taken as 3 by the SC evaluation method. As a result, the number of clustering selected in this paper is 3. Combined with the actual driving scenarios, these 3 patterns are defined as urban, suburban, and highway.

The clustering results are shown in Figure 4 on training data and under the three-dimensional coordinates of the average velocity, idle ratio, and maximum velocity.

Figure 4a shows the prediction results from standard cycles as training data. The driving-pattern-recognition model can match the patterns represented by the different time axes of each standard cycle. Further analysis of the clustering results in Figure 4b shows that the K-means clustering center of red points is (18.10, 0.31, 38.00), which has a low average velocity, a high idle ratio, and a low maximum velocity, so the red points represent a congested urban driving condition. The K-means clustering center of the blue points is (100.60, 0, 111.94); the average velocity of this cluster is high, the idle ratio is 0, and the maximum velocity is also high, so the blue points represent smooth highway-driving conditions. The K-means clustering center of the green points is (48.80, 0.04, 65.74), and the average velocity, idle ratio, and maximum velocity of this cluster are between the congested urban driving condition and the smooth highway driving condition. As a result, these points represent the suburban driving conditions. Thus, the driving conditions can be broadly classified into three typical patterns, namely, urban, suburban, and highway. The classification results are consistent with the actual driving conditions mentioned above, so the recognition results of the model can be considered reasonable.

2.3. Driving Pattern Recognition

Since only three features are used for driving pattern recognition in this paper, after ensuring the real-time model calculation and the low compilation difficulty of the algorithm, in order to further illustrate the accuracy of the driving pattern recognition, actual road spectrum data are collected as a test dataset to verify the recognition accuracy of the model based on the K-means clustering algorithm. The driving-pattern-recognition model will calculate the features in real time and output the recognition results based on the historical 60 s data during the driving process. Due to some abnormal data on velocity caused by the driver's operation and GPS device in the actual road spectrum data (as is the case for many real-world applications), the impulse noise will affect the characteristic of v_max, which may cause an instance of incorrect recognition by the model. When applying the driving-pattern-recognition model to practical predictions, the measured velocity signal should be smoothed using the moving average-filtering method to reduce the random impulse noise [38]. As shown in Formula (5), the basic idea of the moving average filtering method is to perform local averaging on a small number of points along the data of length N, thereby filtering out random noise.

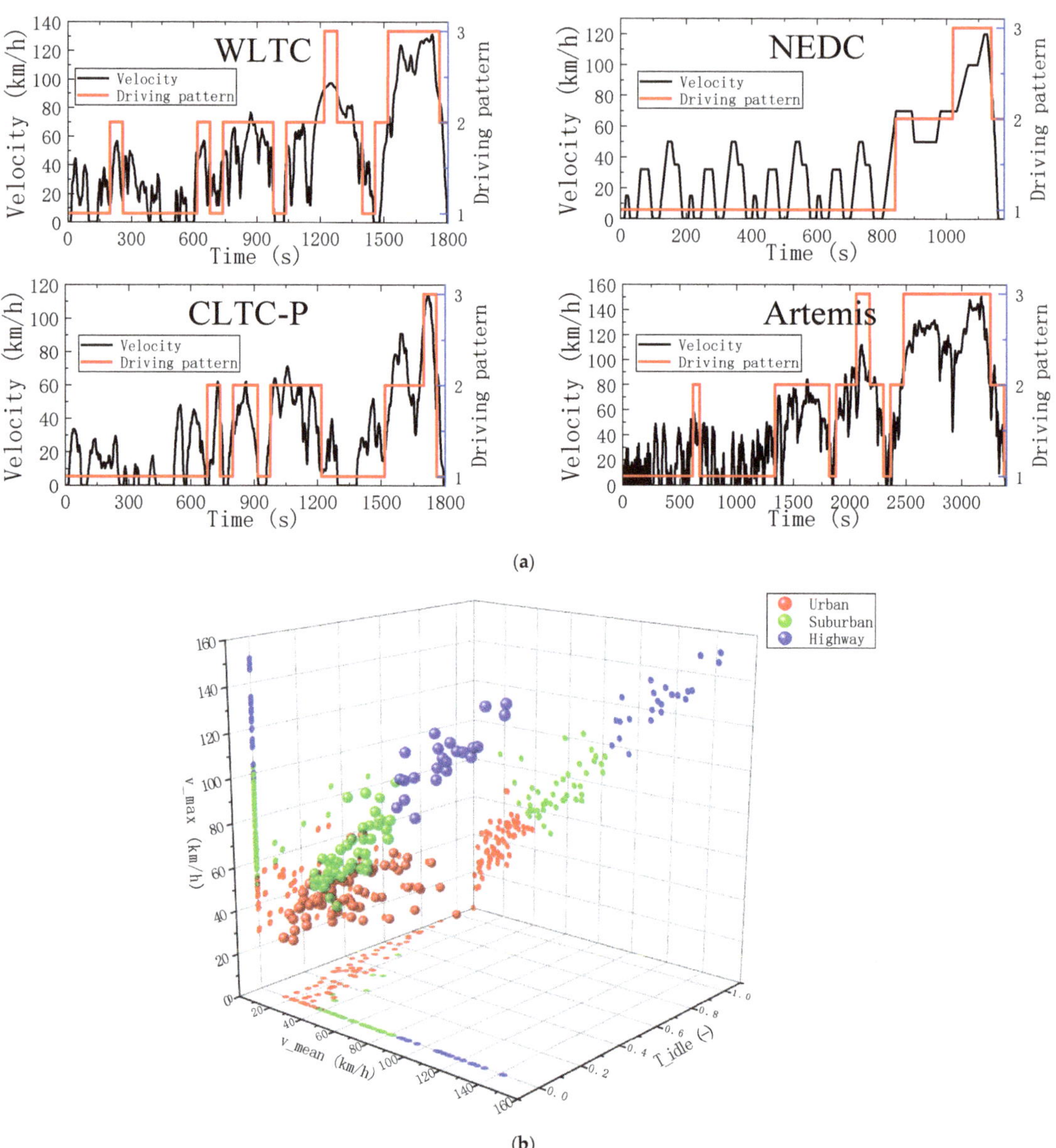

Figure 4. Clustering results (**a**) shown on training data and (**b**) under the three-dimensional coordinates.

$$y_f(k) = \frac{1}{m} \sum_{k-(m-1)/2}^{k+(m-1)/2} y(i) \tag{5}$$

where $y(\cdot)$ represents the original data; $y_f(\cdot)$ represents the filtered data; m is the number of data points used for filtering, which is odd; and i and k represent the data point number. In this paper, m is selected as 5, which represents a 5-second moving average.

Figure 5 represents the prediction results of the driving-pattern-recognition model on the filtered test data.

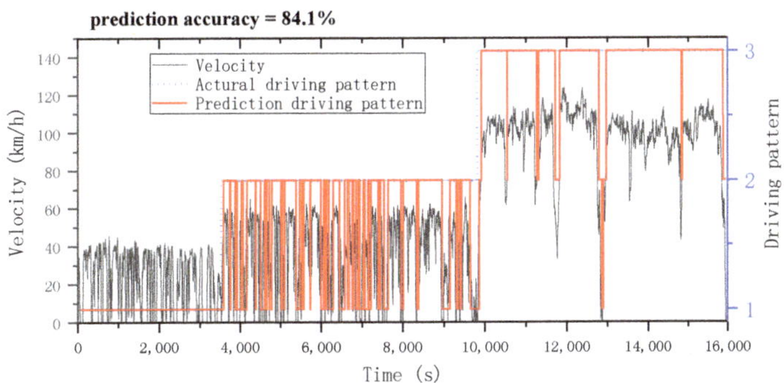

Figure 5. Prediction results of driving-pattern-recognition model based on actual road spectrum data after filtering method.

As shown in Figure 5, the total length of the actual road spectrum data is about 16,000 s, which includes an about 3500 s length of the urban pattern, a 6250 s length of the suburban pattern, and a 6250 s length of highway pattern. Three patterns are shown in the figure as driving patterns one, two, and three, which are indicated by the blue dashed line. The red solid line indicates the prediction result of the driving-pattern-recognition model based on the K-means algorithm. The comparison shows that the driving-pattern-recognition model can identify urban patterns completely and accurately, and the recognition accuracy rate reaches 100%. The suburban pattern is more complicated due to the intermingling of vehicles driving at medium and low velocity or even idling, so the model identifies some of the segments with lower average velocities (e.g., 9000~9100 s segments) or longer idling ratios (e.g., 9600~9800 s segments) as an urban pattern. The prediction result of the highway patterns is also accurate, among which the velocity in the segment 12,750–12,950 s has a sudden drop, which may be due to unexpected traffic accidents or other unexpected situations in the highway pattern at the time of driving, so it cannot be used as a typical feature of a highway pattern; instead, it is more reasonable for it to be identified as a suburban or urban (congested) pattern. In summary, the driving-pattern-recognition model in this paper ensures the real-time calculation speed and low compilation difficulty of the algorithm, while its recognition accuracy is satisfactory, reaching 84.1%, which can be used as the basis for subsequent research.

3. Velocity Prediction Based on the Encoder–Decoder Model

3.1. Basic Conception of LSTM

The LSTM network is an improvement of RNN with a unique store-and-forget function compared to the traditional RNN. By learning the sequence input and extracting the hidden sequence features, it can obtain the dependency relationship between sequences accurately, and overcome the complications of gradient vanishing, gradient explosion, and long-term memory disappearance that occur in RNN during training [39].

There are three inputs to the LSTM network at moment t: the input value x_t at the current moment, the output value h_{t-1} at the previous moment, and the cell state c_{t-1}. The input gate i_t, the output gate o_t, and the forget gate f_t receive the same inputs $[h_{t-1}, x_t]$, which are used to control the update process of the cell state c_t after the activation function σ. The basic LSTM network is shown in Figure 6.

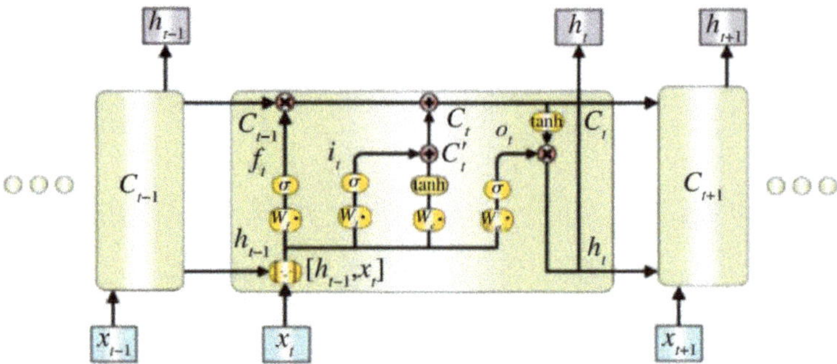

Figure 6. Basic LSTM Network.

A gate is a fully connected layer that receives a vector as an input, while the output is a vector of real numbers between 0 and 1. W is the weight matrix of the gate and b is the bias; then, the gate can be represented as:

$$g(x) = \sigma(Wx + b) \tag{6}$$

where σ denotes the activation function. In this paper, the Sigmoid function is used as the activation function to adjust the input $[h_{t-1}, x_t]$ to the (0,1) interval.

The input gate i_t controls how much of the input x_t is saved to the cell state c_t at the current moment, and is defined as:

$$i_t = \sigma(W_i[h_{t-1}, x_t] + b_i) \tag{7}$$

The forget gate f_t controls how much of the cell state c_{t-1} of the previous moment is saved to the current moment state c_t, and is defined as:

$$f_t = \sigma\left(W_f[h_{t-1}, x_t] + b_i\right) \tag{8}$$

The output gate o_t controls how much of the state c_t at the current moment is output to the current output value h_t, and is defined as:

$$o_t = \sigma(W_o[h_{t-1}, x_t] + b_i) \tag{9}$$

The cell state $\tilde{c}_t$ of the current input is described according to h_{t-1} and the current input x_t, and is defined as:

$$\tilde{c}_t = tanh(W_c[h_{t-1}, x_t] + b_c) \tag{10}$$

The cell state c_t is adjusted by the input gate i_t and the forget gate f_t and is defined as:

$$c_t = f_t c_{t-1} + i_t \tilde{c}_t \tag{11}$$

The final output h_t of the network is determined by both the output gate o_t and the cell state c_t and is defined as:

$$h_t = o_t tanh(c_t) \tag{12}$$

In Formulas (7)–(10), the matrices W_i, W_f, W_o, and W_c are the gate weight matrices and the vectors b_i, b_f, b_o, and b_c are the bias terms of the gates.

The information used for updating the cell state c_t, f_t, and o_t are determined by the gating vectors in Formulas (7)–(10). The cell state and output are updated by Formulas (11) and (12). The cell state is reset or restored by f_t and the new state c_t is obtained by adding partial information through the input gate i_t, while the hidden state h_t is controlled and updated by the output gate o_t.

3.2. An Encoder–Decoder Structure Coupled with Driving Pattern Recognition

The encoder–decoder framework is a structure for time series analysis. In this paper, LSTM neurons are used for establishing the encoder–decoder model and coupled with the driving pattern recognition results obtained above. This multi-input model inputs both the vehicle velocity and driving pattern recognition sequences and the constructed input sequences are encoded and decoded to achieve vehicle velocity prediction, as shown in Figure 7. The driving pattern recognition-encoder decoder (DPR-ED) model mainly contains two parts: the encoder side and the decoder side.

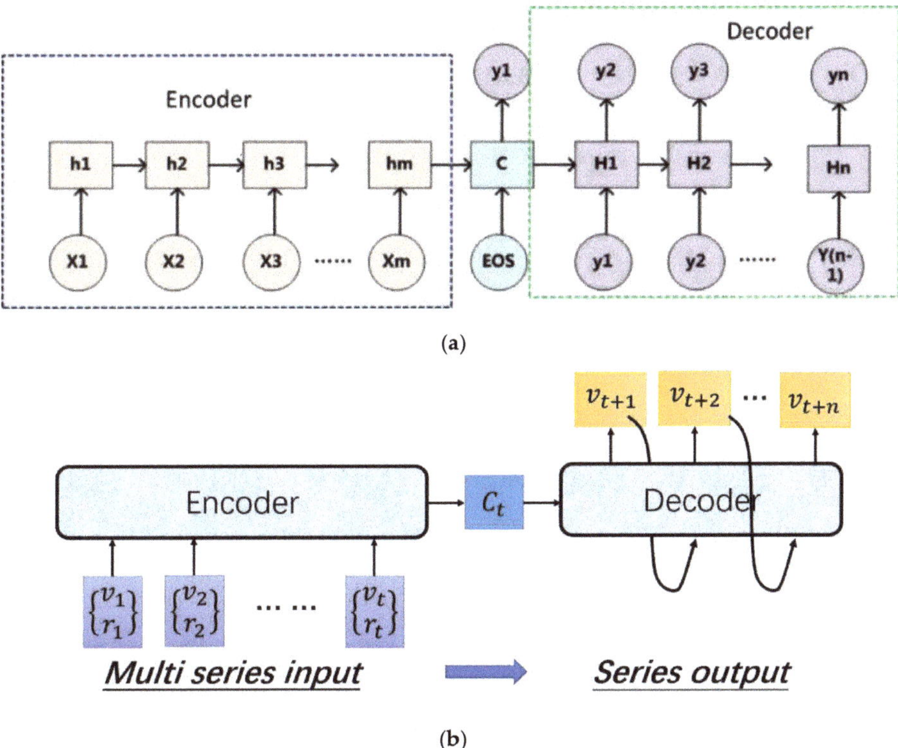

Figure 7. The diagram of the encoder–decoder framework: (**a**) Traditional encoder–decoder framework; (**b**) DPR-ED model.

The operation of the ED-based model consists of two phases: training and testing. The training phase is for learning the implied knowledge and experience from the training data, and the weight matrix of the model is trained for the testing phase; the testing phase applies the trained model and the input data for testing to calculate the corresponding output data. The training and prediction samples are the preparatory work for running the model. All the above-mentioned training and prediction datasets of the DPR-ED model are closely related to the parameters of the history window (W_t), which represents the length of the history series, and the prediction window (W_n), which represents the length of the prediction series. For the traditional ED framework shown in Figure 7a, the input of the training data is a $W_t \times M$ matrix and the output of the training data is a $W_n \times M$ matrix; the input of the prediction data is a $W_t \times 1$ vector and the output of the prediction data is a $W_n \times 1$ vector, where M is the number of training samples. For the DPR-ED model proposed in this paper, shown in Figure 7b, the training data input is a $C \times W_t \times M$ tensor and the training data output is a $W_n \times M$ matrix; the prediction data input is a $C \times W_t$

matrix and the prediction data output is a $W_n \times 1$ vector, where C is the number of features. The number of features in this paper is two, namely, vehicle velocity and driving pattern.

The operation process of the DPR-ED model is as follows. Using the time window sliding method, M training input samples are intercepted from the sequence of typical driving cycles and the corresponding driving pattern sequence, each of which is in the form of $\begin{bmatrix} v_1\ v_2\ \dots\ v_t \\ r_1\ r_2\ \dots\ r_t \end{bmatrix}$, forming the training input data tensor $X_{train} = 2 \times W_t \times M$. According to the same method, M training output samples are intercepted from the sequence of typical driving cycles and the corresponding driving pattern sequence. Each of them is in the form of $[v_{t+1}\ v_{t+2}\ \dots\ v_{t+n}]$, forming the training output data matrix $Y_{train} = W_n \times M$. The number of training samples obtained by the time window-sliding method in this paper is 7914, that is, $M = 7914$. Similarly, the validation input data tensor $X_{val} = 2 \times W_t \times N$ and the validation output data matrix $Y_{val} = W_n \times N$ are obtained. In this paper, the time window-sliding method is used to randomly select one-tenth of the actual road spectrum data as the validation set. Such a small amount of extraction does not affect the persuasiveness of the model analysis on the test set, and even improves the performance of the model by using the complete input of typical driving cycles as training. The number of validation samples in this paper is 1594, that is, $N = 1594$. After the model is trained, the test data input matrix $X_{test} = 2 \times W_t$ is intercepted from the test dataset using the time window-sliding method. The prediction result will be calculated by the trained model as the data output vector $Y_{test} = W_n \times 1$, which is the predicted vehicle velocity sequence.

In terms of the structure of the model, the number of layers of the DPR-ED model in this paper is fixed to a single layer. Setting the number of nodes in the input layer—which represents the length of the historical velocity sequence, W_t—too small may decrease prediction accuracy for a lack of historical information, while setting it too large may increase the complexity of the model structure and thus decrease the computational speed. Regarding the number of nodes in the output layer, which represents the length of the predicted vehicle velocity sequence, W_n, it is indicated in the literature [22] that a prediction sight distance between 1~10 s is beneficial for the PEMS effect of HEVs, while [40] determines it as 5 s and obtains the best results. Considering the above, in the subsequent study of this paper, the range of W_n is set as 1~5 s, and the range of W_t is set as twice the length of W_n, which is 2~10 s. A detailed parameter study and optimization of W_n and W_t are carried out in the subsequent section. The number of neurons in the hidden layer determines the performance of the neural network. Too few neurons will result in the model's inability to obtain enough fitting features and too many will cause the model to run slower and be prone to overfitting, so both conditions will lead to poor prediction results of the trained neural network. The number of hidden-layer neurons is also one of the important parameters to be studied and optimized in detail in this paper.

In order to evaluate the prediction accuracy, different evaluation methods have been used in the literature, the most commonly used of which is the root mean square error (*RMSE*), which reflects the error between the predicted vehicle velocity and the actual vehicle velocity at each step of the predicted sight distance, as shown in Formula (13):

$$RMSE = \sqrt{\frac{1}{n}\sum_{k=1}^{n}\left(Y_P^k - Y_A^k\right)^2} \tag{13}$$

where n is the length of predicted sight distance, which represents the length of the predicted vehicle velocity. It is equal to the number of output nodes of the DPR-ED model. Y_P^k is the predicted vehicle velocity value corresponding to the kth predicted sight distance ($k \in [1, n]$); Y_A^k is the actual vehicle velocity value corresponding to the kth predicted sight distance point.

Since the *RMSE* can only reflect the prediction accuracy at each forecasting step and cannot be used to evaluate the average prediction accuracy of the whole forecasting process, the introduction of an average root mean square error ($RMSE_A$) is proposed as shown in Formula (14):

$$RMSE_A = \frac{1}{M} \sum_{step=1}^{M} RMSE \tag{14}$$

where M is the number of samples. For the sake of the conciseness of expression, the $RMSE$ is used to denote the average root mean square error in all the subsequent research in this paper.

In this paper, to manifest the performance enhancement effect of the DPR-ED model, the traditional MLP model is also established for comparison. The specific method can be found in the literature [41] and the method for comparison draws on the experience of [42].

4. Results and Discussion

Firstly, in order to obtain the optimal model performance, the relationship between the number of neurons of the traditional MLP model and the number of LSTM neurons of the ED-based models versus the fitting accuracy needs to be investigated. In the training process of the models, both the training and validation datasets were normalized. The $RMSE$ was used as the loss. The Adam algorithm was used to update the neural network weights and the early stopping mechanism was introduced to optimize the epochs of training to prevent the overfitting of the model. The parameter of early stopping was defined as patience = 10, which means the training is stopped when the performance of the loss in the validation set is not further improved after ten consecutive weight updates. Finally, in order to better evaluate the compatibility, the accuracy and stability of the model with respect to the parameter length, the history window, and the prediction window were selected as the maximum values (10 s and 5 s) in this study, so that the trained model would have better compatibility with simpler cases. Each model was randomly assigned initial weights, which was repeated five times during training, and the results were analyzed by box plots. The influences of the different numbers of neurons on the learning performance of the traditional MLP model, Basic ED model, and DPR-ED model for the validation set are shown in Figure 8.

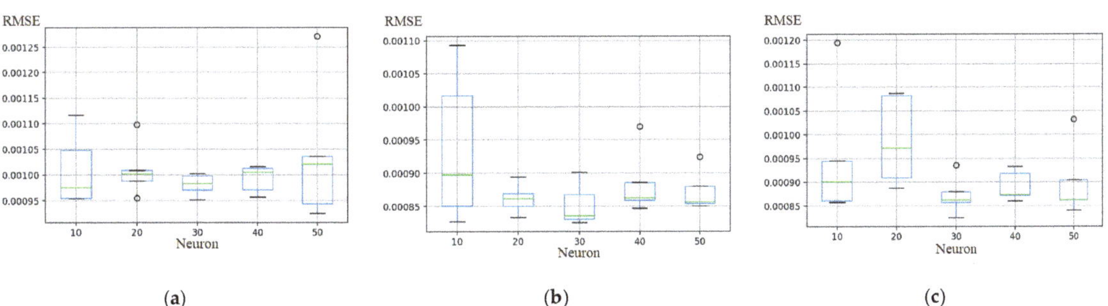

(a)   (b)   (c)

Figure 8. The learning outcome for validation data: (**a**) MLP model; (**b**) Basic ED model; (**c**) DPR-ED model.

Figure 8a shows the accuracy of the traditional MLP model versus different numbers of hidden-layer neurons in the validation set, where each neuron is a perceptron. The mean and standard deviation of the $RMSE$ of the MLP model in the validation set show a trend of decreasing and then increasing as the number of hidden-layer neurons of the MLP model increases, and reach the minimum values of 0.000981 and 0.000021, respectively, when the number of hidden-layer neurons is selected as 30. The accuracy of the basic ED model and DPR-ED model versus different numbers of hidden-layer neurons in the validation set are shown in Figure 8b,c. The neurons in both models are LSTM units. The result indicates that when the number of hidden-layer neurons is taken as 10 for the basic ED model and less than 20 for the DPR-ED model, the performance of the models still cannot be improved (poor stability and accuracy) even after the iterations are completed, which is due to the fact that a simple network is unable to present the characteristics of the historical velocity and the impact of historical information on the future velocity. As the number of hidden-layer neurons increase, the model performance approaches excellence and stability. Therefore, the optimal numbers of hidden-layer neurons of the basic ED model and DPR-ED model are 20 and 30, respectively. The mean values of the *RMSE* are 0.000872 and 0.000862 and the standard deviations of the *RMSE* reach 0.000023 and 0.000032, respectively.

Through the above analysis, the structure of the traditional MLP model is determined as a network with 30 hidden-layer neurons in a single layer, and the structure of the basic ED model and the DPR-ED model is determined as a network with 20 and 30 hidden-layer neurons (LSTM units) in a single layer in the subsequent study. The reason why the DPR-ED model has more hidden-layer neurons than the Basic ED model is probably because the DPR-ED model accepts a multi series input and therefore requires more LSTM units to accurately render the corresponding features.

Figure 9 shows the accuracy of each model versus the history window and prediction window on the test set and comparison.

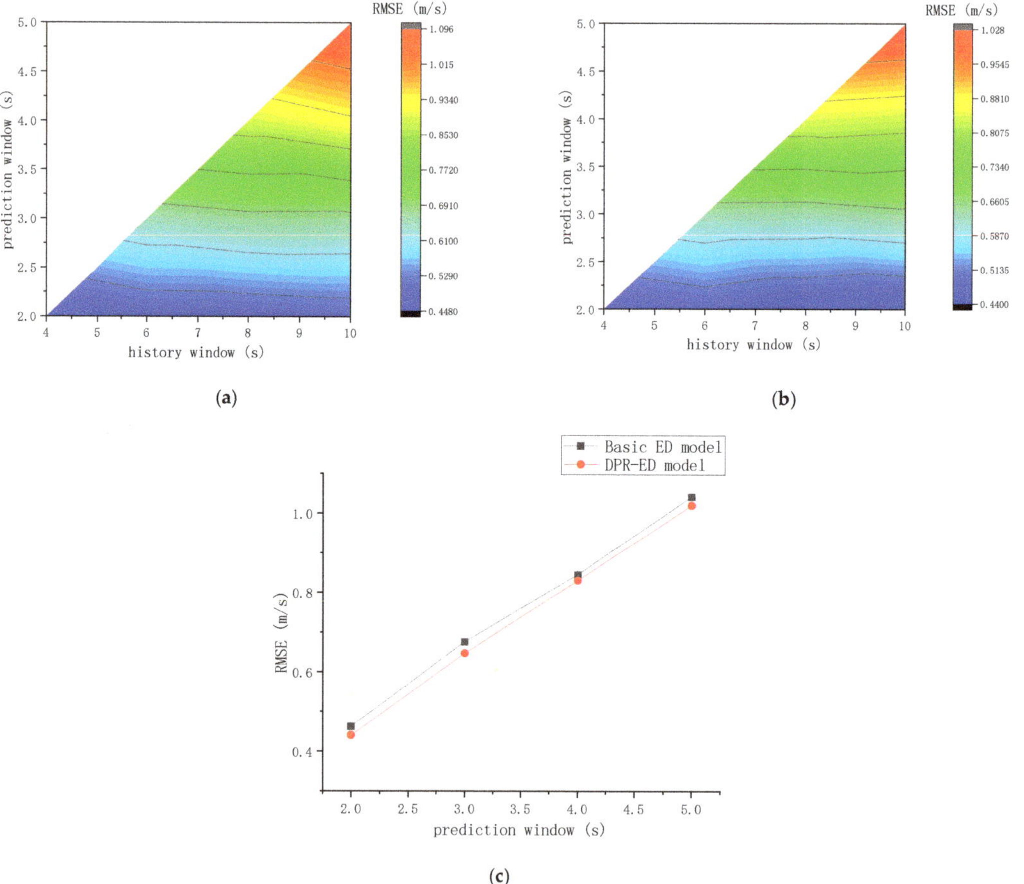

Figure 9. The accuracy of each model versus the history window and prediction window on test set and comparison: (**a**) MLP model; (**b**) DPR-ED model; (**c**) The comparison of basic ED model and DPR-ED model (history window = 10 s).

Figure 9a,b show the impact of different history and prediction windows on the accuracy of the traditional MLP model and the DPR-ED model on the test set. From the results of these two figures, it can be clearly seen that both the history and prediction window directly affect the prediction accuracy of the model. For the MLP model, the history window has a slight effect on the prediction accuracy of the model (as seen from the equipotential line); that is, the larger the history window for the same prediction window case, the more the *RMSE* decreases. For the DPR-ED model, the change of the history window has little effect on the prediction accuracy of the model, so the input sequence length of the model is also less sensitive. The prediction accuracy decreases significantly as the prediction window increases, so in the process of its application in a real vehicle, an excessively large prediction window is generally not used. In addition, it is obvious from the figure that the accuracy of the DPR-ED model is higher than that of the MLP model to different degrees under the same value of the history window and the prediction window; the performance of the DPR-ED model is slightly higher than that of the MLP model when the prediction window is small (e.g., 2 s) For instance, the *RMSE* of the MLP model at a history window of 6 s is 0.483 m/s, while the *RMSE* of the DPR-ED model is 0.476 m/s, which demonstrates a slight improvement of 1.4%. When both the history and prediction window become larger, the DPR-ED model shows a greater advantage: the *RMSE* of the DPR-ED model is 1.028 m/s while the *RMSE* of the MLP model is 1.096 m/s at the history and prediction windows of 10 s and 5 s, respectively, which is a 6.6% deterioration in performance compared to the former. Figure 9c shows the performance improvement of the DPR-ED model compared to the basic ED model in the case of long prediction sequences. In addition to the role of the ED frame, which can better handle time series problems, when the prediction window is large, the LSTM neurons of the DPR-ED model receiving multidimensional inputs are also able to parse more complex information, so the accuracy is further improved by about 2.5% compared to the basic ED model.

Figure 10 shows the prediction results of the DPR-ED model from the test set. Since the total simulation time is too long and too many data points are not conducive to graphical analysis, the results of the urban, suburban, and highway patterns are intercepted to 1000 s each. In order to analyze the prediction performance for the five steps more clearly, the local velocity prediction trajectories for two random time points with different driving pattern characteristics are compared in Figure 11.

The red curve in Figure 10 represents the velocity of the five future-prediction steps, and the blue curve represents the actual velocity data. Under the different driving patterns, the DPR-ED model can basically respond to the transient process and adjust the predicted velocity within 5 s, which shows that the DPR-ED model proposed in this paper is effective. Figure 11a shows the typical characteristics of the urban driving pattern and its prediction results. It is shown from Figure 11a that the urban driving pattern contains frequent accelerations and decelerations, and the predicted velocity is close to the actual velocity when the vehicle experiences low deceleration or a smooth large deceleration at 767 s and 774 s. If a sudden change in acceleration or deceleration occurs (reflecting the various complex road conditions that occur in the urban driving pattern during actual driving), (i.e., 757 s), the predicted velocity deviates from the actual velocity, but still follows the trajectory of the trend. Figure 11b shows the typical characteristics of the highway driving pattern and its prediction results. Figure 11b indicates that the highway driving pattern shows a relatively gentle high-speed cruise or small speed changes at most of the time. The predicted velocity of the DPR-ED model also shows a relatively gentle change trend in the highway driving pattern, which is close to the actual speed. It is evident that since the DPR-ED model additionally accepts the driving pattern sequence information, it can better predict the typical characteristics under different driving patterns.

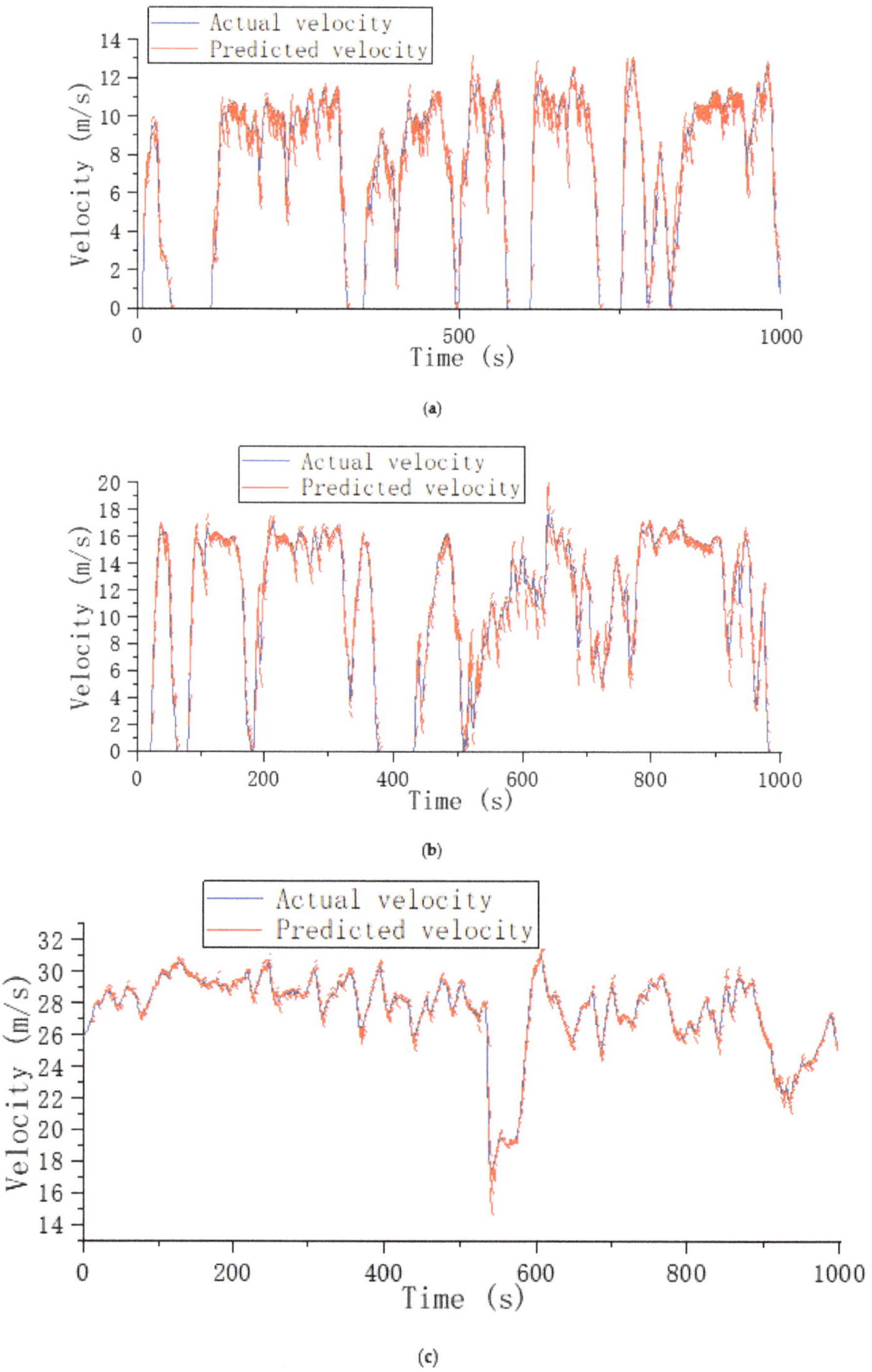

Figure 10. Velocity prediction result of the DPR-ED model: (**a**) City pattern; (**b**) Suburb pattern; (**c**) Highway pattern.

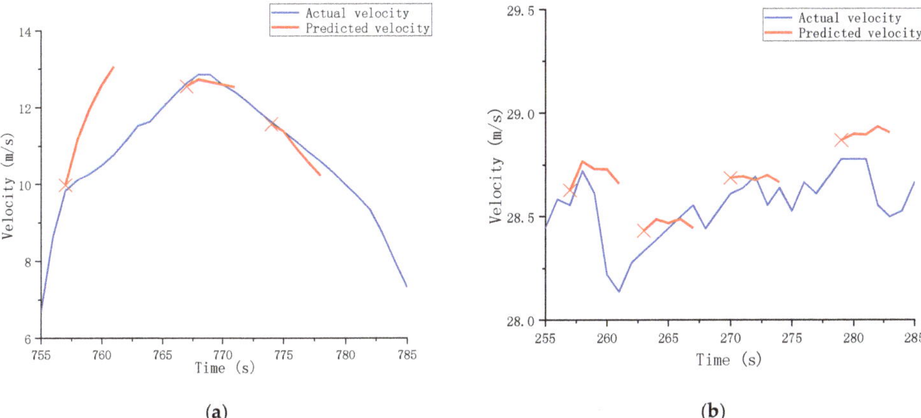

(a)   (b)

Figure 11. Local velocity prediction of DPR-ED model for 5 prediction steps: (**a**) City pattern; (**b**) Highway pattern.

It is worth noting that even though the predicted velocity corresponds well to the actual velocity, the predicted velocity seems quite noisy, especially in the city pattern. Firstly, the unfiltered velocity could be one of the reasons for this phenomenon, as the impulse noise of the data itself will definitely affect the prediction result. Figure 12 shows the prediction results of the DPR-ED model on the city part of the filtered test set. The filtering method is five-second moving average, which has been mentioned above. Secondly, the ability to change the speed of the applied vehicle should be examined and limited while the torque limitation exists for the power system of a certain vehicle. Figure 13 shows the acceleration distribution of the prediction results by the DPR-ED model on the raw test set.

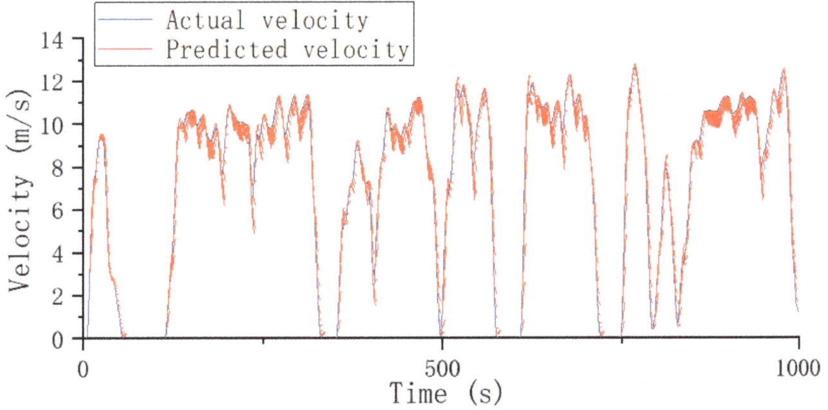

Figure 12. The prediction results of the DPR-ED model on the city part of the filtered test set.

As shown in Figure 12, after filtering the impulse and high frequency noise in the raw test data, the velocity prediction stability is improved significantly in the area where the sudden change is large. The *RMSE* of the prediction results from the filtered test set is an exceptional 0.79 m/s. It indicates that a low-noise vehicle speed signal plays a crucial role in the efficient operation of the model. Achieving signal noise reduction is not within the scope of this paper, so the following analysis is based on raw data. However, for real-world applications, high-precision GPS and advanced signal-processing techniques should be coupled as much as possible. Figure 13 shows that the acceleration distribution is an approximate normal distribution, while the acceleration proportion is slightly more than the deceleration proportion. A small number of data points exhibit a changing rate of velocity greater than 1 m/s, and there are individual points with a changing rate greater than 1.5 m/s, which may also add to the noise of the model. When the power limit or driving style does not allow for such an acceleration, a clipping method can be applied during the real-world application of the model to correct the prediction results within the allowed acceleration range.

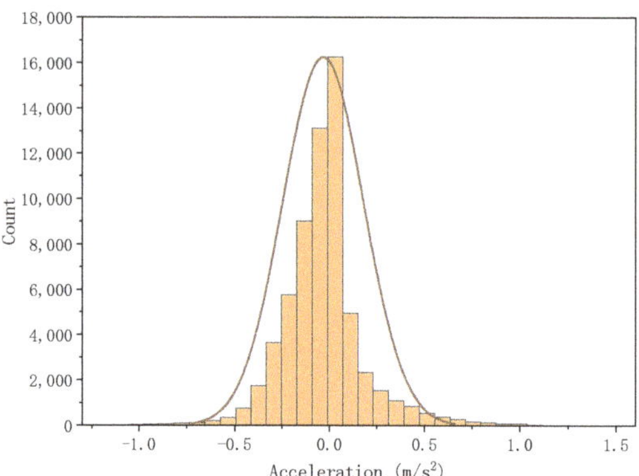

Figure 13. The acceleration distribution of the prediction results by DPR-ED model on the raw test set.

Figure 14 shows the local *RMSE* of the MLP model and the DPR-ED model during the process of the velocity prediction for the whole test set.

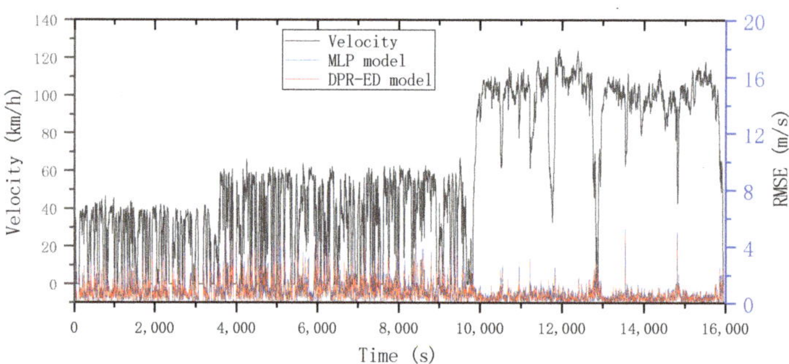

Figure 14. Local *RMSE* comparison between MLP and DPR-ED on test set.

From Figure 14, it is evident that the local *RMSE* has the lowest average value in the highway driving pattern and the highest average value in the suburban driving pattern, which is due to the fact that the velocity changes in the highway driving pattern are smoother, while the complex suburban driving pattern is full of predictive uncertainties. The local *RMSE* is larger when the vehicle is operating under a large rate of change in acceleration, which is due to the fact that large prediction errors usually occur when the vehicle suddenly starts to accelerate after deceleration or in a situation of emergency deceleration after acceleration; such driving conditions are challenging to predict and should be addressed with the involvement of the driver. In addition, it can be found that the DPR-ED model improves the accuracy compared to the MLP model when the vehicles are working in a large rate of change of acceleration, although the local *RMSE* shows a peak (especially for the highway driving pattern). For example, the local *RMSE* of the MLP model's predicted velocities at times 13,538 s and 14,805 s are 5.21 m/s and 5.05 m/s, respectively, while the results predicted by the DPR-ED model at the same time are 4.88 m/s and 4.61 m/s, respectively, which shows an improvement of 6.3% and 8.7%.

Figure 15 shows the comparison of the error distribution between the MLP model and the DPR-ED model for the test data. Table 1 shows the proportion of the prediction results for the test set contained in each error segment.

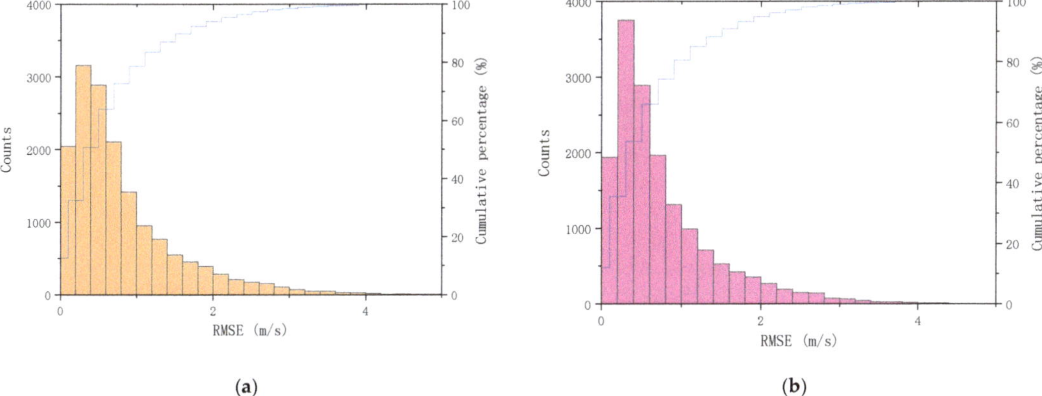

Figure 15. The comparison of the error distribution between the MLP model and the DPR-ED model for test data: (**a**) MLP model; (**b**) DPR-ED model.

Table 1. The proportion of the prediction results of the test set contained in each error segment.

RMSE Segment (m/s)	Proportion (MLP Model)	Proportion (DPR-ED Model)
0~0.1	12.81%	12.15%
0.1~0.3	19.81%	23.55%
0.3~0.5	18.15%	18.13%
0.5~0.7	13.22%	12.34%
0.7~1.1	14.85%	14.5%
1.1~2.1	15.35%	14.37%
>2.1	5.81%	4.96%

As shown in Table 1, the results predicted by the DPR-ED model and the MLP model from the test data with an error of 0~0.1 m/s have similar proportions. However, the results predicted by the DPR-ED model from the test data with an error of 0.1~0.3 m/s have a proportion of 23.55%, which is about 4% higher than the result obtained by the MLP model. Therefore, more prediction results are concentrated on segments with small errors. On the other hand, the DPR-ED model has also eliminated the data with an error of more than 0.5 m/s. It reduced the proportion with an error of 0.5~0.7 m/s and over 1.1 m/s by about 1% and 1.9%, respectively; thus, the DPR-ED model filters out more of the results with larger errors. As a result, the DPR-ED model shows an excellent prediction robustness over different driving patterns.

Consequently, compared with the MLP model's poor prediction accuracy on certain loads due to the discrete characteristics of the training data load, the optimal partially shared network shows an excellent prediction robustness over the full load range.

As shown in Figure 15, it can be found that the average error of the MLP model and the DPR-ED model prediction result is 0.81 m/s and 0.77 m/s, respectively. The DPR-ED model performs 5.2% better than the MLP model on the average prediction accuracy. Moreover, the variance is decreased greatly from 0.52 to 0.45, which is an improvement of 15.6%. This result shows that the DPR-ED model significantly improves both the accuracy and the stability.

5. Conclusions

In this paper, based on the samples from typical driving cycles, a high-precision velocity prediction model was established through the encoder–decoder framework coupled with driving pattern recognition. The main achievements and conclusions are as follows:

(1) The driving-pattern-recognition model was established by a K-means clustering algorithm and validated based on the test data; the driving patterns were identified as urban, suburban, and highway patterns. The model achieved a satisfactory recognition accuracy of 84.1% on the total length of 16,000 s of real road spectrum data, achieving results that can be used as the basis for subsequent studies.

(2) The MLP, basic ED, and DPR-ED models, trained using the early stopping method, were developed. The effect of different numbers of neurons on the prediction accuracy and stability of each model was investigated and the optimization of the models was completed. The results show that the DPR-ED model with 30 LSTM hidden neurons can achieve the optimal overall performance for velocity prediction, which obtains an average *RMSE* of 0.000862 and a standard deviation of 0.000032 after the dataset's normalization.

(3) Compared with the MLP model, the DPR-ED model is designed to improve the performance by implementing multidimensional inputs and applying time series analysis. In the long-time prediction series case, the DPR-ED model shows a significant advantage over the MLP model: the *RMSE* of the DPR-ED model applied to the validation set was 1.028 m/s, while the *RMSE* of the MLP model was 1.096 m/s, with a 6.6% deterioration in performance compared to the former. When the two models were applied to 16,000 s road spectrum data for testing, the proportion with a low error of 0.1~0.3 m/s was improved by 4% and a larger error proportion was filtered for the results predicted by the DPR-ED model. The DPR-ED model performed 5.2% better than the MLP model with respect to the average prediction accuracy. Meanwhile, the variance was decreased by 15.6%. This novel framework enables the processing of long-time sequences with multiple input dimensions, which improves the prediction accuracy under complicated driving patterns and enhances the generalization performance and robustness of the model.

Author Contributions: Conceptualization, D.L. and Y.Z.; methodology, Y.Z. and C.Z.; software, Y.Z.; validation, Y.Z. and L.F.; formal analysis, Y.T.; investigation, D.L. and L.F.; resources, D.L.; data curation, Y.Z.; writing—original draft preparation, D.L. and Y.Z.; writing—review and editing, Y.Z. and L.F.; visualization, D.L. and Y.T.; supervision, D.L.; project administration, D.L.; funding acquisition, D.L. and L.F. All authors have read and agreed to the published version of the manuscript.

Funding: This research was supported by the National Key R&D Program of China (2021YFB2500800).

Conflicts of Interest: The authors declare no conflict of interest.

Definitions/Abbreviations

v_mean	Average velocity
v_max	Maximum velocity
T_idle	idle ratio
WLTC	World Light-duty vehicle Test Cycle
NEDC	New European Driving Cycle
CLTC-P	China Light-duty vehicle Test Cycle-Passenger
SSE	sum of squared errors
SC	Silhouette coefficient
DPR	Driving pattern recognition
LSTM	Long short-term memory
ED	Encoder–decoder
MLP	Multilayer perceptron
RMSE	Root mean square error

References

1. Wang, Y.; Lou, D.; Xu, N.; Fang, L.; Tan, P. Energy management and emission control for range extended electric vehicles. *Energy* **2021**, *236*, 121370. [CrossRef]
2. Karbaschian, M.A.; Söffker, D. Review and Comparison of Power Management Approaches for Hybrid Vehicles with Focus on Hydraulic Drives. *Energies* **2014**, *7*, 3512–3536. [CrossRef]
3. Biswas, A.; Emadi, A. Energy management systems for electrified powertrains: State-of-the-art review and future trends. *IEEE Trans. Veh. Technol.* **2019**, *68*, 6453–6467. [CrossRef]
4. Rizzoni, G.; Onori, S. Energy Management of Hybrid Electric Vehicles: 15 years of development at the Ohio State University. *Oil Gas Sci. Technol. Rev. D'Ifp Energ. Nouv.* **2014**, *70*, 41–54. [CrossRef]
5. Tran, D.D.; Vafaeipour, M.; El Baghdadi, M.; Barrero, R.; Van Mierlo, J.; Hegazy, O. Thorough state-of-the-art analysis of electric and hybrid vehicle powertrains: Topologies and integrated energy management strategies. *Renew. Sustain. Energy Rev.* **2020**, *119*, 109596. [CrossRef]
6. Wu, J.; Zhangc, H.; Cuin, X. Fuzzy energy management strategy for a hybrid electric vehicle based on driving cycle recognition. *Int. J. Automot. Technol.* **2012**, *13*, 1159–1167. [CrossRef]
7. Chen, B.C.; Guan, J.C.; Li, J.H. Adaptive Power Management Control of Range Extended Electric Vehicle. *Energy Procedia* **2014**, *61*, 67–70. [CrossRef]
8. Skugor, B.; Deur, J. Instantaneous Optimization-based Energy Management Control Strategy for Extended Range Electric Vehicle. In *SAE Technical Paper Series*; SAE International: Warrendale, PA, USA, 2013.
9. Rezaei, A.; Burl, J.B.; Zhou, B.; Rezaei, M. A New Real-Time Optimal Energy Management Strategy for Parallel Hybrid Electric Vehicles. *IEEE Trans. Control. Syst. Technol.* **2019**, *27*, 830–837. [CrossRef]
10. Zeng, Y.; Cai, Y.; Kou, G.; Gao, W.; Qin, D. Energy Management for Plug-In Hybrid Electric Vehicle Based on Adaptive Simplified-ECMS. *Sustainability* **2018**, *10*, 2060. [CrossRef]
11. Peng, J.; He, H.; Xiong, R. Rule based energy management strategy for a series–parallel plug-in hybrid electric bus optimized by dynamic programming. *Appl. Energy* **2017**, *185*, 1633–1643. [CrossRef]
12. Xi, L.; Zhang, X.; Sun, C.; Wang, Z.; Hou, X.; Zhang, J. Intelligent Energy Management Control for Extended Range Electric Vehicles Based on Dynamic Programming and Neural Network. *Energies* **2017**, *10*, 1871. [CrossRef]
13. Zhang, P.; Yan, F.; Du, C. A comprehensive analysis of energy management strategies for hybrid electric vehicles based on bibliometrics. *Renew. Sustain. Energy Rev.* **2015**, *48*, 88–104. [CrossRef]
14. Wu, J.; He, H.; Peng, J.; Li, Y.; Li, Z. Continuous reinforcement learning of energy management with deep Q network for a power split hybrid electric bus. *Appl. Energy* **2018**, *222*, 799–811. [CrossRef]
15. Li, Y.; He, H.; Khajepour, A.; Wang, H.; Peng, J. Energy management for a power-split hybrid electric bus via deep reinforcement learning with terrain information. *Appl. Energy* **2019**, *255*, 113762.1–113762.13. [CrossRef]
16. Li, G.; Görges, D. Energy management strategy for parallel hybrid electric vehicles based on approximate dynamic programming and velocity forecast. *J. Frankl. Inst.* **2019**, *356*, 9502–9523. [CrossRef]
17. Li, Y.; He, H.; Peng, J.; Zhang, H. Power management for a plug-in hybrid electric vehicle based on reinforcement learning with continuous state and action spaces. *Energy Procedia* **2017**, *142*, 2270–2275. [CrossRef]
18. Tan, H.; Zhang, H.; Peng, J.; Jiang, Z.; Wu, Y. Energy management of hybrid electric bus based on deep reinforcement learning in continuous state and action space. *Energy Convers. Manag.* **2019**, *195*, 548–560. [CrossRef]
19. Olia, K.; Shahverdi, M.; Mazzola, M.; Sherif, A. *Developing a Model Predictive Control-Based Algorithm for Energy Management System of the Catenary-Based Electric Truck*; SAE International: Warrendale, PA, USA, 2016.
20. Zhang, S.; Xiong, R.; Sun, F. Model predictive control for power management in a plug-in hybrid electric vehicle with a hybrid energy storage system. *Appl. Energy* **2017**, *185*, 1654–1662. [CrossRef]
21. Rezaei, A.; Burl, J.B. Prediction of Vehicle Velocity for Model Predictive Control. *IFAC-Paper* **2015**, *48*, 257–262. [CrossRef]
22. Liu, K.; Asher, Z.; Gong, X.; Huang, M.; Kolmanovsky, I. *Vehicle Velocity Prediction and Energy Management Strategy Part 1: Deterministic and Stochastic Vehicle Velocity Prediction Using Machine Learning*; SAE International: Warrendale, PA, USA, 2019.
23. Shin, J.; Sunwoo, M. Vehicle Speed Prediction Using a Markov Chain with Speed Constraints. *IEEE Trans. Intell. Transp. Syst.* **2019**, *20*, 3201–3211. [CrossRef]
24. Wang, F.; Luo, Y.T. A Research on Power Splitting Strategy for Hybrid Energy Storage System Based on Driving Condition Prediction. *Automot. Eng.* **2019**, *41*, 1251–1258, 1264.
25. Sun, C.; Hu, X.; Moura, S.J.; Sun, F. Velocity Predictors for Predictive Energy Management in Hybrid Electric Vehicles. *IEEE Trans. Control. Syst. Technol.* **2015**, *23*, 1197–1204.
26. Guo, J.; He, H.; Sun, C. ARIMA-Based Road Gradient and Vehicle Velocity Prediction for Hybrid Electric Vehicle Energy Management. *IEEE Trans. Veh. Technol.* **2019**, *68*, 5309–5320. [CrossRef]
27. Jing, J.; Filev, D.; Kurt, A.; Özatay, E.; Michelini, J.; Özgüner, Ü. Vehicle speed prediction using a cooperative method of fuzzy Markov model and auto-regressive model. In Proceedings of the 2017 IEEE Intelligent Vehicles Symposium (IV), Los Angeles, CA, USA, 11–14 June 2017; pp. 881–886.
28. Rosolia, U.; Zhang, X.; Borrelli, F. Data-Driven Predictive Control for Autonomous Systems. *Annu. Rev. Control. Robot. Auton. Syst.* **2018**, *1*, 259–286. [CrossRef]

29. Fengqi, Z.; Xiaosong, H.; Kanghui, X.; Xiaolin, T.; Yahui, C. Current Status and Prospects for Model Predictive Energy Management in Hybrid Electric Vehicles. *J. Mech. Eng.* **2019**, *55*, 86–108.

30. Choudhury, A.; Shanmugavadivelu, S.; Velpuri, B. Real Time Speed Trend Analysis and Hours of Service Forecasting Using LSTM Network. In Proceedings of the 2019 IEEE 5th International Conference for Convergence in Technology (I2CT), Bombay, India, 29–31 March 2019; pp. 1–6.

31. Shi, F.; Wang, Y.; Chen, J.; Wang, J.; Hao, Y.; He, Y. Short-term Vehicle Speed Prediction by Time Series Neural Network in High Altitude Areas. *IOP Conf. Ser. Earth Environ. Sci.* **2019**, *304*, 032072. [CrossRef]

32. Wang, Y.Q. Traffic Flow Prediction Based on Combined Model of ARIMA and RBF Neural Network. In Proceedings of the 2017 2nd International Conference on Machinery, Electronics and Control Simulation (Mecs 2017), Taiyuan, China, 24–25 June 2017; Volume 138, pp. 82–86.

33. Sutskever, I.; Vinyals, O.; Le, Q.V. Sequence to sequence learning with neural networks. *Adv. Neural Inf. Process. Syst.* **2014**, *27*. [CrossRef]

34. Sun, C.; Moura, S.J.; Hu, X.; Hedrick, J.K.; Sun, F. Dynamic traffic feedback data enabled energy management in plug-in hybrid electric vehicles. *IEEE Trans. Control. Syst. Technol.* **2014**, *23*, 1075–1086.

35. Liu, G.C. *Study of Energy Management Strategy of Range-Extended Electric Vehicle*; Shanghai Jiao Tong University: Shanghai, China, 2017.

36. Hu, J.; Liu, D.; Du, C.; Yan, F.; Lv, C. Intelligent energy management strategy of hybrid energy storage system for electric vehicle based on driving pattern recognition—ScienceDirect. *Energy* **2020**, *198*, 17298. [CrossRef]

37. Rousseeuw, P. Silhouettes: A graphical aid to the interpretation and validation of cluster analysis. *J. Comput. Appl. Math.* **1999**, *20*, 53–65. [CrossRef]

38. Cao, M.; Xu, Q.; Bian, H.; Yuan, X.; Du, P. Research on configuration strategy for regional energy storage system based on three typical filtering methods. *IET Gener. Transm. Distrib.* **2016**, *10*, 2360–2366. [CrossRef]

39. Li, M.; Lu, F.; Zhang, H.; Chen, J. Predicting Future Locations of Moving Objects with Deep Fuzzy-LSTM Networks. *Transp. A Transp. Sci.* **2020**, *16*, 119–136. [CrossRef]

40. Xiang, C.; Ding, F.; Wang, W.; He, W. Energy management of a dual-mode power-split hybrid electric vehicle based on velocity prediction and nonlinear model predictive control. *Appl. Energy* **2017**, *189*, 640–653. [CrossRef]

41. Wu, Y.; Zhang, Y.; Li, G.; Shen, J.; Chen, Z.; Liu, Y. A predictive energy management strategy for multi-mode plug-in hybrid electric vehicles based on multi neural networks. *Energy* **2020**, *208*, 118366. [CrossRef]

42. Shi, C.; Zhang, Z.; Ji, C.; Li, X.; Di, L.; Wu, Z. Potential improvement in combustion and pollutant emissions of a hydrogen-enriched rotary engine by using novel recess configuration. *Chemosphere* **2022**, *299*, 134491. [CrossRef] [PubMed]

Article

A Study of Evaluation Method for Turbocharger Turbine Based on Joint Operation Curve

Sheng Yin *, Jimin Ni, Houchuan Fan, Xiuyong Shi and Rong Huang

School of Automotive Studies, Tongji University, Shanghai 201804, China
* Correspondence: 1811475@tongji.edu.cn; Tel.: +86-021-6958-9980

Abstract: Turbochargers have evolved with the advancement of engine technology. In this study, we pro-posed a concept of joint operation, based on the operating characteristics of the compressor and turbine. Furthermore, a turbine evaluation method was proposed based on this concept, and an optimization application study of the turbine impeller blade number and turbine casing was con-ducted and verified. The results showed that the performance evaluation method based on the joint point could predict the optimization trend of turbine performance more accurately, the turbine output power optimized based on our new method evidently had advantages over the original turbine, and the joint point showed better overall performance. The original single-entry turbine could be optimized into a 9-blade twin-entry turbine having better response characteristics. The maximum torque of the optimized engine was 5.4% higher than that of the original engine, and the minimum brake specific fuel consumption (BSFC) was reduced by 2.1%. In the low and medium speed operating region, engine torque was increased by up to 3.2% and BSFC was reduced by up to 1.1% compared to the turbine optimized by conventional methods. Hence, the optimization effect of our new method was proven.

Keywords: engine; turbocharger; turbine; internal joint operation curve; joint point

Citation: Yin, S.; Ni, J.; Fan, H.; Shi, X.; Huang, R. A Study of Evaluation Method for Turbocharger Turbine Based on Joint Operation Curve. *Sustainability* **2022**, *14*, 9952. https://doi.org/10.3390/su14169952

Academic Editors: Cheng Shi, Jinxin Yang, Jianbing Gao and Peng Zhang

Received: 10 July 2022
Accepted: 10 August 2022
Published: 11 August 2022

Publisher's Note: MDPI stays neutral with regard to jurisdictional claims in published maps and institutional affiliations.

1. Introduction

Due to rising fuel prices and stricter greenhouse gas emission regulations, reducing fuel consumption and emissions has become an endeavor for the internal combustion engine industry [1–4]. For example, the European Union has set a 37.5% reduction in fuel consumption for passenger cars by 2030 compared to 2021. For the internal combustion engine itself, stringent emission regulations and rising fuel control requirements have prompted the internal combustion engine industry to develop more advanced technologies [5–8]. To achieve this goal, exhaust gas turbochargers, a key component of the internal combustion engine's air exchange system, could play a particularly important role. Similarly, the studies of alternative liquid fuels also need to consider the effect of the turbocharger [9]. The turbocharger and the engine match and interact with each other, and the operating characteristics of the engine must be considered while studying the turbocharger [10].

On account of the exhaust characteristics of the engine, the turbocharger turbine operates in an exhaust gas pulse environment where the turbine inlet pressure and temperature change periodically [11,12]. Under pulsed, unsteady conditions, the turbine mostly operates in deviation from the design point condition and even in a zero-admission condition [13]. Copeland et al. [14] studied the difference between the unsteady and steady flow performance of a separated twin-entry turbine, and the steady flow efficiency was found to be higher than the unsteady flow efficiency. Moreover, Rajoo [15] showed in his study that the pulsed, unsteady average efficiency of the turbine deviated from the corresponding quasi-steady efficiency value by up to 32%, indicating that the steady approach was inaccurate.

The automotive turbocharger pulse boost system sets the turbine to work in the engine's periodic pulse, unsteady exhaust environment; the admission condition at the turbine end changes continuously, and the performance of the turbine further changes periodically with the change in the intake state at the turbine end. Multi-fluid turbine structures and various boost systems have emerged to improve the utilization of pulse energy [16,17]. Currently, the swallowing capacity and efficiency of a turbine can be easily improved by using a twin-entry turbine instead of a single-entry turbine [15]. Other research studies that have focused on the use of single-entry turbine are include [18–20]. Cerdoun [21] investigated the unsteady flow process in an asymmetric parallel twin-entry turbine and concluded that increasing the radius of the volute shape could enhance the centrifugal force of the airflow. In addition, Chiong MS [22,23] studied a one-dimensional (1D) numerical model of a twin-entry turbine under full admission pulse flow conditions and predicted the corresponding performance. The full admission condition of the twin-entry turbine occurs only momentarily, and the turbine operates with unequal admission [24–26].

In terms of unequal admission conditions, Newton [27] employed a computational fluid dynamics (CFD) software to investigate the flow loss of a twin-entry turbine under full and partial admissions. Costall [28] developed a twin-entry turbine model, which could be used as a simplified single-entry model under full admission conditions, while a more complex model is required for unequal admission conditions. Hajilouy [29] investigated the performance and internal flow field characteristics of a twin-entry turbine under full and extreme admission conditions. Other studies [30,31] have also performed the same experiment, and revealed the mechanism of the influence of the volute structure on the turbine performance.

The compressor is also the key component of the turbocharger, which directly determines the turbocharger performance, as well as the matching performance of the turbocharger and the engine. The compressor is affected by the fluctuation of the engine admission, which makes the flow rate and pressure of the compressor periodically change with the engine operating process. This could further affect the engine performance; for example, the compressed air supercharging system could improve the driving force during the phase of the engine's increasing crankshaft rotational speed [32]. The works of [33] and [34] presented the periodic changes in air flow at the inlet and outlet of the compressor. Other studies related to the compressor of a turbocharger can be found in [35–37].

The performance of turbochargers in a harsh working environment of high temperature and high pressure for a long time will directly affect the performance of an engine. Fan proposed a novel Adaptive Local Maximum-Entropy Surrogate Model, carried out a turbine disk reliability analysis under geometrical uncertainty, and achieved a desirable result [38]. Meng constructed a smooth response surface of the turbine performance by the saddlepoint approximation reliability analysis method and solved a turbine blade design problem [39].

As can be seen from the brief description above, the actual operation of the booster has a non-constant flow at both ends, and the difference between the non-constant flow performance and the corresponding constant flow value is obvious. The performance at both ends of the turbocharger is limited by the structural properties of another end, and a complete performance map is usually unavailable [40]. Meanwhile, the current conventional method of calculating the total efficiency of a turbocharger is to multiply the turbine efficiency, the compressor efficiency, and the turbocharger mechanical efficiency [41]. This calculation method is not objective enough, because the maximum efficiency points of the turbine and compressor are often not in the same operating condition. Moreover, the operation of the turbocharger is a dynamic process of speed change, and the actual efficiency of both the turbine and the compressor is dynamically changing, therefore there are different total efficiencies of the turbocharger for different operating conditions. The internal matching of the compressor to the turbine fundamentally determines the actual operating performance of the turbocharger and the matching performance between the turbocharger and the engine.

There is no in-depth and systematic research on the internal matching between the compressor and turbine, and the important role of the internal matching of their joint, in the design of turbochargers, is not fully reflected. In this study, a combination of experimental and numerical simulations was used to study different turbine structures, and GT-Power and ANSYS CFX were used as the simulation platforms. First, an automotive engine turbocharger turbine was studied and verified. Second, different volutes and turbine impellers with different numbers of blades were designed to produce different turbine structures, and simulations were performed to compare the new turbine structure with the original turbine and determine the differences between the conventional evaluation method and new method. Lastly, different turbine structures were evaluated using our new method, and the best turbine structure was identified. Following this, the corresponding superposition of the compressor power consumption curve and turbine effective power curve were coupled. A new method for optimizing the actual operating performance of the turbine from the perspective of joint matching between the compressor and turbine was proposed. The turbine performance optimized by the two methods was compared, and the new optimization method was proven to be effective and feasible when using equal and unequal admission experiments. The evaluation process and correction method used in this study can provide a future reference for the research and optimal design of turbocharger turbines. Figure 1 shows the framework structure of this paper.

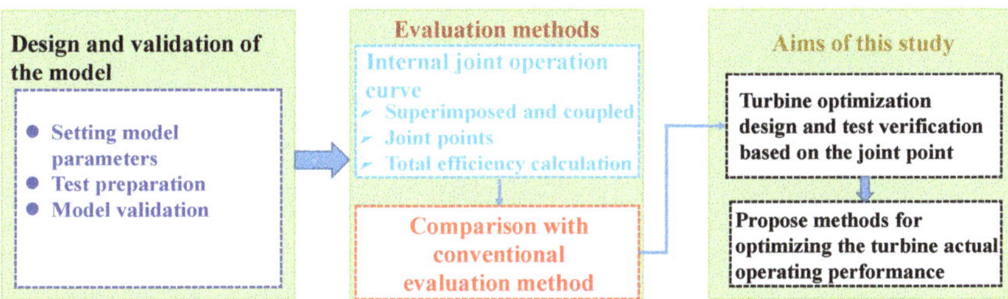

Figure 1. Flow chart of study framework.

2. Materials and Methods

2.1. Research Methodology and Research Objects

A 4-cylinder automotive engine turbocharger turbine was used as the research object. Figure 2 shows the three-dimensional (3D) structure of the physical prototype of the turbocharger. There were five turbine impeller structures of 8-blade, 9-blade, 10-blade, 11-blade, and 12-blade in the optimized design scheme. The volute structures were single-entry and twin-entry, and the two types of volutes were matched with 11-blade and 9-blade impellers, respectively. A total of four turbine structures were combined, as shown in Table 1. Additionally, other structural parameters remained unchanged.

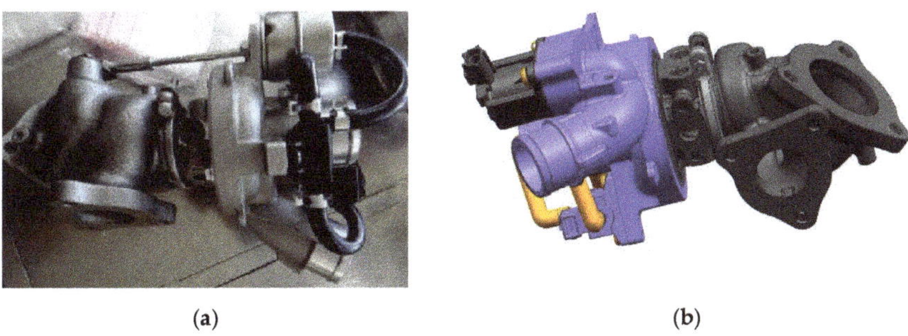

(a)   (b)

Figure 2. (**a**) Physical prototype; (**b**) three-dimensional structure model.

Table 1. Optimization design schemes.

Scheme	Volute Structure	Impeller Blades
1 (Original Turbine)	Single-entry	11-blade
2	Twin-entry	11-blade
3	Single-entry	9-blade
4	Twin-entry	9-blade

2.2. Setting Model Parameters

The turbine used a full-wheel disk turbine impeller, whose volute was a single-entry volute without a leaf nozzle, and the number of blades of the turbine impeller was 11. Table 2 lists turbine parameters.

Table 2. Turbine parameters.

No.	Parameter	Profile
1	Volute inlet diameter/mm	36.0
2	Volute nozzle width/mm	5.0
3	Impeller inlet diameter/mm	37.6
4	Impeller outlet diameter/mm	33.1
5	Impeller axial length/mm	18.9
6	Impeller inlet blade height/mm	5.1
7	Impeller inlet blade angle/deg	0
8	Impeller exit mean blade angle/deg	56.4
9	Radial and axial tip clearance/mm	0.5
10	Number of blades	11
11	Flow range/(kg/s)	0.02–0.13

The frozen rotor interface was selected as the interface type between the volute and rotation domains and between the rotation and outlet transition domains. The turbulence was modeled in the Navier–Stokes equation solver, using the k-epsilon equation [42]. The walls of the model were set to be smooth, nonslip, and adiabatic [43]. Ideal gas was used as the working fluid. The flow rate and total temperature were set as turbine inlet conditions. At the turbine outlet, where the flow is considered to be subsonic, the static pressure is imposed.

The complex structure of the separate turbine volute required the unstructured grid. The rotational domain of the turbine impeller was arranged with a perfectly matched circumferential grid, and an encrypted O-grid was set around the impeller blade to increase the number of grid layers in the blade top clearance region to obtain a more accurate top clearance flow. The turbine inlet duct and outlet duct were added for better computational convergence. Surface layer meshes were added at each working wall to increase the accuracy of the near-wall flow simulation. Table 3 lists the details for each domain of the turbine model.

Table 3. Details for each domain.

Fluid Domain	Mesh Type	Mesh Numbers
Inlet duct	Structured	73,219
Separate turbine volute	Unstructured	620,189
Impeller single blade channel	Structured	203,038
Wheel back clearance	Unstructured	242,283
Outlet transition	Unstructured	527,878
Outlet duct	Structured	169,575
Total		3,866,562

2.3. Test Preparation

In the test, the twin-entry volute was interchangeable with the single-entry volute of the original turbine, and the 9-blade turbine rotor assembly was interchangeable with the 11-blade turbine rotor assembly of the original turbine. The turbine performance tests were conducted on a Kratzer turbocharger test stand, and Figure 3 shows the schematic diagram of the bench arrangement.

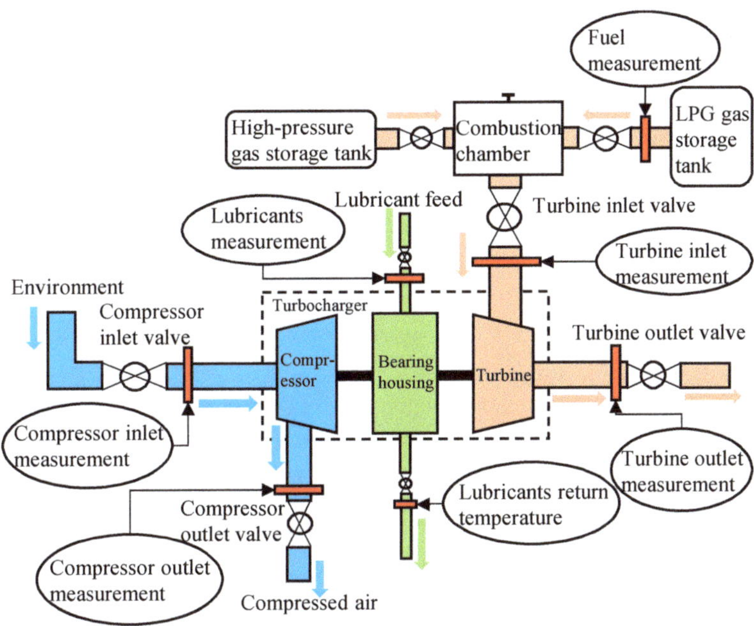

Figure 3. Schematic diagram of the bench arrangement.

2.4. Model Validation

The validation test of the single-entry prototype was carried out on the Kratzer turbocharger test stand, and the following three speed conditions of 110,000 rpm, 150,000 rpm, and 190,000 rpm were selected to represent the low, medium, and high speeds of the turbocharger operation, respectively. In this study, we used the validation method published in previous papers [16,43,44].

Figure 4 shows the comparison between the test and simulation of the swallowing capacity of the original turbocharger turbine. Overall, the test and simulation of the swallowing capacity were in good agreement with each other and had a maximum error value of 2.62% at the speed of 150,000 rpm, which was within the acceptable range. This discrepancy was attributed to the simplification of the secondary feature in the geometry, the parameter settings for heat transfer in the simulation, and a manufacturing error [45,46]; moreover, it could be influenced by the optimal application scope of the SST model [47].

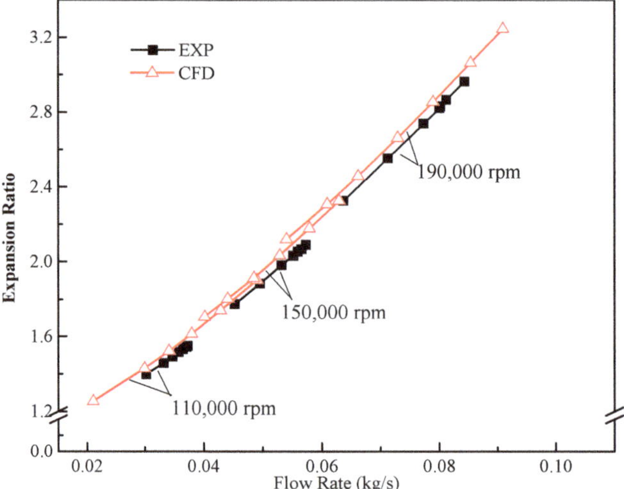

Figure 4. Experimental verification of the original turbine model.

3. Results

3.1. Coupling of Internal Joint Operating Curves

The compressor and turbine are rigidly connected by the rotor shaft. They have the same speed, and their working states are related to the operation of the engine. The following three are the basic conditions for the operation of the booster:

1. Speed balance: the compressor, turbine, and rotor shaft have the same speed.

$$n_C = n_T = n_R,\tag{1}$$

where n_C is the compressor speed (rpm), n_T is the turbine speed (rpm), and n_R is the rotor shaft speed (rpm);

2. Flow rate balance: the compressor working mass flow rate plus the engine fuel mass flow rate is equal to the turbine working mass flow rate, and this flow rate balance condition is also followed when the turbine bypass valve is opened.

$$m_T = m_C + m_F,\tag{2}$$

where m_T is the turbine operating flow rate (kg/s), m_C is the compressor operating flow rate (kg/s), and m_F is the fuel flow rate (kg/s);

3. Power balance: The power transfer from the turbine side to the compressor side will lose part of the power, which includes turbine rotor lubrication and cooling losses, as well as the heat radiation of the components and other losses. The turbine power output minus this loss is equal to the compressor power consumption.

$$P_T = P_C + P_{Loss},\tag{3}$$

where P_T is the turbine output power (kW), P_C is the compressor power consumption (kW), and P_{Loss} is the power loss during energy transfer (kW).

The turbocharger operation must comply with Equations (1)–(3). This shows that the matching of the turbocharger to the engine contains three aspects: the matching of the compressor to the engine, the matching of the turbine to the engine, and the internal matching between the compressor and turbine of the turbocharger [48].

In the fixed-speed line, the compressor power consumption curve and turbine effective power curve were superimposed and coupled. The intersection point of both the compressor and turbine at the time of coupling was the compressor flow rate, and the intersection point of the power balance was obtained, as shown in Figure 5.

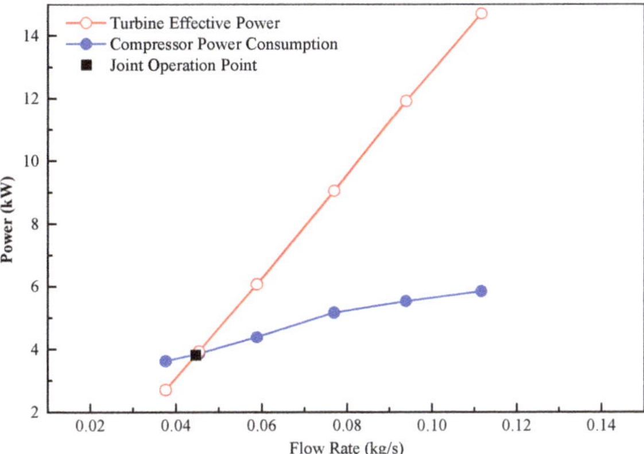

Figure 5. Schematic diagram of the power balance at both ends of the turbocharger at 150,000 rpm.

In comparison, the compressor power consumption curves and turbine effective power curves for all speeds were superimposed and coupled to obtain the joint operation points of the turbocharger at each speed, as shown in Figure 6a. Connecting each joint operation point in turn formed the complete turbocharger joint operation curve for the condition of the turbine bypass valve being closed, as shown in Figure 6b. The parameters in Figures 5 and 6 are for the common operating conditions of the matched engine, and the parameter ranges are from the engine calibration and turbocharger map.

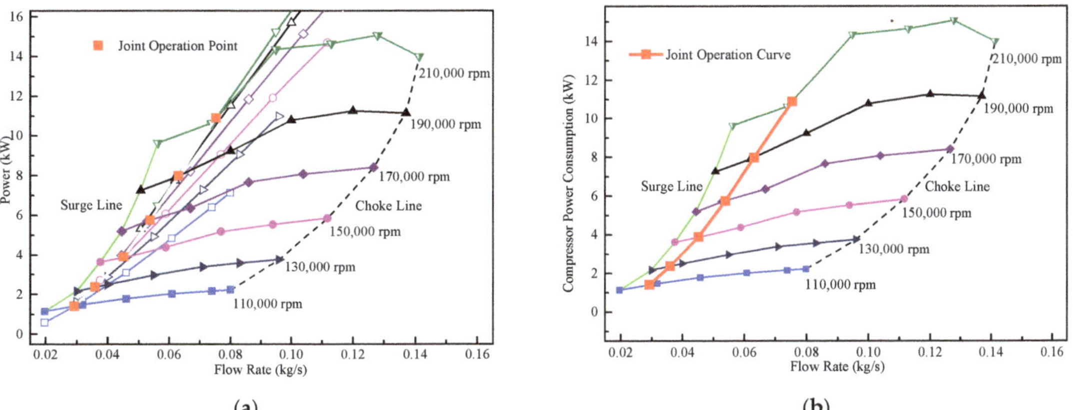

Figure 6. (a) Distribution of the joint operation points at each speed; (b) distribution of the joint operation curve on the compressor power consumption map.

Furthermore, the turbocharger joint operation curve was the joint operation matching curve between the compressor and the turbine. When matching a turbocharger with an engine, following the turbocharger joint operating curve must be prioritized. The joint operation curve reflects the matching performance between the turbocharger and the engine, which helps to simplify matching the work between the turbocharger and the engine and improves the matching accuracy.

3.2. Turbine Optimization Evaluation Method

The performance parameters were described and compared in the research. The data related to these parameters were obtained from experiments and CFD simulations. The turbine performance parameters mainly include the expansion ratio, efficiency, output power, etc. The expansion ratio reflects turbine swallowing capacity. Different parameters have different effects on turbine performance, and the calculation formula is shown in Equation (4), where variable parameters B_1, B_2, and B_3 are the corresponding performance weights, and the calculated turbine performance index, Φ_T, is a dimensionless value.

$$\Phi_T = B_1 \times \frac{P_{T2}}{P_{T1}} + B_2 \times \frac{\pi_{T1}}{\pi_{T2}} + B_3 \times \frac{\eta_{T2}}{\eta_{T2}}, \tag{4}$$

here, P is the joint point output power, π is the joint point expansion ratio, η is the joint point efficiency, T_1 refers to the original turbine, and T_2 refers to the modified turbine.

For the weight distribution of each parameter, the role of the turbine is to extract more engine exhaust gas energy and transfer it to the coaxial compressor; moreover, the turbine output power directly determines the distribution of the joint operating curve. Therefore, the weight distribution of the turbine output power is the largest. The swallowing capacity is the parameter that must be guaranteed by the turbine with the second highest weight, while the turbine efficiency weight is the last consideration. Table 4 lists the weight assignment of each performance parameter.

Table 4. Weight assignment for each performance parameter.

Joint Point Performance Parameters	Output Power	Expansion Ratio	Efficiency
Weight assignment	60	30	10

3.3. Analysis of Turbine Impeller Blade Number Based on Joint Point

Five groups of turbine impellers with blade numbers of 8, 9, 10, 11, and 12 were designed, and the 3D modeling of each group was kept consistent with that of the original turbine. The same topology and meshing method were used for the impeller models, and the optimization study was carried out at 150,000 rpm.

The turbine was evaluated based on the conventional method, which compares the turbine performance curves as one component rather than favoring the joint points. Figure 7 shows the effect of different blade numbers on the turbine performance, and all the turbine joint points are marked in the figure.

The output power, expansion ratio, and efficiency of the turbine decreased significantly at 150,000 rpm when the number of blades was lower than nine, which indicated that the number of turbine impeller blades should not be too small. When the number of blades was greater than nine, the turbine performances were very close.

This indicated that there was no significant difference in performance between the turbine structures for blade numbers greater than 9. Therefore, based on the conventional evaluation method, the number of turbine impeller blades could be reduced from 11 to 9 blades to reduce the turbine rotor mass and improve the turbocharger response characteristics without any significant reduction in the overall turbine performance.

The turbine was evaluated based on the joint point, and the output power and efficiency near the joint point of the nine-blade turbine were found to be lower than those of the original turbine at all speeds. Table 5 shows the joint point performance comparison of turbines with different numbers of blades based on a single-entry volute. With reference to the performance of the original turbine, the turbine performance changed with the number of blades. Whether the number of impeller blades was greater than 11 or less than 11, the turbine performance index was less than that of the original turbine. When the number of impeller blades was less than nine, the turbine performance index decreased significantly, in terms of turbine output power and efficiency. The nine-blade turbine performance was different from the 11-blade original turbine performance based on the joint point performance evaluation, and the turbine impeller blade number could not be reduced. For a single-entry turbine, the number of the original turbine blades was reasonable. This differed from the conclusions of the conventional method evaluation.

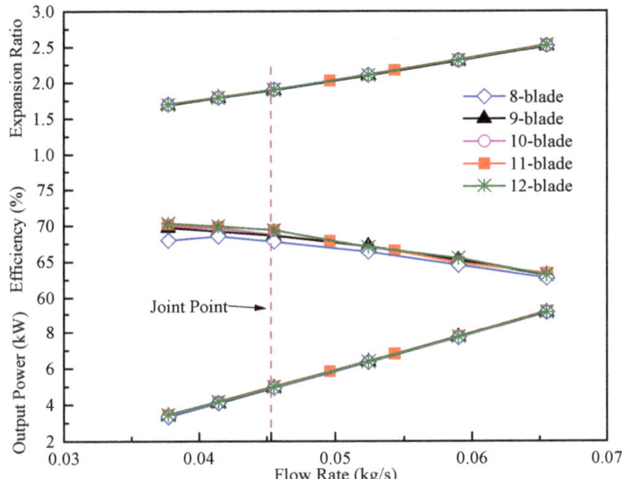

Figure 7. The effect of the number of blades on performance at 150,000 rpm.

Table 5. Comparison of turbines with different number of blades.

Number of Blades	8	9	10	11
Joint point output power (kW)	3.58	3.60	3.62	3.63
Joint point expansion ratio	1.72	1.71	1.71	1.72
Joint point efficiency (%)	66.14	67.3	67.25	67.61
Turbine Performance Index	98.9	99.7	99.9	100.0

Flow field analysis was performed for the mid-speed joint point, and Figure 8 shows the entropy increase distribution at 80% leaf spread height of the impeller channel, with different numbers of blades at the mid-speed joint point. The lower the number of blades, the higher the entropy downstream of the blade channel was, and the larger the region of high entropy. The high entropy area near the suction surface of the blades increased significantly compared with the original turbine when the number of impeller blades was nine. The circumferential flow field of the impeller was not uniformly distributed, indicating that the internal flow evidently became turbulent, and that the flow loss downstream of the blade channel increased significantly, thus reducing the efficiency of the turbine.

According to the turbine working principle, the flow efficiency is high in the blade channel, and the Mach number or relative velocity of the flow in the airflow direction increases. The greater the relative Mach number and the lower the absolute velocity, the smaller the turbine outlet residual velocity loss and the higher the efficiency at the exit of the blade.

The Mach number of each scheme at the inlet of the blade channel was equal in the middle and downstream position of the blade channel. The greater the number of blades, the greater the Mach number was, which implied that the better the utilization of exhaust gas energy, the smaller the corresponding absolute speed at the exit of the impeller, as shown in Table 6. As seen in the following table, the difference in the absolute outlet velocity between blades 8 and 9 was large. Therefore, the number of turbine impeller blades could be reduced within a reasonable range to reduce the rotor mass.

Figure 8. Entropy increase distribution at 80% leaf spreading height of impeller channels.

Table 6. Turbine outlet residual velocity loss.

Number of Blades	8	9	10	11	12
Outlet Absolute Speed (m/s)	201.42	196.87	190.68	190.50	190.11

3.4. Comparative Analysis of Engine Performance with Optimized Structure of Two Methods

The difference between the conventional method and the joint point-based turbine optimization method was verified using the GT model. Figure 9 shows the 1D model constructed using GT-Power. The basic assumptions [49] in this simulation are as follows:

1. The working fluid is a uniform state, and the air entering the cylinder and the residual exhaust gas are completely mixed instantaneously;
2. Air and mixed gas are considered ideal gases, and their thermodynamic parameters are affected by the temperature and composition of the gas;
3. A steady flow process has been regarded for the process of working fluid;
4. The import and export kinetic energy of the working fluid is negligible, and there is no leakage during the combustion process;
5. The combustion heat release process is regarded as a thermodynamic process in which the external heats the working fluid inside the system in accordance with the established heat release law.

Keeping the compressor of the original turbine unchanged, the number of turbine impeller blades was varied to obtain the external characteristic torque and BSFC distribution of the engine, as shown in Figure 10. In the low- and medium-speed ranges of the engine, the output torque of the engine was significantly lower when matched with a nine-blade, single-entry turbine than when matched with the original turbine. Furthermore, the overall BSFC of the nine-blade, single-entry turbine was higher than that of the original turbine. Its maximum torque was 3.5% lower than that of the original turbine, and the minimum BSFC was 3.7% higher than that of the original turbine.

Reducing the number of turbine impeller blades reduced the engine power and fuel economy performance of the single-entry turbine. Therefore, the conclusion of reducing the number of turbine impeller blades obtained based on the conventional method was not reasonable, while the evaluation method based on the joint point performance could accurately predict the optimization trend of the number of turbine impeller blades.

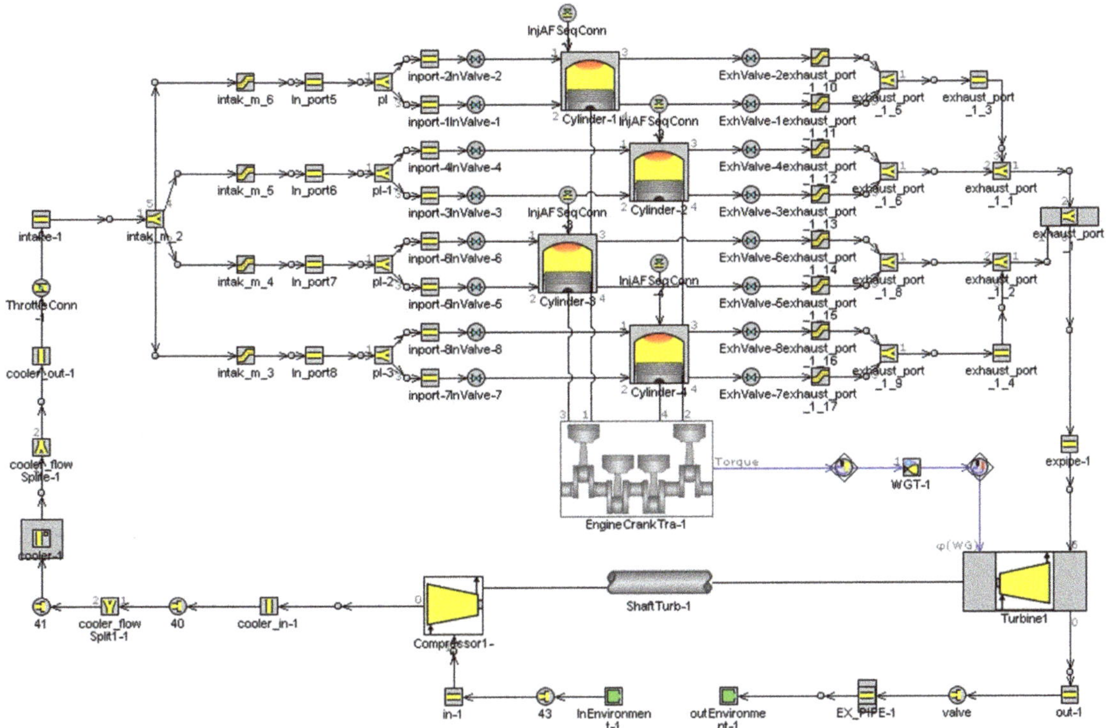

Figure 9. Engine 1D model.

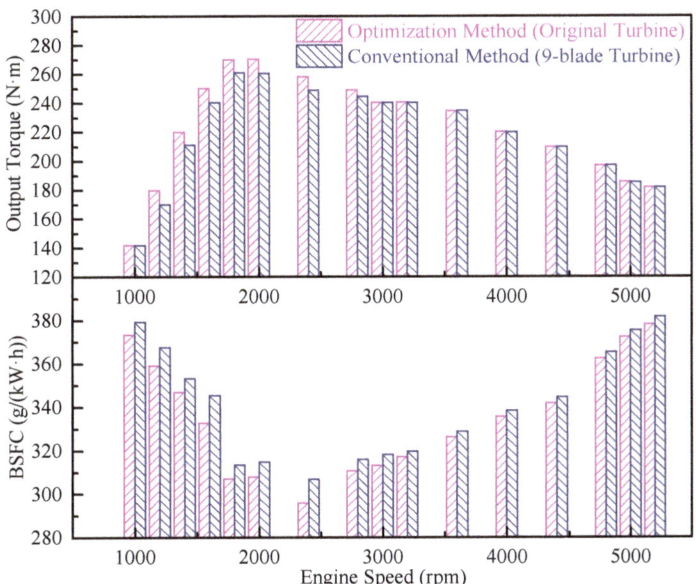

Figure 10. Effect of turbine structure on engine performance.

3.5. Optimized Design of Twin-Entry Turbine Structure under the Equal Admission Conditions

A twin-entry volute was designed based on the single-entry volute of the original turbine to compare its performance with a single-entry turbine under the same conditions. The two air inlets of the twin-entry turbine used equal-flow balanced admission. Figure 11 shows the performance comparison of two volute structures at 150,000 rpm. Compared to the single-entry turbine, the overall output of the twin-entry turbine was higher at the joint point. However, its expansion ratio was also higher, thus making the twin-entry turbine performance index only slightly greater than that of the original turbine. The performance indices of the 9-blade and 11-blade twin-entry turbines were equal at the joint point, indicating that the number of impeller blades could be reduced from 11 to 9 for the twin-entry turbine.

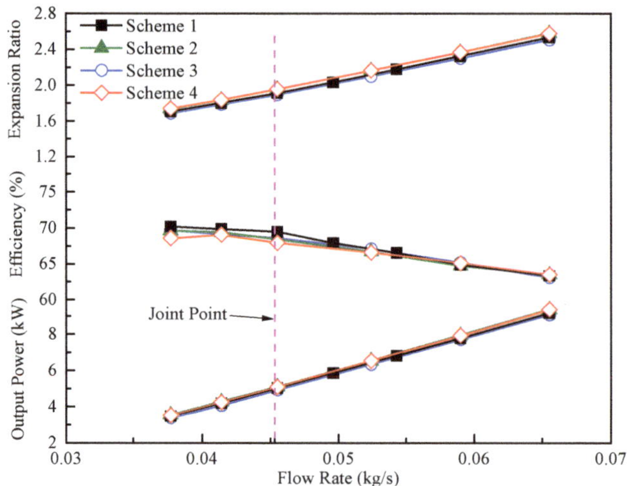

Figure 11. The effect of different turbine structures on performance at 150,000 rpm.

The output power of the twin-entry turbine was slightly higher than that of the single-entry turbine at the joint operating point, and there was no significant difference in the overall swallowing capacity. Furthermore, the efficiency of the joint point of the twin-entry turbine was slightly lower; however, the flow inside the impeller was more uniform. The overall performance of the joint point of the twin-entry turbine was slightly higher than that of the original turbine, and the joint point performance indices of the 9-blade and 11-blade, twin-entry turbines were equal. Therefore, the original 11-blade, single-entry turbine could be optimized to a 9-blade, twin-entry turbine, and the 9-blade, twin-entry turbine exhibited better response characteristics and a more stable operation.

Equal admission performance tests of the twin-entry turbine with different numbers of blades were conducted to illustrate the feasibility of optimizing the number of blades of the twin-entry turbine impeller. Figure 12 shows the equal admission experimental comparison of the performance of the 11-blade and 9-blade, twin-entry turbines. The maximum difference between the two expansion ratios (of 2.0%) was located at the high evolution speed joint point. The maximum difference in the relative efficiency at the joint point was 1.5%. The output powers of the two were close to each other near the joint working condition; the maximum difference between them was 2.1% at the joint point of 150,000 rpm, and the absolute difference was less than 0.2 kW.

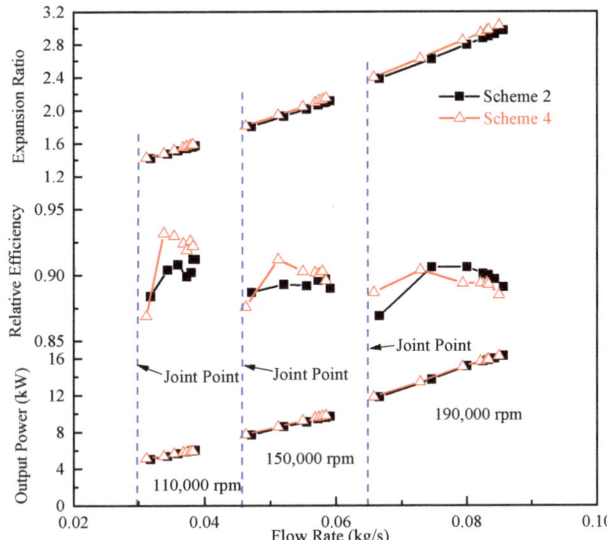

Figure 12. The equal admission experimental comparison of twin-entry turbines with a different number of blades.

From the analysis of the equal admission test, it was found that the swallowing capacity, efficiency, and output power of the 11-blade and 9-blade, twin-entry turbines were relatively close near the joint operating conditions, and there was no obvious difference between the two performances, indicating that the number of blades of the twin-entry turbine impeller could be optimized from 11 to 9 blades, which was consistent with the optimization conclusion from the simulation.

3.6. Analysis of the Unequal Admission Performance of the Twin-Entry Turbine

The exhaust pipes of cylinders 1 and 2 were connected to the outer runner of the twin-entry turbine, and the exhaust pipes of cylinders 3 and 4 were connected to the inner runner. The comprehensive performance comparisons of the turbine in different states under unequal admission condition were carried out at 150,000 rpm, as shown in Table 7. The output power of the twin-entry turbine had a significant advantage over the single-entry turbine, and its performance index was higher than that of the single-entry turbine. For the twin-entry turbine, the performance index was slightly larger for the outer runner.

Table 7. Performance comparison under unequal admission conditions at 150,000 rpm.

Turbine State	Scheme 1	11-Blade-Outer	11-Blade-Inner	9-Blade-Outer	9-Blade-Inner
Joint point output power (kW)	8.08	10.07	10.03	10.07	10.07
Joint point expansion ratio	2.41	3.94	3.96	3.95	3.98
Joint point efficiency (%)	62.59	53.65	52.54	53.79	52.23
Turbine Performance Index	100	101.6	101.1	101.7	101.3

Under unequal admission conditions, there was no significant difference in expansion ratio, efficiency, and output power overall between the 11-blade and 9-blade, twin-entry turbines with a fixed inlet. Therefore, the original 11-blade, single-entry turbine can be optimized to a 9-blade, twin-entry turbine with better performance.

The device for an unequal admission turbine inlet performance test was used to add a gasket at the inlet of the twin-entry turbine housing to cut off the flow rate to one inlet while the other inlet was normal. The unequal admission experiment of the twin-entry turbine swallowing capacity was carried out, as shown in Figure 13. When the same runner was closed, the expansion ratios of the 11-blade and 9-blade turbines were very close to each other, which was the same as the simulated trend. For the same number of blades, the turbine expansion ratio of the inner runner admission was slightly smaller than that of the outer runner intake; however, the expansion ratio error of the inner and outer runner admission was within the allowable range, indicating that there was no significant difference in the swallowing capacity of the inner and outer runners, and the experimental trend agreed well with the simulation conclusion.

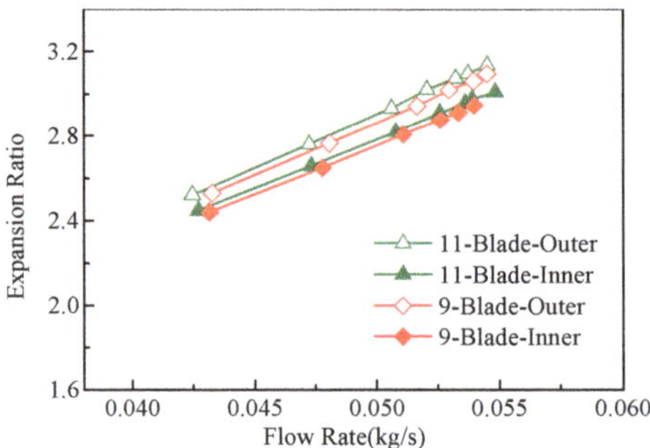

Figure 13. The unequal admission experimental comparison of the twin-entry turbine swallowing capacity.

4. Discussion

Based on the above method, all the performance points of engine meet the joint operating point requirements. Keeping the compressor of the original structure unchanged, the turbine structure was varied to obtain the external characteristic torque and the BSFC distribution of the engine based on the joint operating points, as shown in Figure 14. In the low- and middle-speed range of the engine, the output torque of the engine was significantly higher when matched with a twin-entry turbine than when matched with a single-entry turbine. However, its BSFC was lower than that of the single-entry turbine.

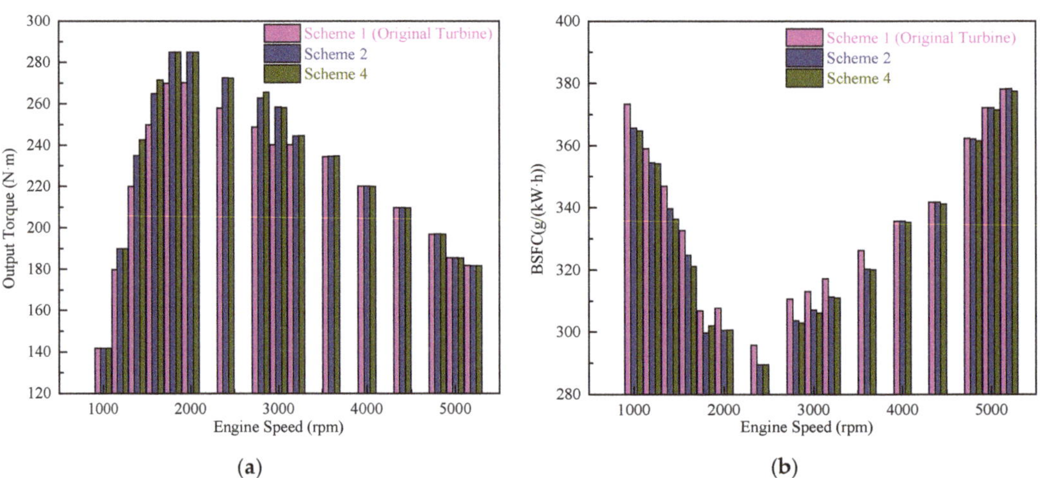

Figure 14. (**a**) The effect of different turbine structures on engine torque; (**b**) the effect of different turbine structures on engine BSFC.

For the twin-entry volute, the torque and BSFC curves of the 11-blade and 9-blade turbines overlapped, showing no significant difference. Hence, the 9-blade impeller could replace the 11-blade impeller and significantly reduce the turbine rotor mass and improve the response characteristics of the turbocharger and engine. The maximum engine torque of the 9-blade, twin-entry turbine was 5.4% higher than that of the original engine, and its lowest BSFC was 2.1% lower than that of the original engine. It was seen that the 9-blade, twin-entry turbine could effectively improve the engine performance at low- and medium-speed, and that the turbine performance was significantly improved.

Figure 14 shows the results of the 1D engine performance simulation, which shows that the torque and BSFC are basically nonlinear. This conclusion has also been confirmed by Samuel et al. [50] and Yusaf et al. [51]. On the other hand, the turbine obtained by the optimization method based on the joint operating curve has no significant difference at high speed compared to the conventional method. However, the turbine with the new optimized method has improved torque and BSFC in the low and medium speed. In this operating region, engine torque was increased by up to 3.2% and BSFC was reduced by up to 1.1% compared to the turbine optimized by conventional methods.

5. Conclusions

In this study, according to the basic conditions of turbocharger internal operation, the turbocharger joint point was determined, and the internal joint operation curve was obtained. A turbine evaluation method was proposed based on the internal joint operation, and the number of impeller blades and turbine housing were selected for optimal application design. The following conclusions were obtained within the scope of this study:

(1) Based on the joint operating characteristics of the two ends of the turbocharger compressor and turbine, the internal joint operating curve of the turbocharger was coupled using the performance distribution of the compressor and turbine, which was closer to the actual situation and was more practical;

(2) The evaluation method based on the joint point performance could predict the optimization trend of the turbine more accurately. Based on the internal joint operation curve of the turbocharger, the number of turbine impeller blades was optimized compared with the conventional evaluation method. The 1D engine performance simulation showed that the engine power and fuel economy of the turbine structure optimized by the conventional method were worse than those of the original engine;

(3) The turbine based on joint point optimization has better overall performance under the unequal admission condition. The output power of the twin-entry turbine has a significant advantage over the original turbine. Under the unequal admission condition, the 11-blade and 9-blade, twin-entry turbines have a similar performance, and the optimized turbine with 9-blade outer runner has the best overall performance, which is 1.7% higher than the original turbine;

(4) The turbine structure determined using the optimization method showed clear advantages. The findings of the study were verified using performance tests. The engine 1D performance simulation further showed that the maximum torque of the engine with the 9-blade, twin-entry turbine was 5.4% higher than that of the original engine and that the minimum BSFC was reduced by 2.1%.

The analysis determined that the optimal turbocharger was a 9-blade, twin-entry turbine. The performance of the optimized turbocharger was verified using the turbocharger test and the engine 1D performance simulation results. In the low- and medium-speed operating regions, the engine torque was increased by up to 3.2%, and BSFC was reduced by up to 1.1% compared to the turbine optimized by conventional methods.

In this study, from the perspective of internal joint matching between the compressor and turbine, and by coupling the internal joint operation curve of the turbocharger, the optimization application study of the turbine was carried out with the objective of improving the performance of the turbocharger turbine under internal joint operation conditions. This optimization method can provide future reference for the research and design optimization of the turbocharger turbine.

Author Contributions: Conceptualization, S.Y. and H.F.; methodology, S.Y.; software, H.F.; validation, S.Y., H.F. and R.H.; formal analysis, H.F.; writing—original draft preparation, S.Y.; writing—review and editing, S.Y.; project administration, J.N.; funding acquisition, J.N. and X.S. All authors have read and agreed to the published version of the manuscript.

Funding: This research was funded by the National Natural Science Foundation of China youth Science Foundation project (Grant No. 22102116) and Nanchang Intelligent New Energy Vehicle Research Institute Foresight Project (TPD-TC202110-12).

Institutional Review Board Statement: Not applicable.

Informed Consent Statement: Not applicable.

Data Availability Statement: The data presented in this study are available in the main text of the article.

Acknowledgments: The team of authors acknowledges anonymous reviewers for their feedback, which certainly improved the clarity and quality of this paper.

Conflicts of Interest: The authors declare no conflict of interest.

References

1. Helmers, E.; Leitao, J.; Tietge, U.; Butler, T. CO_2-equivalent emissions from European passenger vehicles in the years 1995–2015 based on real-world use: Assessing the climate benefit of the European "diesel boom". *Atmos. Environ.* **2019**, *198*, 122–132. [CrossRef]
2. Chen, Y.S.; Ma, J.Q.; Han, B.; Zhang, P.; Hua, H.N.; Chen, H.; Su, X. Emissions of automobiles fueled with alternative fuels based on engine technology: A review. *J. Traffic Transp. Eng.-Engl. Ed.* **2018**, *5*, 318–334. [CrossRef]
3. Cha, J.; Lee, J.; Chon, M.S. Evaluation of real driving emissions for Euro 6 light-duty diesel vehicles equipped with LNT and SCR on domestic sales in Korea. *Atmos. Environ.* **2019**, *196*, 133–142. [CrossRef]
4. Ko, J.; Jin, D.; Jang, W.; Myung, C.L.; Kwon, S.; Park, S. Comparative investigation of NOx emission characteristics from a Euro 6-compliant diesel passenger car over the NEDC and WLTC at various ambient temperatures. *J. Appl. Energy* **2017**, *187*, 652–662. [CrossRef]
5. Liang, X.Y.; Wang, Y.S.; Chen, Y.; Deng, S. Advances in emission regulations and emission control technologies for internal combustion engines. *SAE J. Steep* **2021**, *2*, 101–119. [CrossRef]
6. Krishnamoorthi, M.; Malayalamurthi, R.; He, Z.X.; Kandasamy, S. A review on low temperature combustion engines: Performance, combustion and emission characteristics. *Renew. Sustain. Energy Rev.* **2019**, *116*, 109404. [CrossRef]
7. Shi, C.; Zhang, Z.; Ji, C.W.; Li, X.Y.; Wu, Z.R. Potential improvement in combustion and pollutant emissions of a hydrogen-enriched rotary engine by using novel recess configuration. *Chemosphere* **2022**, *299*, 134491. [CrossRef] [PubMed]
8. Shi, X.Y.; Qian, W.W.; Wang, H.Y.; Pan, M.Z.; Wang, Q.W.; Ni, J.M. Development and verification of a reduced dimethoxymethane/ n-heptane/toluene kinetic mechanism and modelling for CI engines. *Appl. Therm. Eng.* **2022**, *214*, 118855. [CrossRef]
9. Rosdi, S.M.; Mamat, R.; Azri, A.; Sudhakar, K. Effects of lean combustion on Bioethanol-Gasoline blends using turbocharged spark ignition engine. *Int. J. Automot. Mech. Eng.* **2021**, *18*, 9140–9148. [CrossRef]
10. Lujan, J.M.; Serrano, J.R.; Piqueras, P.; Diesel, B. Turbine and exhaust ports thermal insulation impact on the engine efficiency and aftertreatment inlet temperature. *J. Appl. Energy* **2019**, *240*, 409–423. [CrossRef]
11. Lavagnoli, S.; De Maesschalck, C.; Paniagua, G. Uncertainty analysis of adiabatic wall temperature measurements in turbine experiments. *Appl. Therm. Eng.* **2015**, *82*, 170–181. [CrossRef]
12. Burke, R.D.; Vagg, C.R.M.; Chalet, D.; Chesse, P. Heat transfer in turbocharger turbines under steady, pulsating and transient conditions. *Int. J. Heat Fluid Flow.* **2015**, *52*, 185–197. [CrossRef]
13. Schorn, N. The radial turbine for small turbocharger applications: Evolution and analytical methods for twin-entry turbine turbochargers. *SAE Int. J. Engines* **2014**, *7*, 1422–1442. [CrossRef]
14. Copeland, C.D.; Martinez-Botas, R.; Seiler, M. Unsteady performance of a double entry turbocharger turbine with a comparison to steady flow conditions. *J. Turbomach.* **2012**, *134*, 021022. [CrossRef]
15. Rajoo, S.; Romagnoli, A.; Martinez-Botas, R.F. Unsteady performance analysis of a twin-entry variable geometry turbocharger turbine. *Energy* **2012**, *38*, 176–189. [CrossRef]
16. Fan, H.C.; Ni, J.M.; Wang, H.; Zhu, Z.F.; Liu, Y.G. Numerical study of unsteady performance of a double-Entry turbocharger turbine under different A/R value conditions. In Proceedings of the SAE 2016 World Congress and Exhibition, Detroit, MI, USA, 5 April 2016. [CrossRef]
17. Romagnoli, A.; Copeland, C.D.; Martinez-Botas, R.; Seiler, M.; Rajoo, S.; Costall, A.W. Comparison between the steady performance of double-entry and twin-entry turbocharger turbines. *J. Turbomach.* **2013**, *135*, 011042. [CrossRef]
18. Romagnoli, A.; Martinez-Botas, R.F.; Rajoo, S. Steady state performance evaluation of variable geometry twin-entry turbine. *Int. J. Heat Fluid Flow* **2011**, *32*, 477–489. [CrossRef]
19. Chiong, M.S.; Rajoo, S.; Romagnoli, A.; Costall, A.W.; Martinez-Botas, R.F. Integration of meanline and one-dimensional methods for prediction of pulsating performance of a turbocharger turbine. *Energy Convers. Manag.* **2014**, *81*, 270–281. [CrossRef]

20. Romagnoli, A.; Martinez-Botas, R.F. Performance prediction of a nozzled and nozzleless mixed-flow turbine in steady conditions. *Int. J. Mech. Sci.* **2011**, *53*, 557–574. [CrossRef]
21. Cerdoun, M.S.; Ghenaiet, A. Unsteady behaviour of a twin entry radial turbine under engine like inlet flow conditions. *Appl. Therm. Eng.* **2018**, *130*, 93–111. [CrossRef]
22. Chiong, M.S.; Rajoo, S.; Martinez-Botas, R.F.; Costall, A.W. Engine turbocharger performance prediction: One-dimensional modeling of a twin entry turbine. *Energy Convers. Manag.* **2012**, *57*, 68–78. [CrossRef]
23. Chiong, M.S.; Rajoo, S.; Romagnoli, A.; Costall, A.W.; Martinez-Botas, R.F. One-dimensional pulse-flow modeling of a twin-scroll turbine. *Energy* **2016**, *115*, 1291–1304. [CrossRef]
24. Padzillah, M.H.; Rajoo, S.; Martinez-Botas, R.F. Influence of speed and frequency towards the automotive turbocharger turbine performance under pulsating flow conditions. *Energy Convers. Manag.* **2014**, *80*, 416–428. [CrossRef]
25. Ding, Z.; Zhuge, W.; Zhang, Y.; Chen, H.; Martinez-Botas, R.; Yang, M. A one-dimensional unsteady performance model for turbocharger turbines. *Energy* **2017**, *132*, 341–355. [CrossRef]
26. Galindo, J.; Tiseira, A.; Fajardo, P.; Garcıa-Cuevas, L.M. Development and validation of a radial variable geometry turbine model for transient pulsating flow applications. *Energy Convers. Manag.* **2014**, *85*, 190–203. [CrossRef]
27. Newton, P.; Copeland, C.; Martinez-Botas, R.; Seiler, M. An audit of aerodynamic loss in a double entry turbine under full and partial admission. *Int. J. Heat Fluid Flow* **2012**, *33*, 70–80. [CrossRef]
28. Costall, A.W.; McDavid, R.M.; Martinez-Botas, R.F.; Baines, N.C. Pulse performance modeling of a twin entry turbocharger turbine under full and unequal admission. *J. Turbomach.* **2011**, *133*, 021005. [CrossRef]
29. Hajilouy-Benisi, A.; Rad, M.; Shahhosseini, M.R. Flow and performance characteristics of twin-entry radial turbine under full and extreme partial admission conditions. *Arch. Appl. Mech.* **2009**, *79*, 1127–1143. [CrossRef]
30. Cai, L.X.; Xiao, J.F.; Wang, S.S.; Gao, S.; Duan, J.Y.; Mao, J.R. Gas-particle flows and erosion characteristic of large capacity dry top gas pressure recovery turbine. *Energy* **2017**, *120*, 498–506. [CrossRef]
31. Xue, Y.X.; Yang, M.Y.; Martinez-Botas, R.F.; Romagnoli, A.; Deng, K.Y. Loss analysis of a mix-flow turbine with nozzled twin-entry volute at different admissions. *Energy* **2019**, *166*, 775–788. [CrossRef]
32. Mamala, J.J.; Praznowski, K.; Kołodziej, S.; Ligus, G. The Use of Short-Term Compressed Air Supercharging in a Combustion Engine with Spark Ignition. *Int. J. Automot. Mech. Eng.* **2021**, *18*, 8704–8713. [CrossRef]
33. Galindo, J.; Serrano, J.R.; Climent, H.; Tiseira, A. Experiments and modelling of surge in small centrifugal compressor for automotive engines. *Exp. Thermal. Fluid Sci.* **2008**, *32*, 818–826. [CrossRef]
34. Bellis, V.D.; Marelli, S.; Bozza, F.; Capobianco, M. 1d simulation and experimental analysis of a turbocharger turbine for automotive engines under steady and unsteady flow conditions. *Energy Procedia* **2014**, *45*, 909–918. [CrossRef]
35. Fatsis, A.; Vlachakis, N.; Leontis, G. A centrifugal compressor performance Map empirical prediction method for automotive turbochargers. *Int. J. Turbo. Jet. Eng.* **2019**, *38*, 411–420. [CrossRef]
36. Kim, S.; Park, J.; Ahn, K.; Baek, J. Improvement of the performance of a centrifugal compressor by modifying the volute inlet. *J. Fluids Eng.* **2010**, *132*, 091101. [CrossRef]
37. Chung, J.; Lee, S.; Kim, N.; Lee, B.; Kim, D.; Choi, S.; Kim, G. Study on the effect of turbine inlet temperature and backpressure conditions on reduced turbine flow rate performance characteristics and Correction Method for Automotive Turbocharger. *Energies* **2019**, *12*, 3934. [CrossRef]
38. Fan, J.; Yuan, Q.H.; Jing, F.L.; Xu, H.B.; Wang, H.; Meng, Q.Z. Adaptive local maximum-entropy surrogate model and its application to turbine disk reliability analysis. *Aerospace* **2022**, *9*, 353. [CrossRef]
39. Meng, D.B.; Yang, S.Q.; Zhang, Y.; Zhu, S.P. Structural reliability analysis and uncertainties-based collaborative design and optimization of turbine blades using surrogate model. *Fatigue Fract. Eng. Mater. Struct.* **2019**, *42*, 1219–1227. [CrossRef]
40. Payri, F.; Serrano, J.R.; Fajardo, P.; Reyes-Belmonte, M.A.; Gozalbo-Belles, R. A physically based methodology to extrapolate performance maps of radial turbines. *Energy Convers. Manag.* **2012**, *55*, 149–163. [CrossRef]
41. Cheng, L.; Dimitriou, P.; Wang, W.; Peng, J.; Aitouche, A. A novel fuzzy logic variable geometry turbocharger and exhaust gas recirculation control scheme for optimizing the performance and emissions of a diesel engine. *Int. J. Engine Res.* **2020**, *21*, 1298–1313. [CrossRef]
42. Xu, C.; Wang, S.; Mao, Y. Numerical study of porous treatments on controlling flow around a circular cylinder. *Energies* **2022**, *15*, 1981. [CrossRef]
43. Fan, H.C.; Ni, J.M.; Shi, X.Y.; Jiang, N.; Qu, D.Y.; Zheng, Y.; Zheng, Y.H. Unsteady performance simulation analysis of a waste-gated turbocharger turbine under different valve opening conditions. In Proceedings of the SAE 2017 International Powertrains, Fuels and Lubricants Meeting, Detroit, MI, USA, 8 October 2017. [CrossRef]
44. Fan, H.C.; Ni, J.M.; Shi, X.Y.; Qu, D.Y.; Zheng, Y.; Zheng, Y.H. Simulation of a combined nozzled and nozzleless twin-entry turbine for improved efficiency. *J. Eng. Gas Turbines Power* **2019**, *141*, 051019. [CrossRef]
45. Romagnoli, A.; Martinez-Botas, R. Heat Transfer Analysis in a turbocharger turbine: An experimental and computational evaluation. *Appl. Therm. Eng.* **2012**, *38*, 58–77. [CrossRef]
46. Bohn, D.; Heuer, T.; Kusterer, K. Conjugate flow and heat transfer investigation of a turbo charger. *J. Eng. Gas Turbines Power* **2005**, *127*, 663–669. [CrossRef]
47. Wang, Z.H.; Ma, C.C.; Zhang, H.; Zhu, F. A novel pulse-adaption flow control method for a turbocharger turbine: Elastically restrained guide vane. *Proc. Inst. Mech. Eng. Part C* **2020**, *234*, 2581–2594. [CrossRef]

48. Galindo, J.; Fajardo, P.; Navarro, R.; Garcia-Cuevas, L.M. Characterization of a radial turbocharger turbine in pulsating flow by means of CFD and its application to engine modeling. *Appl. Energy* **2013**, *103*, 116–127. [CrossRef]
49. Duan, Y.F.; Shi, X.Y.; Kang, Y.; Liao, Y.; Duan, L.S. Effect of Hydrous Ethanol Combined with EGR on Performance of GDI Engine. In Proceedings of the SAE 2020 World Congress and Exhibition, Detroit, MI, USA, 14 April 2016. [CrossRef]
50. Samuel, O.D.; Waheed, M.A.; Taheri-Garavand, A.; Verma, T.N.; Dairo, O.U.; Bolaji, B.O.; Afzal, A. Prandtl number of optimum biodiesel from food industrial waste oil and diesel fuel blend for diesel engine. *Fuel* **2021**, *285*, 119049. [CrossRef]
51. Yusaf, T.F.; Buttsworth, D.R.; Saleh, K.H.; Yousif, B.F. CNG-diesel engine performance and exhaust emission analysis with the aid of artificial neural network. *Appl. Energy* **2009**, *87*, 1661–1669. [CrossRef]

MDPI

St. Alban-Anlage 66

4052 Basel

Switzerland

www.mdpi.com

Sustainability Editorial Office

E-mail: sustainability@mdpi.com

www.mdpi.com/journal/sustainability